MECHANICS OF CELLULAR PLASTICS

MECHANICS OF CELLULAR PLASTICS

Edited by
N. C. HILYARD
Department of Applied Physics, Sheffield City Polytechnic, Sheffield, UK

APPLIED SCIENCE PUBLISHERS LTD
LONDON

APPLIED SCIENCE PUBLISHERS LTD
Ripple Road, Barking, Essex, England

British Library Cataloguing in Publication Data

Mechanics of cellular plastics.
1. Polymers and polymerization
I. Hilyard, N.C.
620.1'92   TA455.P8

ISBN 0-85334-982-7

WITH 31 TABLES AND 251 ILLUSTRATIONS

© APPLIED SCIENCE PUBLISHERS LTD 1982

All rights reserved. No part of this publication may be reproduced, stored in a retrieval system, or transmitted in any form or by any means, electronic, mechanical, photocopying, recording, or otherwise, without the prior written permission of the publishers, Applied Science Publishers Ltd, Ripple Road, Barking, Essex, England

Photoset in Malta by Interprint Ltd
Printed in Great Britain by Galliard (Printers) Ltd,
Great Yarmouth

PREFACE

Cellular plastics have been available commercially for many years but still have considerable potential for further development in terms of material properties, conversion routes and product applications. The main reasons for this are (i) cellular plastics can exhibit unique mechanical properties, (ii) they provide a highly efficient use of polymer raw material, which is likely to be a decreasing resource in the future, and (iii) they provide a material system with a high stiffness to weight ratio, a factor which is of considerable benefit in many actual or potential markets.

This book considers both the theoretical aspects of the mechanical behaviour of rigid and flexible cellular polymers and the results of recent systematic investigations into the relationship between the mechanical properties, composition and structure of these materials. As such, it will be of interest to material scientists concerned with the development of new and improved material systems and engineers involved in material specification and designing for cellular plastics. For the most part, the cellular polymers considered previously in the literature have been complex material systems and often have been inadequately characterised from a microstructural point of view. In addition, material data have been presented which were obtained using test procedures which were not completely appropriate to these materials. It is hoped that to a large extent these deficiencies have been overcome in the present work. Although emphasis is placed on the presentation of information in graphical and tabular form, this publication is not intended to be a handbook on materials engineering or product design; its main purpose is to state and analyse our current knowledge and understanding of the mechanics of cellular plastics and indicate areas for future investigation.

The topics covered include an introduction to the structure of cellular plastics and structural model representations, the stiffness and strength of homogeneous rigid and flexible foams, resiliency and fatigue in flexible foams with particular attention being given to materials for comfort cushioning, the analysis of material behaviour under oscillatory and impact loading conditions, the design and performance of cellular shock mitigation systems and the strength and stiffness of structural, fibre reinforced and syntactic foams.

It is inevitable in a book of this nature that there will be differences of style and approach in the contributions from different authors. Although some degree of uniformity has been achieved, the amount of editorial change has been kept to a minimum in order to maintain the essential character of the individual contributions.

I wish to thank the authors for their patience and understanding during the period in which the manuscript has been under preparation. I also wish to thank Dr. A. N. Gent, University of Akron, who first introduced me to the science of cellular polymers and gave useful advice during the initial planning stages of the book.

N. C. HILYARD

CONTENTS

Preface v

List of Contributors ix

1. Introduction 1
   N. C. Hilyard and J. Young

2A. Stiffness and Strength—Rigid Plastic Foams 27
   G. Menges and F. Knipschild

2B. Stiffness and Strength—Flexible Polymer Foams 73
   N. C. Hilyard

3. Cushioning and Fatigue 99
   H. W. Wolfe

4. Dynamic Mechanical Behaviour 143
   N. C. Hilyard

5. Shock Mitigation—Material Behaviour 179
   N. C. Hilyard

6. Shock Mitigating Systems 207
   M. A. Mendelsohn

7. Structural Foams 263
   J. L. Throne

8. Reinforced Foams 323
   J. METHVEN and J. R. DAWSON

9. Syntactic Foams 359
   A. R. LUXMOORE and D. R. J. OWEN

Index 393

LIST OF CONTRIBUTORS

J. R. Dawson
  Department of Metallurgy and Materials Science, University of Liverpool, P.O. Box 147, Liverpool L69 3BX, UK

N. C. Hilyard
  Department of Applied Physics, Sheffield City Polytechnic, Pond Street, Sheffield S1 1WB, UK

F. Knipschild
  Institut für Kunststoffverarbeitung, Technischen Hochschule Aachen, Pontstrasse 49, D5100 Aachen, West Germany

A. R. Luxmoore
  Department of Civil Engineering, University of Wales, Singleton Park, Swansea SA2 8PP, UK

M. A. Mendelsohn
  Research and Development Center, Westinghouse Electric Corporation, 1310 Beulah Road, Pittsburgh, Pennsylvania 15235, USA

G. Menges
  Institut für Kunststoffverarbeitung, Technischen Hochschule Aachen, Pontstrasse 49, D5100 Aachen, West Germany

J. Methven
  Fillite Limited, 12 Arkwright Road, Astmoor Industrial Estate, Runcorn, Cheshire WA7 INU, UK

D. R. J. Owen
  Department of Civil Engineering, University of Wales, Singleton Park, Swansea SA2 8PP, UK

J. L. Throne
  Research and Development Department, Amoco Chemicals Corporation, P.O. Box 400, Naperville, Illinois 60540, USA

H. W. WOLFE
   Elastomers Laboratory, E.I. du Pont de Nemours and Company, Chestnut Run, Wilmington, Delaware 19898, USA

J. YOUNG
   Department of Applied Physics, Sheffield City Polytechnic, Pond Street, Sheffield S1 1WB, UK

Chapter 1

INTRODUCTION

N. C. HILYARD and J. YOUNG
Department of Applied Physics, Sheffield City Polytechnic, Sheffield, UK

1.1 THE NATURE AND CLASSIFICATION OF CELLULAR POLYMERS

Cellular polymers are multiphase material systems that consist of a polymer matrix and a fluid phase, the fluid usually being a gas. Most polymers can be expanded into cellular product, but only a small number have been exploited commercially.[1] In terms of volume consumed, polyurethane, polystyrene, poly(vinyl chloride) and the polyolefins have dominated the markets. However, over the last few years there has been increased utilisation of engineering structural foams for load-bearing applications. The polymers used include high-density polyethylene, modified polyphenylene oxide, polycarbonate and ABS. In some cases, additional solid phases may be present in the cellular system. Examples include fibre-reinforced foams and syntactic foams (which are composite materials consisting of hollow glass, ceramic or plastic microspheres dispersed throughout a polymer matrix).

Because of their complex nature, polymer foams have been classified in a variety of ways. The most usual are; the cellular morphology, the mechanical behaviour and the composition. Structurally the material system can be described as open or closed cell. With closed-cell materials the gas is dispersed in the form of discrete gas bubbles and the polymer matrix forms a continuous phase. In open-cell foams the voids coalesce so that both the solid and the fluid phases are continuous. Schematic representations of the different physical forms of cellular polymers are given in Fig. 1.1.

With open-cell structures, as depicted in Fig. 1.1(a), the fluid is able to flow through the polymer matrix under the action of some driving poten-

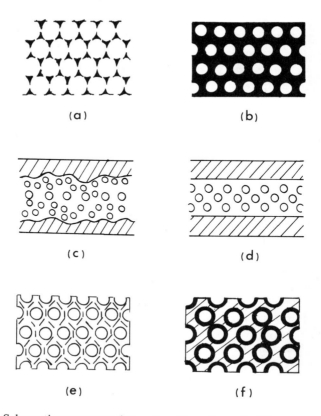

FIG. 1.1. Schematic representations of sections through different types of cellular polymer. (a) Low-density open-cell foam, (b) High-density closed-cell foam, (c) single-component structural foam with cellular core and integral solid skin, (d) multicomponent structural foam, (e) fibre-reinforced closed cell foam and (f) syntactic foam.

tial, whereas in closed-cell structures (Fig. 1.1b), gas transport takes place by diffusion through the cell walls. The ease of movement of the fluid phase through the matrix is one of the factors governing the physical and mechanical properties of the cellular polymer. In practice, the two cellular morphologies can co-exist so that a polymer foam is not always completely open or closed cell. The volume fraction of closed cells has a considerable influence on the mechanical behaviour of these systems so it is an important structural characteristic.

These materials are also classified according to their stiffness, the two extremes being rigid and flexible. Skochdopole and Rubens[2] have defined a rigid foam as one in which the polymer matrix exists in the crystalline state, or if amorphous, is below its glass transition temperature. Following from this, a flexible foam is a system in which the matrix polymer is above the crystalline melting-point or above its glass transition temperature. According to this classification most polyolefin, polystyrene, phenolic, polycarbonate, polyphenylene oxide and some polyurethane foams are rigid, whereas rubber foams, elastomeric polyurethanes, certain polyolefins and plasticised poly(vinyl chloride) are flexible. Intermediate between these two extremes is a class of cellular polymer known as semi-rigid. Although these materials have an elastic modulus higher than that of flexible foams, their stress–strain behaviour is closer to that of flexible systems than that exhibited by rigid foams.

Rigid cellular polymers can be further subdivided according to whether they are used for non-load-bearing applications (such as thermal insulation) or as load-bearing structural materials (which require high stiffness, strength and impact resistance).

A particular class of flexible foam is known as high resiliency (also termed cold-cured PUR foam). This classification refers to the mechanical hysteresis exhibited by the material when taken through a compression–deformation cycle. High resiliency (HR) polyurethane (PUR) foam has a stress–strain diagram in compression intermediate between that of conventional PUR flexible foam and rubber latex foam. The mechanical hysteresis of this material is much smaller than that of open cell flexible PUR foam and is close to that of rubber latex foam (see Chapter 3).

Cellular polymers exist in a wide range of different structural forms. They may be homogeneous with a uniform cellular morphology throughout or they may be structurally anisotropic. They may have an integral solid polymer skin or they may be multicomponent in which the polymer skin is of different composition to the polymeric cellular core. Some of these structures are shown schematically in Fig. 1.1.

1.2 FORMATION OF CELLULAR POLYMERS

1.2.1 The Expansion Process

Chemical and technological aspects of the manufacture of polymer foams have been described in detail in the publications edited by Frisch and Saunders.[3] Although foams can be produced in a variety of ways, the most commonly employed method is the expansion process. In principle, this consists of (i) the nucleation of gas bubbles in a liquid polymer system, either a suspension, liquid mixture or melt, (ii) the growth and stabilisation of these bubbles and (iii) the solidification of the polymeric phase by crosslinking or cooling to give a structurally stable cellular system. The details of the process depend on the nature of the starting materials, i.e. thermoplastic, thermosetting plastic or latex suspension. The structures resulting from the expansion process may be open cell or closed cell. In some cases, such as with some flexible cellular polyurethanes, the green foam undergoes a post-expansion treatment, called reticulation, which ruptures the membranes between adjacent cells to produce an almost completely open-cell structure.

Several techniques are used to entrain the gas in the liquid polymer system before expansion. These include the mechanical frothing of suspensions and the incorporation of blowing agents, either physical or chemical, in the starting formulation.[4-6] There are two important methods of physical blowing: (a) the injection of an inert gas, such as nitrogen or carbon dioxide, into the polymer melt at high pressure which subsequently expands when the melt is extruded from a die or injected into a mould; and (b) the inclusion of low-boiling-point liquids, such as chlorofluorocarbons and methylene chloride, in the starting materials which volatilize on heating, resulting in the formation of gas bubbles when the pressure is released. Chemical blowing is achieved by (a) dry blending the starting polymer with finely powdered solid agents, such as azodicarbonamide, which decompose over a limited temperature range with the evolution of gas and (b) the interaction between two reactive chemicals. A particularly important example of the last of these processes is found in the manufacture of PUR foams.

Polyurethane foams have been described as being unique in the family of cellular polymers. They exist in a wide variety of physical forms, i.e. open cell, closed cell and microcellular, and can be rigid, semi-rigid, flexible or highly resilient. They are manufactured by the very accurate metering and mixing of two catalysed liquid components, a polyol and an isocyanate. The polyol component has either a polyether or a polyester backbone chain and contains either a chain extender or crosslinking agent and a catalyst. Once the components have been mixed the polymerisation and expansion take place simultaneously. If the reactants are bifunctional, a linear polymer product results. Higher functionality leads to the formation of branched-chain materials or crosslinked polymer networks. When water is present in the mixture the H_2O–isocynate reaction causes the generation of CO_2 which blows the foam. However, there are limits to the amount of water that can be employed because of the excessive heat generated by the exothermic reaction which damages the cellular product. As a consequence, physical blowing agents are often included in the formulation which are completely responsible for, or assist in, the blowing of the foam.

The mechanical properties of cellular polymers are related in a complex way to a wide range of variables. Included in these are structural parameters such as the cell structure geometry, the cell size, the orientation of the cells with respect to the direction of loading, the uniformity in the size of the cell elements and the volume fraction of open or closed cells. Although the structures of different types of polymer foam have been described on many occasions in the literature, it is worthwhile to consider briefly the features that are particularly relevant to our understanding of the mechanical behaviour of these materials.

1.2.2 Bubble Growth and Idealised Cellular Morphologies

The principles of bubble nucleation and growth in polymer melts and liquid starting mixtures have been described by a number of authors.[7-11] According to Saunders and Hansen,[7] self nucleation of bubbles occurs in the absence of nucleating agents when the gas–polymer solution becomes supersaturated. For example in water-blown PUR foams, sufficiently rapid gas generation can be achieved by the catalysis of the water–isocyanate reaction. As soon as bubble nucleation relieves the supersaturation, no new bubbles are formed and the concentration of gas in solution decreases because of diffusion into existing bubbles. After this, the bubbles grow by coalescence, thermal expansion of the gas within the bubble and diffusion of gas from the smaller bubbles into the larger ones. The dynamics of bubble growth have been discussed by Burt[10] and Throne.[11] With systems containing bubble-nucleating agents, the behaviour is expected to be much the same as that of systems without these agents except that nucleation occurs at lower gas concentrations.

The cellular structures that result from the expansion of the bubbles depend on factors such as the supersaturation temperature, the pressure of the dissolved gas, the rate of gas generation and the processing conditions, e.g. the injection speed in moulded foam parts.[9] Of particular importance in this respect are the rheological and surface properties of the expanding polymer, e.g. viscosity, retarded elastic behaviour and surface energy.[7, 9, 10] A further complication is that foam expansion is a dynamic process in which the mechanical properties of the polymer, which are influenced by factors such as the degree of crystallisation, molecular orientation and crosslinking, vary as a function of time.

Two-dimensional representations of idealised cellular structures at different stages of bubble expansion are shown in Fig. 1.2. These are based

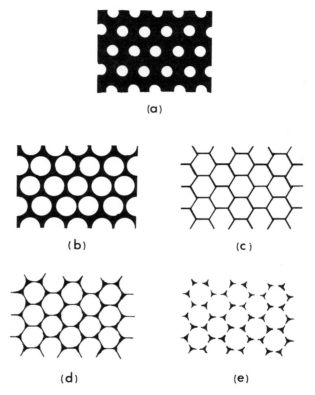

FIG. 1.2. Two-dimensional representations of idealised cellular structures at different stages of bubble expansion (based on the description of Harding[8]).

on the description of Harding.[8] Initially the small bubbles are dispersed throughout the polymer melt or liquid system. If stabilised in this state, the resulting foam would be of high density. The lowest density that can be achieved with spherical bubbles of uniform size is when the bubbles are close packed as shown in Fig. 1.2(b). Although small bubbles can exist in the material interstices between the close packed spherical voids, as shown by Menges and Knipschild in Chapter 2A, Harding[8] has indicated that this configuration is unstable and is not typical of polymer foam structures. Further expansion, and hence a lowering of the foam density, causes the large bubbles to distort such that cells bounded by polyhedra with plane surfaces of uniform thickness are formed (Fig. 1.2c). These surfaces are referred to as the cell walls, windows or membranes.

The stability of the cellular structure of a liquid foam is influenced by several processes. One of these is the drainage of the liquid in the cell walls towards the cell junctions under the action of gravitational and capillary forces.[7, 8] This movement of liquid polymer results in a thinning of the cell walls and the formation of ribs (or struts) at the lines of intersection between the cells, as shown in Fig. 1.2(d). Further drainage will cause the rupture of the membranes and if stabilised in this state, the resulting foam will be open cell and will consist of a three dimensional array of interconnecting polymer struts (Fig. 1.2e). Excessive drainage of the liquid polymer sometimes causes the collapse of the cellular structure during foam manufacture.

In low-density foams, as depicted in Fig. 1.2(c), (d) and (e), the cell morphology is influenced by energy and structural symmetry considerations.[12] Although a tetrakaidecahedron cell, having fourteen membranes, minimises the surface area of the cell, and hence the surface energy, the angular symmetry associated with the pentagonal dodecahedron cell, which has twelve five sided membranes, means that this type of cell is commonly found in cellular polymers. This is particularly true of flexible and rigid PUR foams. A schematic representation of a single cell of this type is shown in Fig. 1.3, together with a photomicrograph of a single cell excised from a specimen of reticulated open cell flexible PUR foam. The cell struts, or junctions, have a cross-section that is approximately triangular in shape, as shown in Fig. 1.4, but can be better described as a hypocycloid with three cusps.[13] In some foams, the section of the strut has uniform dimensions over a considerable proportion of its length, i.e. between the strut junctions, or nodes. This is shown for a reticulated PUR flexible foam in Fig. 1.9. However, in other materials, it is very non-uniform as can be seen from Fig. 2A.4.

Invariably there is some distribution of cell morphology within a foam specimen. For example, with PUR foams four- and six-sided membranes are observed alongside the pentagonal membranes. The presence of cell elements with these configurations is necessary to achieve complete packing. In addition, anisotropy in the dimensions of the cells often

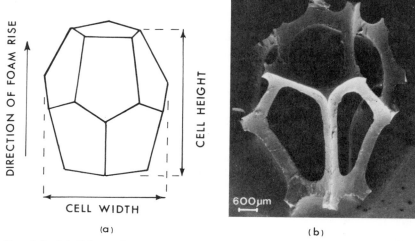

FIG. 1.3. (a) Schematic representation of a pentagonal dodecahedron cell. (b) SEM micrograph of a single cell excised from a low-density PUR flexible foam.

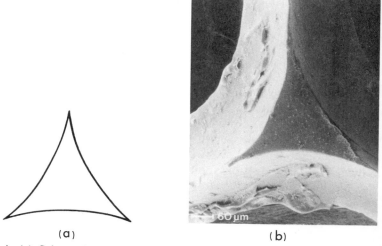

FIG. 1.4. (a) Schematic representation of cross-section through a cell strut. (b) SEM micrograph of a cell strut in a low-density PUR flexible foam.

exists.[12] This structural anisotropy is related to the expansion process and the maximum cell dimension is in the direction of material movement. For free-rising foams, the major axis of the cell is, generally speaking, in the direction of foam rise. However, as described by Hermanson,[14] Doherty *et al.*,[15] and other workers, it can vary from place to place within a foam slab. Anisotropy in the cellular morphology of free-rising foams is discussed further in Section 2A.3.2. In moulded foam parts, the direction of material movement, and hence cell orientation, is influenced by factors such as part configuration, the degree of reaction during the distribution of the material and the thermal input from the mould.[14]

1.3 CELLULAR STRUCTURES

1.3.1 Typical Cellular Morphologies

The stabilisation of a cellular structure can be brought about after different degrees of bubble expansion and all of the idealised morphologies depicted in Fig. 1.2 are found in practice. However, it should be recognised that particular morphologies are often characteristic of particular categories, or types, of cellular polymer and that within a specimen the structure is not completely uniform. This does not mean that different categories of cellular polymer cannot have the same structure. For example, it is shown throughout this book that the pentagonal dodecahedron cell structure (Fig. 1.3) is found in both rigid and flexible foams and foams with different base polymer. In this section we will consider briefly the structures of some typical low-density polymer foams corresponding to the idealised structures shown in Fig. 1.2(c), (d) and (e). Micrographs of high-density foams, corresponding to Fig. 1.2(a) and (b), and other low-density materials, are given in the following chapters.

SEM micrographs of three closed-cell foams exhibiting a structure corresponding to Fig. 1.2(c) are shown in Fig. 1.5, 1.6 and 1.7. The base polymers of these foams are SBR, polystyrene and polyethylene, respectively. It is seen that the predominant structure is a five-sided cell with uniform wall thickness. However, it can also be seen that there are some four- and six-sided cells within these materials. It is interesting to note the wrinkling of the cell walls in Fig. 1.7. This can be attributed to the relaxation of 'frozen-in' stresses caused by the sectioning and/or the heating of the specimen during microscopy. This point is discussed further below.

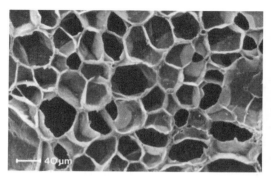

FIG. 1.5. SEM micrograph of a closed-cell rubber foam with density $\rho_f \approx$ 154 kg m^{-3}.

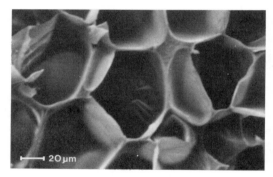

FIG. 1.6. SEM micrograph of a closed-cell polystyrene foam with density $\rho_f \approx$ 115 kg m^{-3}.

FIG. 1.7. SEM micrograph of a closed-cell polyolefin foam with density $\rho_f \approx$ 31 kg m^{-3}.

Figure 1.8 shows a cell structure intermediate between the structures of Fig. 1.2(d) and 1.2(e). This is an SEM micrograph of a reticulated PUR flexible foam in which some of the cell membranes have not been removed, i.e. it is a partly closed cell. Again it is interesting to note the wrinkling of the membranes indicating some degree of stress relaxation.

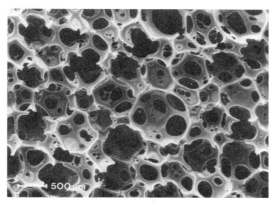

FIG. 1.8. SEM micrograph of a low-density partially reticulated PUR flexible foam of density $\rho_f \approx 30$ kg m^{-3}.

The classic example of a low density open cell foam is reticulated PUR and micrographs of this material have been given many times in the literature. An example is presented in Fig. 1.9. It is seen that this cellular

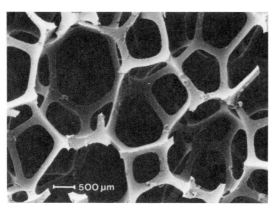

FIG. 1.9. SEM micrograph of a reticulated open cell PUR flexible foam of density $\rho_f \approx 36$ kg m^{-3}.

polymer consists of interconnecting struts, with approximately triangular cross-section, and that the predominant configuration is a pentagonal array of cell elements. For a given density the side dimension and the length of the struts can vary over a considerable range according to the processing conditions. A similar morphology is found in the rubber latex foam shown in Fig. 1.10. However, some rubber latex foams exhibit a

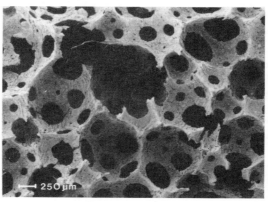

FIG. 1.10. SEM micrograph of an open-cell SBR/NR rubber latex flexible foam of density $\rho_f \approx 78$ kg m^{-3}.

much more irregular structure. An example of a three-phase cellular system is given in Fig. 1.11. This shows the fracture surface through a

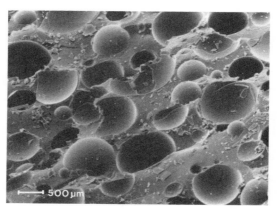

FIG. 1.11. SEM micrograph of a syntactic foam of 0·3 g cm^{-3} glass spheres in an epoxy resin. Packing fraction of spheres approximately 62%.

syntactic foam consisting of hollow glass microspheres distributed throughout an epoxy matrix. The debonding and the fracture of the microspheres is clearly seen together with the distribution in the size of the spheres. Packing fractions up to 75–80% can be achieved using a bimodal sphere size distribution.

1.3.2 Dimensional Description of Cellular Structures

The dimensions of the repeating structural unit in a cellular polymer have been described by means of the pore count, which is the number of cells per linear distance in the foam, the average cell diameter and the width or side length of the polyhedral cell membranes. From a fundamental point of view it is the dimensions of the basic cell elements such as the spherical voids, the cell membranes or the cell struts which are of importance. However, the dimensions of the membranes and struts are not easily measured, and although the cells associated with these structural units are not spherical it has been common practice to take the average cell diameter d_{av} as the characteristic dimension. This is effectively the average value of the maximum dimension of the cell projected onto a plane perpendicular to the direction of measurement.[16] Although this quantity is easily measured using a magnifier or microscope the values obtained are not always directly related to the actual dimensions of the cell, see for example Section 2A.3.2. Typically for commercially manufactured low density PUR flexible foams of density about 32 kg m^{-3} d_{av} lies in the range 0·25 mm to 2·5 mm and for microcellular PUR foams ($\rho_f \approx$ 600 kg m^{-3}) the cell-size range is approximately 0·01 mm to 0·1 mm.

As explained above, and as depicted in Fig. 1.3(a), a single cell may be anisotropic in size. This structural anisotropy has been characterised using the ratio of the dimensions of the cell parallel and perpendicular to the direction of foam rise, i.e. the height/width ratio. Although the cell dimensions in the plane perpendicular to the rise direction are usually considered to be the same this is not always the case and a better description of the cell dimensions is obtained by taking the cell size in three mutually perpendicular directions (1,2,3). As far as the mechanical properties are concerned it is also necessary to take account of the anisotropy in the cell-size distribution. This is considered in Chapter 2B.

Examples of the cell-size distributions and the anisotropy in the distributions to be expected in commercially manufactured PUR flexible foams are given in Fig. 1.12, 1.13 and 1.14. The three foams, A, B and C were supplied by different firms. The data were obtained from the

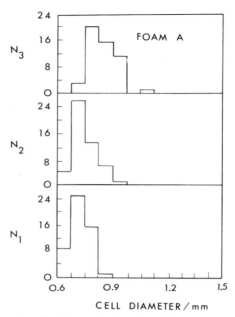

FIG. 1.12. The number distribution in the size of the cell diameter in three orthogonal directions; Foam A.

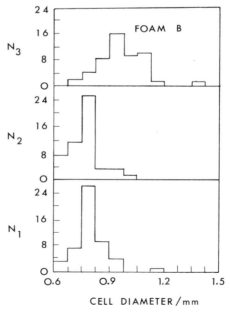

FIG. 1.13. The number distribution in the size of the cell diameter in three orthogonal directions; Foam B.

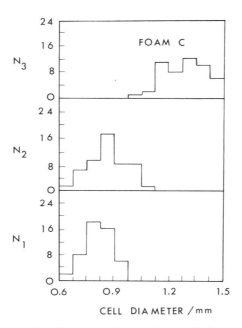

FIG. 1.14. The number distribution in the size of the cell diameter in three orthogonal directions; Foam C.

measurement of about one hundred cells in each direction and direction 3 was taken as the direction for maximum average cell diameter. It can be seen that with these specimens not only does the anisotropy in the average cell dimension vary from foam to foam, but in two of the samples the width of the cell-size distribution was dependent on direction also.

1.3.3 Structural Models

The approaches used in the analysis of the mechanical behaviour of cellular materials and for the development of equations that can be employed to predict their behaviour can be broadly divided into three categories. These are; (i) the use of phenomenological models, such as that employed by Skochdopole and Rubens[2] to explain the influence of closed-cell structures on the compressive stress–strain behaviour of a semi-rigid foam; (ii) the application of equations developed for composite materials, such as the Kerner, Halpin–Tsai and Nielsen equations, an approach that has been widely used in the analysis of the elastic properties of rigid foams;[17, 18] and (iii) the analysis of the load-deformation behaviour of models representative of the physical structure of the cellular material. Many of these models have been reviewed by Meinecke and Clark[19] so all that is necessary here is to consider how some of the model representations cited in the following chapters conform to the structures that are likely to be found in practice.

A particular type of model that has been widely used is that based on the pentagonal dodecahedron cell structure shown in Figure 1.3. One example of this is the model representation given by Chan and Nakamura[20] which is shown in Fig. 1.15. This can be applied to both open- and closed-cell structures. In the latter case the foam is considered to be

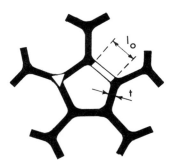

FIG. 1.15. The closed-cell pentagonal dodecahedron structural model of Chan and Nakamura[20] showing in section the pentagonal cell membranes of side l_0 and thickness t and the triangular prismatic struts of length l_0 and base t.

composed of dodecahedral voids which are bounded by pentagonal interfaces of matrix polymer. The material surrounding the voids is separated into elements having a regular geometry. These include pentagonal plates of uniform thickness t and side length l_0, triangular prismatic bars of length l_0 and side t and tetrahedra of side t. A modified form of this model is employed by Menges and Knipschild in Chapter 2A in their analysis of the elastic and strength properties of PUR and PVC rigid foams. Clearly this model conforms closely to the ideal structure of a broad class of cellular polymer. However, as pointed out by these authors, simplifications and assumptions have to be made when it is applied to a real material. Regardless of this, the model has been quite successful in predicting the relationship between structure and properties and the resulting equations contain few fitting parameters.

Another complex type of model, used by Ko[21] in his study of the material properties of open-cell elastomeric foams, is based on the cellular network formed when the interstices between close-packed spherical voids are filled with polymer. He investigated the networks formed by two types of packing, hcp and fcp, and demonstrated that a free-standing unit cell excised from the network by a hexagonal cylinder could be taken as being representative of the foam specimen. The unit cell resulting from the hexagonal close packing of voids is shown in Fig. 1.16. The important structural parameters of the unit cell are the length l_0 and cross-

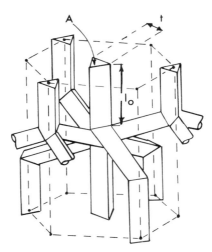

FIG. 1.16. The unit cell of an idealised open cell structure formed by filling the interstices between hcp spherical voids with polymer (based on the description of Ko,[21] published courtesy of Technomic Publishing Co., Inc., Westport, USA).

section area A of the strut elements and the inclination of the struts with respect to each other. The struts were assumed to have a triangular cross-section. Although this type of model exhibits the three dimensional symmetry found in many cellular polymers, the cell structure formed by the interstices of close-packed spherical voids does not conform to the pentagonal dodecahedron structures found in these materials.

Because of the difficulties associated with the analysis of the load-deformation behaviour of complex models, such as those just described, various simplified structures have been proposed. Two of these are shown in Fig. 1.17 and 1.18. The first is the cubical arrangement of

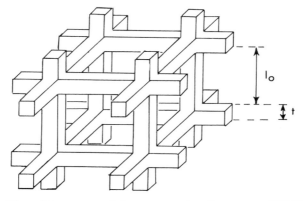

FIG. 1.17. The cubic strut model representation for open cell foams proposed by Gent and Thomas.[22]

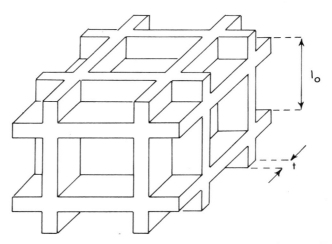

FIG. 1.18. The cubic plate model representation for closed cell foams proposed by Matonis.[23]

square-section struts used by Gent and Thomas[22] in their studies of the elastic properties and the stress–strain behaviour of open-cell flexible foams. The struts have a free length l_0 and side dimension t. In their analysis they assumed that the cubes of material, side length t, at the junctions of the struts are essentially undeformable compared with the struts themselves, and they referred to this as dead volume. An equivalent model for closed-cell systems has been proposed by Matonis.[23] This is shown in Fig. 1.18 and consists of a cubical array of uniform plates. Although these simple models do not conform to the structures actually observed in cellular polymers they have been useful in establishing structure–property relationships in situations where it is difficult to apply the more sophisticated models. For example, the cubical rod model has been used to investigate the influence of cell structure variables on the load bearing capacity of flexible foams[22] and the relationship between structural and mechanical anisotropy.[24]

The first of the models described by Gent and Thomas was, in fact, the ball-and-strut model shown in Fig. 1.19 which they used to represent the cellular morphology of open-cell rubber latex foams (see Fig. 1.10). It

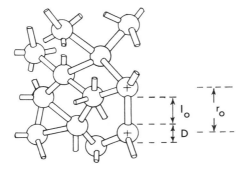

FIG. 1.19. The ball and strut model of Gent and Thomas[25] for open cell foams.

consists of circular struts of cross-section area A and free length l_0 joined at their intersections by polymer spheres of diameter D. As with the cubical strut model these connecting elements were assumed to be undeformable. The orientation distribution of the struts was taken to be random and it was assumed that the entire surface of the spherical dead volumes is covered by strut elements. If n struts enter each sphere then according to this model $nA/\pi D^2 = 1\cdot0$. The model employed by Lederman[26] was essentially the same as that of Gent and Thomas, except that he considered the condition $nA/\pi D^2 = 1\cdot0$ to be a special case and that in general the spherical dead volume need not be entirely covered by strut elements, i.e. $nA/\pi D^2 \leqq 1\cdot0$. He showed that the value of $nA/\pi D^2$ is related to processing variables and took this quantity as a fitting constant. However he assumed it would have the same value in foams manufactured under the same conditions. The important structural parameter in this model is the ratio of the sphere diameter to the free strut length, D/l_0.

1.4 MECHANICAL BEHAVIOUR

1.4.1 General Considerations

In addition to the cellular morphology, the mechanical properties of the matrix polymer exert a very strong influence on the mechanical behaviour of a polymer foam. Although these properties are governed primarily by the composition of the base polymer they are also affected by the processing conditions which can control, for example, the degree of molecular orientation, crystallisation and crosslink density. Other factors influencing the behaviour include the nature of the fluid enclosed by the matrix, the test piece size, the temperature and the strain and strain rate. In most respects we can regard cellular polymers as being non-linear viscoelastic and rheologically complex materials. Because of this wide range of variables, it has been necessary to establish different model approaches to explain the different aspects of their mechanical behaviour.

The different classes of foams are normally operated under different conditions of strain. Thus with rigid structural foams we are concerned primarily with the small strain quasi-static elastic and strength properties for the three modes of loading; tension, compression and shear. The creep and stress relaxation behaviour of these materials are also of importance for design purposes. As pointed out in the following chapter, this information is not always readily available.

With flexible foams we are concerned mainly, but not exclusively, with compressive deformation, and in this case the shape of the stress–strain curve and the size of the stress at large strains, $\varepsilon \gg 25\%$, is of particular significance. These are governed by the elastic modulus of the matrix polymer, the cell size, the spatial distribution in the size of the cell elements and the volume fraction of closed cells. Flexible and semi-rigid foams may also be employed under conditions where they are subjected to dynamic loading, e.g. impact and oscillatory deformations. The small- and large-strain dynamic mechanical behaviour is influenced by the nature of the fluid enclosed by the polymer matrix, i.e. whether it be compressible or incompressible, the resistance to the flow of the fluid, which is related to the mechanical properties of the fluid and the cellular morphology of the foam, and the strain and strain rate.

Although the reported data concerning the influence of cell size on the small-strain static properties are not always in agreement there is much evidence to show that under certain conditions the elastic properties, E_f, of structurally isotropic foams of the same composition are not significantly affected by this parameter. They are governed primarily by the elastic modulus E_p of the matrix polymer and the density ρ_f of the foam. Since ρ_f is an easily measured quantity considerable attention has been given to establishing empirical and theoretical relationships between the three parameters E_f, E_p and ρ_f.

1.4.2 Property–Density Relationships

The classic equation that has been widely used to relate the properties of rigid foams to the foam density is [14, 27–29]

$$E_f = K\rho_f^n \qquad (1.1)$$

where K and n are empirically determined constants and K is related in some way to the properties of the matrix polymer. This expression has been shown to be valid for both the elastic and strength properties under tensile and compressive loading. The exponent n lies in the range $1\cdot0 \leq n \leq 2\cdot0$. The reported data indicate that the value of n depends on the mode of deformation; the value for tensile loading being less than that for compression.[27] It has been found for PUR rigid foams that the parameter K is related to the test temperature T by an equation of the form [28]

$$K = (a - bT)^m \qquad (1.2)$$

where a and b are numerical coefficients. According to De Gisi and Neet,[28] the values of a and b depend on the property of interest, i.e. elastic modulus or strength, and $m = 1\cdot0$. Phillips and Waterman[30] have suggested that it is difficult to compare directly the elastic properties of PUR rigid foams having a wide range of densities because of differences in the properties of the matrix polymer. As explained in Chapter 3, the mechanical properties of polyurethanes are controlled to a large extent by hydrogen bonding. At elevated temperatures a transition occurs which can be attributed to the breaking up of the hydrogen bonded structures. These workers found that when the elastic properties of PUR rigid foams of different density are compared at a constant temperature difference, $T_s - T$, relative to the temperature T_s for this transition, a direct proportionality between modulus and density is obtained. Their results indicated that for a given cellular polymer the value of the exponent in eqn (1.2) is about 0·58, and that the constants a and b are related to T_s.

An alternative way of expressing the relationship between foam properties and density is in terms of the density ratio, or volume fraction of polymer, ϕ. The mass M_f of a foam specimen of volume V_f is $M_f = V_f \rho_f$. Since the enclosed gas phase makes little contribution to the mass, $M_f = V_p \rho_p$ where V_p is the volume of polymer of density ρ_p in the specimen. Consequently the volume fraction of polymer ϕ is equal to the density ratio ρ_f/ρ_p. The relevant property relationship is[17,18,31]

$$E_f = E_p \phi^n \tag{1.3}$$

It has been usual practice to take E_p and ρ_p as the properties of the base polymer since, as with the foam density, these are easily measured quantities. Throne has argued (see for example Chapter 7) that for rigid foams the value of n is 2·0, as found by Moore et al.[18]

It is shown below, in Chapter 2B, that data for the initial tensile elastic modulus of PUR and rubber latex flexible foams can be fitted quite well by eqn.(1.3) using values of n between about 1·4 and 1·7.

There are limitations to the applicability of eqns.(1.1) and (1.3). Hermanson[14] has pointed out that in moulded foam parts the density varies across the section thickness so that the strength and elastic properties are not necessarily directly related to the overall density of the foam specimen. These density variations are influenced by processing parameters such as the mould temperature and heat capacity, packing ratio and gel conditions. Similar considerations apply to free-rising foams. The importance of the degree of density variation is discussed in more detail in Section 3.2 of Chapter 2A. It is also known that structural anisotropy, which is commonly found in cellular polymers, leads to anisotropy in their mechanical properties. Although the influence of cell orientation may be less than that of density variations[14] it has been demonstrated in many publications (see Section 2B.5) that it does produce measurable effects.

Another factor that has been widely discussed is the effect of the expansion process and the processing variables on the mechanical properties of the matrix polymer. One aspect of this is the possibility of 'frozen-in' molecular orientation. Benning[32] has shown that at elevated temperatures an isolated membrane from a thermoplastic foam behaves in much the same way as shrink film. From this he concluded that these membranes must possess a certain degree of molecular orientation. However he found that the amount of orientation in the matrix polymer of a foam was less than that of a free-blown film of the same material. This he attributed to stress relaxation during the foaming process.

1.4.3 Structure–Density Relationships

Since the density plays a central role in determining the small-strain quasi-static mechanical behaviour, it is useful to examine how this parameter has been related to the cell-structure geometry. This can be done using the models described in the literature. These model approaches place particular emphasis on the ratio of the thickness to length dimensions of the basic cell elements. As examples of open-cell structures we can take the models of Gent and Thomas (Fig. 1.17 and 1.19), and that of Ko (Fig. 1.16). For the cubical-rod model (Fig. 1.17), there are twelve polymer rods, of length l_0 and side t, each of which are shared by four adjacent cells, and eight cubes of side t shared between eight cells. Consequently the volume of polymer in the repeating structural unit is $V_p = 3\,l_0\,t^2 + t^3$. Since the overall volume of the repeating unit is $V_f = (l_0 + t)^3$, the volume fraction of polymer is given by

$$\phi = (3\beta^2 + \beta^3)/(1 + \beta)^3 \qquad (1.4)$$

where the cell structure parameter $\beta = t/l_0$. The corresponding expression for the ball-and-strut model (Fig. 1.19), is the same as eqn. (1.4) with the parameter $\beta = D/l_0$, where D is the diameter of the spherical dead volume. In this particular model it is assumed that the total surface area of the spheres is covered by cell elements. If this is not the case, then the appropriate expression is, according to Lederman,[26]

$$\phi = 3\beta^2\{(nA/\pi D^2) + (\beta/3)\}/(1 + \beta)^3 \qquad (1.5)$$

where $nA/\pi D^2$ is the fraction of surface area covered by the strut elements. In his original publication, Ko expressed the density of voids, d, in terms of the dimensions of the cell elements. This is related to the volume fraction of polymer by $d = V_g/V_f = 1 - \phi$, where V_g is the volume of gas in a foam specimen of volume V_f. For an open-cell polymer network formed by the filling of the interstices between hexagonal close packed spheres (Fig. 1.16), the volume fraction of polymer is given by

$$\phi = (9/8)(t/l_0)^2$$

where t is the side dimension and l_0 the length of the cell struts which are assumed to have an equilateral triangular shape.

The variation of the density ratio ϕ as a function of the cell structure parameter D/l_0, or t/l_0 predicted by eqns. (1.4) and (1.5) are shown as the broken lines (c) and (d), respectively, in Fig. 1.20. The relationship given by the Lederman model is for a system in which only half the area

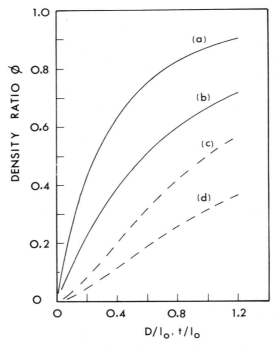

FIG. 1.20. Variation of the density ratio $\phi = \rho_f/\rho_p$ as a function of the cell structural parameter D/l_0, or t/l_0, predicted by different structural models; continuous lines, closed cell structures, broken lines, open cell structures. (a) Matonis,[23] equation 1.6, (b) Chan and Nakamura,[20] equation 1.7, (c) Gent and Thomas,[22,25] eqn. (1.4) and (d) Lederman,[26] eqn. (1.5) (with $nA/\pi D^2 = 0.5$).

of the spherical dead volume is covered by strut elements, i.e. $nA/\pi D^2 = 0.5$. These expressions are intended to apply primarily to low-density open-cell foams. In the limiting condition $\phi \to 0$, both the Gent–Thomas and the Ko models predict that $\phi \approx 3\beta^2$.

The Chan and Nakamura and the Matonis models, which are shown in Fig. 1.15 and 1.18, respectively, may be taken as representative of two forms of closed-cell structures. In the Matonis cubical-plate model, the repeating structural unit is a cube of side length $l_0 + 2(t/2)$, enclosing a cubic volume of gas having side length l_0. Thus for this structure the volume fraction of polymer is given by

$$\phi = 1 - 1/(1 + \beta)^3 \quad (1.6)$$

where $\beta = t/l_0$. In the more complex pentagonal dodecahedron closed-cell model described by Chan and Nakamura, the density ratio is related to the side length l_0 of the undeformed pentagonal cell elements by $\phi = 1 \cdot 0 - 7 \cdot 663 \, l_0^3 N$, where N is the number of cells per unit volume. Using the equation based on a material balance given by these workers it can be shown for this structural model that

$$\phi \approx \beta(10 \cdot 32 + 4 \cdot 33\beta + 0 \cdot 60\beta^2)/(7 \cdot 63 + 10 \cdot 32\beta + 4 \cdot 33\beta^2 + \beta^3) \quad (1.7)$$

where $\beta \doteq t/l_0$ and t is the side length of the triangular struts connecting the pentagonal plates which have side l_0. The relationships between the density ratio and the structure parameter t/l_0 predicted by these closed cell-models are shown as the continuous lines (a) and (b), respectively, in Fig. 1.20.

1.5 CONCLUDING REMARKS

From the preceding discussion it is clear that cellular polymers are complex material systems and that although, in the first instance, the elastic and strength properties are closely related to the overall density, the mechanical behaviour cannot always be explained in such simple terms. In general, the properties are controlled by a large number of chemical, physical and process variables which influence for example, the properties of the matrix polymer, the density distribution and the isotropy of the cellular structure. In addition, under dynamic loading the mechanical behaviour is strongly influenced by the mechanical properties of the fluid enclosed by the polymer matrix, the time rate of compression and the dimensions of the test specimen.

In the remainder of this book, material data for the different categories of cellular polymer obtained under specified test conditions are presented and discussed. The observed mechanical behaviour is compared with that predicted by different model representations.

REFERENCES

1. *Foamed Plastics*, in *1973–1974 Modern Plastics Encyclopedia*, Vol. 50 (No. 10A), McGraw-Hill, October, 1973, p.125.
2. SKOCHDOPOLE, R. E. and RUBENS, L. C. (1965) *J. Cell. Plast.*, **1**, 91.
3. FRISCH, K. C. and SAUNDERS, J. H. (1972) *Plastic Foams*, Parts I and II, Marcel Dekker Inc., New York.
4. COLLINGTON, K. T. (1975) *J. Cell. Plast.*, **11**, 213.
5. HALLAS, R. S. (1977) *Plast. Eng.*, **33**, 17.
6. REES, J. L. and MORNINGSTAR, G. (1978) *Elastomerics*, **110**, 43.
7. SAUNDERS, J. H. and HANSEN, R. H. (1972) *Plastic Foams*, Ed. Frisch, K. C. and Saunders, J. H., Marcel Dekker Inc., New York, Chap. 2.
8. HARDING, R. H. (1965) *J. Cell. Plast.*, **1**, 385.
9. HOBBS, S. Y. (1976) *Polym. Eng. Sci.*, **16**, 270.
10. BURT, J. G. (1978) *J. Cell. Plast.*, **14**, 341.
11. THRONE, J. L. (1976) *J. Cell. Plast.*, **12**, 161.
12. HARDING, R. H. (1967) *Resinography of Cellular Plastics*, ASTM STP 414, Am. Soc. Testing Mats., p3.
13. JONES, R. W. and FESMAN, G. (1965) *J. Cell. Plast.*, **1**, 200.
14. HERMANSON, R. (1968) *J. Cell. Plast.*, **4**, 46.
15. DOHERTY, D. J., HURD, R. and LESTER, G. R. (1962) *Chem. Ind.*, 1340.
16. WATERMAN, N. R. and PHILLIPS, P. J. (1974) *Polym. Eng. Sci.*, **14**, 72.
17. THRONE, J. L. (1978) *J. Cell. Plast.*, **14**, 21.
18. MOORE, D. R., COUZENS, K. H. and IREMONGER, M. J. (1974) *J. Cellular Plast.*, **10**, 135.
19. MEINECKE, E. A. and CLARK, R. C. (1973) *Mechanical Properties of Polymeric Foams*, Technomic Publishing Company Inc., Westport.
20. CHAN, R. and NAKAMURA, M. (1969) *J. Cell. Plast.*, **5**, 112.
21. KO, W.L. (1965) *J. Cell. Plast.*, **1**, 45.
22. GENT, A. N. and THOMAS, A. G. (1963) *Rubber Chem. Tech.*, **36**, 597.
23. MATONIS, V. (1964) *SPE J.*, **20**, 1024.
24. KANAKKANATT, S. V. (1973) *J. Cell. Plast.*, **9**, 50.
25. GENT, A. N. and THOMAS, A. G. (1959) *J. Appl. Polym. Sci.*, **1**, 107.
26. LEDERMAN, J. M. (1971) *J. Appl. Polym. Sci.*, **15**, 693.
27. FERRIGNO, T. H. (1967) *Rigid Plastic Foams*, Reinhold Publishing Corporation, p. 150.
28. DE GISI, S. L. and NEET, T. E. (1976) *J. Appl. Polym. Sci.*, **20**, 2011.
29. BENNING, C. J. (1969) *Plastic Foams: The Physics and Chemistry of Product Performance and Process Technology*, Wiley–Interscience, Vol. 1, p. 93.
30. PHILLIPS, P. J. and WATERMAN, N. R. (1974) *Polym. Eng. Sci.*, **14**, 67.
31. PRAMUK, P. F. (1976) *Polym. Eng. Sci.*, **16**, 559.
32. BENNING, C. J. (1967) *J. Cell. Plast.*, **3**, 125.

Chapter 2A

STIFFNESS AND STRENGTH— RIGID PLASTIC FOAMS

G. Menges and F. Knipschild
*Institut für Kunststoffverarbeitung,
Technischen Hochschule Aachen, Aachen,
West Germany*

Translated by N. C. Hilyard
*Department of Applied Physics, Sheffield City Polytechnic,
Sheffield, UK*

NOTATION

A	Area
$C_{1,2}$	Constants
E	Elastic modulus
E_0	Modulus in first linear region
E_1	Modulus in second linear region
F	Force
$K_{1,2}$	Constants
I	Area moment of inertia
OD	Degree of cell orientation
R	Radius
V	Volume
a	Empirical specimen parameter
b	Width
d	Cell diameter, specimen side dimension
h	Specimen height
j	Two-dimensional structure orientation factor (≈ 0.64)
$\bar{j}$	Three-dimensional structure orientation factor (≈ 0.46)
k	Statistical constant used in the calculation of the mean cell diameter (≈ 1.27)
l	Length of a cell element

l^0 Length between cell nodes
l_0 Gauge length
n_0 Number of cells per unit volume
n_R Number of cell rods per unit volume
r Radius
t Thickness of cell rods

α Clamping factor for cell rods ($1 < \alpha < 4$)
γ Shear strain
δ Deformation
ϵ Tensile, compressive strain
θ Temperature
ν_p Poisson ratio of the base polymer
ρ Density
σ Tensile, compressive stress
τ Shear stress
ϕ Foam density ratio
$\mathcal{X}$ Characteristic angle of cell model
ψ Characteristic angle of cell model

Subscripts

B Breaking, fracture
C Compressive
R Rod
T Tensile
f Foam
k Critical, buckling value
m Mean value
p Polymer

2A.1 INTRODUCTION

The mechanical properties of plastic rigid foams (PRF) can be described using rigid polyurethane (PUR) and poly (vinyl chloride) (PVC) foams as examples. Since the properties of a cellular polymer are strongly dependent on the cell structure it is necessary to describe this structure for both systems.

PVC rigid foam is of closed-cell structure. If consists of an assembly of polyhedra with edges of different shape. The cell walls have constant thickness over their entire length. This is because of the high viscosity of the melt which leads, during foaming, to a uniform expansion of the skeleton. The basic coarse cellular structure, with its polyhedron cells bounded by three- to eight-cornered plates, is interspersed with a network of smaller cells which cause an accumulation of material at the edges and the corners of the cells. These lenticular polyhedron-shaped secondary cells are much smaller than the primary cells. The proportion of the small cell range is dependent on the density. Figures 2A.1 and 2A.2 show the structure of a PVC rigid foam and a single element of this foam.

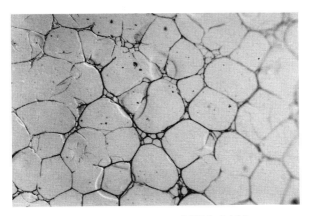

FIG. 2A.1. The structure of PVC rigid foam.

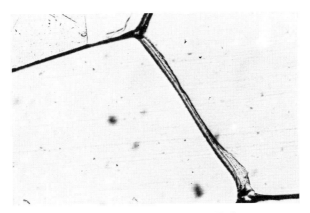

FIG. 2A.2. A PVC rigid-foam cell element.

The PUR rigid foam is a closed-cell structure and for foams in the density range 50–100 kg m^{-3} a well developed, relatively uniform, polyhedron structure is observed[1-4]. The polyhedra have eight to sixteen sides which are bounded by four to six edges, respectively, according to the density or the proportion of gas. Usually five edges are found. The intersection of three cells is formed by rods with concave surfaces, as described in Chapter 1 above. The cell structure of a PUR rigid foam and a single element of a PUR rigid foam are shown in Figs. 2A.3 and 2A.4.

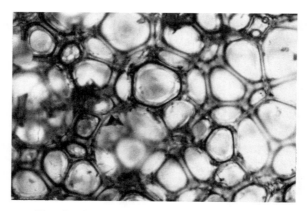

FIG. 2A.3. The structure of PUR rigid foam.

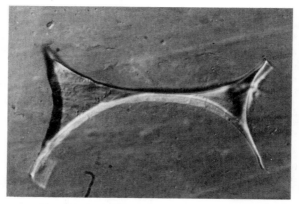

FIG. 2A.4. A PUR rigid-foam cell element.

2A.2 STRUCTURAL MODELS AND THE MECHANICAL BEHAVIOUR OF RIGID FOAMS

When describing multiphase materials by means of models it is necessary to make simplifying assumptions. In foams, where the second phase is a gas, the total deformation is made-up of two parts.[5] In addition to the Hookean deformation of the cellular skeleton, the external load creates bending and shear stresses which give rise to angle changes, bending and the distortion of the cell structure. This is particularly important in low-density foams where it makes a significant contribution to the overall deformation and makes it difficult to describe the deformation behaviour using models. Some of the structural models that have been used previously to estimate the mechanical properties of these materials are described in Chapter 1. In order to predict the mechanical properties of PRF it is necessary to modify these models as explained below.

Structural investigations[4,6,7] have shown that the model for the foam structure can be based on a dodecahedron structure. The polymer volume that encloses the cavities can be divided into three different cell elements:[9] 12 uniform five-cornered plates, 30 triangular prismatic rods and 20 tetrahedrons. The following elements are shared between adjacent unit cells: 2 cells, one plate; 3 cells, one prismatic rod; 4 cells, one tetrahedron. With open-cell foams the uniform five-cornered plates are absent. Figure 2A.5 shows these cell elements together with their dimensions and volume and the way they are combined in a unit cell.

2A.2.1 Open-Cell Foams

It has been established[8] that the structure of an open-cell PUR rigid foam is described well by the pentagonal dodecahedron model. The PUR rigid foam considered here is of closed-cell structure. However the cell walls are thinner than the triangular prismatic rods by a factor of ten to twenty times, so that their influence on the mechanical behaviour can be neglected. Because of this, the PUR rigid foam will be treated as a quasi open-cell foam. This assumption was made on the basis of microscopic investigations related to the stress behaviour.

The inaccuracy introduced by neglecting the supporting effect of the cell walls is compensated for by adding a contribution to the volume of the load bearing elements. Such a simplification seems justifiable for the low-density foams under consideration. Thus in the first instance the load-bearing elements are the prism rods, as is the case with open-cell foams.

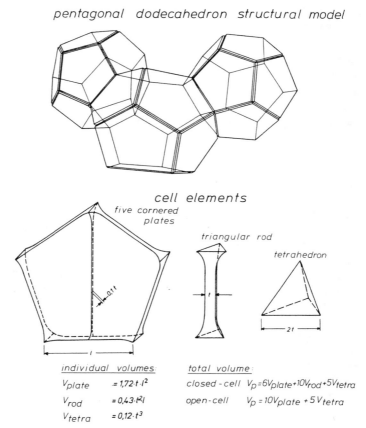

FIG. 2A.5. The pentagonal dodecahedron structural model and cell elements.

Their dimensions were determined as a function of the density and are shown in Fig. 2A.6.

The free-rod length, l, decreases and the thickness, t, increases with increasing foam density. The rod length, l^0, re-defined as the distance between the centres of two tetrahedra,[9] and the number of cells remain constant. Whether this is true for other foamed plastic materials has yet to be investigated. The relationships between the various cell-structure parameters are deduced from Fig. 2A.5 and 2A.6 as[6,9] the number of cells per unit volume $n_0 = 1/\{7\cdot 633\,(l^0)^3\}$, rod length $l^0 = l + 2t$, rod thickness

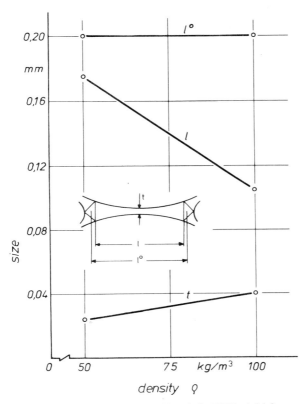

FIG. 2A.6. The size of the cell rods in PUR rigid foam.

$t = 0.661^0(\phi)^{1/2}$, where ϕ is the density ratio defined as ρ_f/ρ_p. All other parameters can be determined if ϕ and l^0 are known.

(a) *Deformation and Failure Mechanisms*

The investigations described here were carried out on a commercial PUR rigid foam produced by the block foaming process. Various foam densities in the range 50–100 kg m^{-3} were studied. The samples were taken from the central homogeneous region of the blocks which had dimensions $0.3 \times 1.0 \times 2.0$ m^3. The cells in this region showed essentially zero flow orientation. For this homogeneous material no directional dependence between the foaming direction and the loading direction could be established.

The deformation behaviour of the load-carrying cell elements has been studied under the microscope for different modes of loading, tension, compression and shear, using small specimens (dimensions, 5 × 10 × 50 mm³).[6,7] It was observed in these studies that the very thin cell walls buckled at the slightest deformation, wrinkled, and under tension fractured. These observations confirmed the assumption made above that even if there is a closed-cell foam the elements that actually support the load are not the cell walls but are the cell rods. These experience different stresses, and consequently deformations, according to the mode of loading and their orientation relative to the direction of external loading. The orientation of the rods in a cross-section is statistically distributed. Between the two extreme positions, parallel and perpendicular to the external load directions, all intermediate orientations are possible. The loading of the individual rods can be related to the external loading, as shown in Fig. 2A.7.

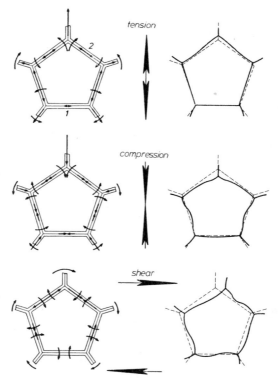

FIG. 2A.7. Qualitative stressing of the cell rods and the resulting deformations.

Under tensile loading, the rod lying in the direction of the external force is stressed in tension. Failure occurs on reaching the limit of the tensile strength of the base material by the fracture of these rods. Rods lying perpendicular to the applied force are either stressed in bending (rod 1, Fig. 2A.7) or remain unstressed, except for a small tensile loading, and consequently are not deformed. Rods in intermediate orientations experience combined normal and bending stresses.

For compressive loading, the stressing of individual rods is similar to the case of tensile loading but of opposite sign. Here also failure occurs in those rods which lie in the direction of the external loading. However, for the low-density foams under investigation this happens before the limit of material strength is reached because of buckling of the cell rods (instability failure). According to the thickness of the node, which determines the degree of clamping, and the slenderness ratio of the rod three modes of deformation corresponding to the different buckling situations described by Euler were observed for rods stressed only in compression.[8]

Whereas for normal stresses the position of the rods has to be considered in one plane only, with shear the spatial orientation of the rods must be taken into account. The rod stresses may be normal stresses or bending stresses or a combination of the two. In this case the critical stress is due to either pure compression or a combination of compression and buckling. Failure then occurs because of buckling of those rods that lie at an angle of about 45° to the direction of the external load. Thus we have instability failure as in compression.

Figure 2A.8 shows a shear-loaded PUR rigid foam at different stages of deformation. It can clearly be seen that the very thin cell walls buckle under small deformations and wrinkle. The series of illustrations show the possible deformations—buckling in compression, bending, tension—and indicate clearly that failure under shear loading occurs in those rods which are loaded in compression and that eventually, at large deformations, the rods stressed in tension will break.

(b) *Prediction of Material Properties*

Theoretical equations for predicting the mechanical properties may be derived from the observations given in the preceding section. The purpose is to predict these properties as a function of the density. It is necessary to distinguish between material parameters associated with strength and those associated with elastic properties. It can be deduced from the behaviour of the cell elements under load that the strength of a rigid foam under normal stress is determined solely by those rods oriented

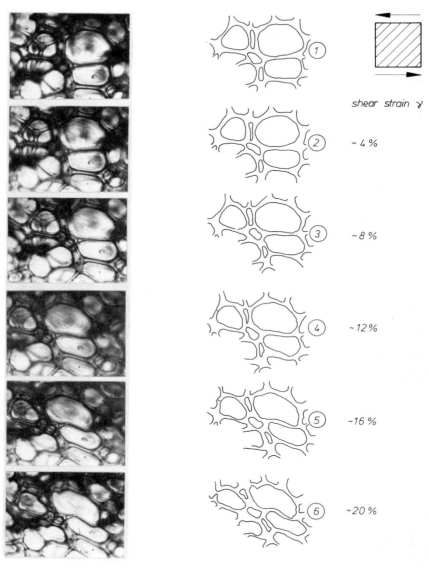

FIG. 2A.8. The deformation of the cell structure of PUR rigid foam in shear.

in the external loading direction. The elastic properties, however, are determined by the deformation behaviour of all rods since the change in length in the loading direction is made up of the buckling of the rods in the direction of the load and the bending of rods perpendicular to the direction of loading.

Shear forces arising in the rods and the resulting shear deformations are not considered in this mathematical analysis. Results described later show that this assumption is justified for low-density foams. The cell stiffening due to the thin cell walls is accounted for in the analysis by an addition of a material allowance to the cell rods.

(i) *Strength*

As far as strength is concerned it is necessary to distinguish between material failure and instability failure. Instability failure occurs under compressive and shear loading, material failure under tensile loading.

Tension: In order to determine tensile strength, the actual foam volume is reduced to the load-bearing polymer volume. This means the load-carrying cell rods. The orientation of the rods relative to the direction of the external load, and consequently the size of the normal stress, is accounted for by an orientation factor j. The volume fraction of polymer in the form of rods, $\phi = v_{rod}/v_{total}$, is related[8] to the foam density ratio $\bar{\phi}$ by $\bar{\phi} \approx 0.5\,\phi$. Taking account of orientation the volume fraction of polymer in rods lying parallel to the direction of loading is $v_{rod\,\|}/v_{total} = \bar{\phi} j$. The ratio of the cross-section area of these rods to the total cross-section of the foam is also given by $A_{rod\,\|}/A_{total} = \bar{\phi} j$. The tensile strength of the foam, σ_{TB}, is related to the tensile strength of the base polymer, σ_{TBp}, and the density ratio, $\bar{\phi}$, by

$$\sigma_{TB} = \sigma_{TBp} \bar{\phi} j a$$

where a is a factor that takes account of the effect of sample preparation. This is determined empirically and has a value between 0·7 and 0·8.

From the statistical distribution of the rods in a cross-section, the normal stress components may be expressed as a sinusoidal distribution. The mean value of this normal stress component, x_m, in a planar cross-section corresponds to the orientation factor j,

$$j = x_m = \int_0^{\pi/2} \frac{\cos x}{\pi/2}\,dx = \frac{2}{\pi} \approx 0.64$$

The tensile strength of a foam is dependent on the tensile strength of the base material and increases in proportion to the density ratio. The individual load carrying rods oriented in the direction of external loading are then stressed to a point of material failure.

Compression: As can be seen under the microscope, unstable failure occurs under compressive loading. This means that failure occurs before the actual ultimate strength of the material is reached. Whereas under tensile stress the entire cross-section of the load-bearing material elements is relevant, in this case one has to start from the load-carrying capacity, i.e. the critical load, of a rod. The critical load of a rod is given, according to the mode of clamping, by Euler as

$$F_k = \alpha(\pi^2 I E_p / l^2)$$

The factor α takes into consideration the manner of clamping. Since neither an ideally pivoted nor ideally rigid clamping is present we assume here a mean clamping factor of $\alpha = 2.5$.

If one multiplies the buckling load of a rod by the number of rods per unit area one obtains the compressive strength, σ_{CB}, of the foam. According to the derivation carried out in Fig. 2A.9 the compressive strength is

$$\sigma_{CB} = a \, \alpha E_p j 0.0425 \, \phi^2$$

The following conversions and simplifications have been made: $l = l^0$, $I = (\sqrt{3}/96)t^4$, $(t/l)^4 = 0.2 \, \phi^2$. The compressive strength is a function of the elastic modulus of the base polymer and increases with the square of the density ratio. The influence of density on the clamping factor was neglected.

Shear: With shear loading, failure also occurs in the compressively stressed rods as shown in Fig. 2A.8. When the present biaxial loading is taken into account using a spatial orientation factor $\bar{j}$, then one obtains the shear stress at fracture as

$$\tau_B = a\alpha E_p \bar{j} 0.0425 \phi^2$$

The orientation factor $\bar{j}$ takes account of the dependence of the normal stress components in the compressively stressed rods on the orientation in space. It corresponds to a mean value $\bar{x}_m$ which is obtained from the ratio of the volume, V, to the base area, A, of the body enclosed by a sine curve; $\bar{j} \approx \bar{x}_m = V/A$, where

$$V = 2\pi \int_0^\pi x \cos x \, dx = \pi^2 - 2\pi$$

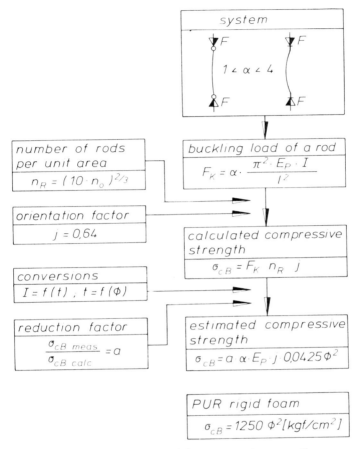

FIG. 2A.9. Estimation of the compressive strength.

and $A = \pi r^2$ with $r = \pi/2$. The orientation factor

$$\bar{j} \approx \bar{x}_m \approx 0.46$$

(ii) *Modulus of elasticity*

In calculating strength only the normal stress components of the load-carrying cell rods are taken into account. The elastic behaviour however is determined by the deformation caused by bending and normal stresses.

The model shown in Fig. 2A.10 is used as the basis for determining the

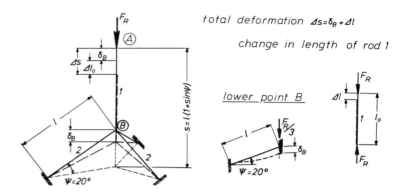

FIG. 2A.10. The rod system for estimating the elastic modulus.

individual components of the deformation. Consider a system of four cell rods which meet in a tetrahedron. The lowering of the point of application of the force A is composed of the change in length $\Delta l^0 = F_R l^0 / A_R E_p$ of rod 1 due to the normal force plus the lowering of point B, which is composed of a bending and a normal force component, $\delta_B = \delta_b + \delta_N$. The total deformation of the system may be described as

$$\Delta s = \frac{F_R l^0}{A_R E_p} + \frac{F_R (l^0)^3}{I_R E_p} K_1 + \frac{F_R l^0}{A_R E_p} K_2$$

where K_1 and K_2 are numerical constants resulting from the conversions. The elastic modulus of the foam emerges from this equation according to the derivation in Fig. 2A.11 as

$$E_f = E_p C_1 \phi^2 / (\phi + C_2)$$

For the PUR rigid foam under consideration here the constants were $C_1 = 0.65$ and $C_2 = 0.23$.

The elastic modulus of the foam can thus be described as a function of the density and the elastic modulus of polymer. According to this analysis the same value is obtained for tensile and compressive loading.

(c) *Comparison of Predicted and Measured Values*

Using the characteristic data of the base material determined from independent measurement, $\sigma_{TBp} = 6 \times 10^2$ kgf cm^{-2} and $E_p = 2.5 \times 10^4$ kgf cm^{-2}, the following relationships are obtained for the PUR rigid foam under investigation:

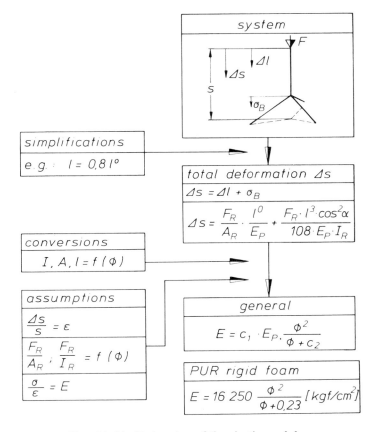

FIG. 2A.11. Estimation of the elastic modulus.

tensile strength, $\sigma_{TB} = 150\,\phi$ kgf cm^{-2}
compressive strength, $\sigma_{CB} = 1250\,\phi^2$ kgf cm^{-2}
shear strength, $\tau_B = 800\,\phi^2$ kgf cm^{-2}
elastic modulus, $E_f = 16250\,\phi^2/(\phi + 0\cdot23)$ kgf cm^{-2}

The predicted and measured characteristic data are shown in Figs. 2A.12 and 2A.13 together with those given in the literature.[10] When the predicted values for the tensile strength are compared with the measured parameter, good agreement is obtained in all cases. For the compressive and shear strengths the calculated function increased somewhat

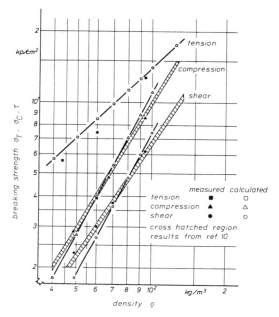

FIG. 2A.12. The strength of PUR rigid foam as a function of the density.

more rapidly. There is also good agreement between the estimated values of the tensile and compressive moduli and the values obtained experimentally. However, the moduli in tension tended to be higher than the moduli in compression (see Section 2A.4.1(b)). Although the characteristic data from the literature[10] run parallel to the predicted function, in general, they lie below the values determined in the investigations described here. Comparable data points in Fig. 2A.13 indicate that these differences can be explained by differences in the sample geometry and the strain measuring procedure (see Section 2A.3).

2A.2.2 Closed-Cell Foams

The structure of closed-cell PVC rigid foam has been shown in Figs. 2A.1 and 2A.2. In comparison with PUR rigid foam, which as far as mechanical behaviour is concerned is quasi open cell, with PVC rigid foam we are dealing with a closed-cell structure. The individual cells form a closed load-bearing system. The fact that in previous model

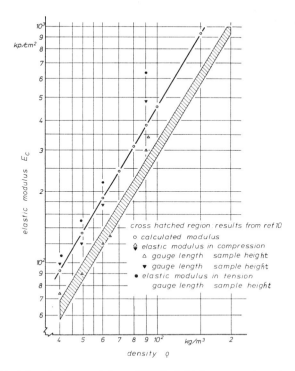

FIG. 2A.13. The elastic modulus of PUR rigid foam as a function of density.

representations[1,11–15] deformations are calculated from forces acting on a single cell element extracted from the closed load-bearing system is a rough simplification and is bound to lead to large deviations from the real material behaviour. Taking the compressive strength as an example the way in which a better estimation can be made is outlined below.

(a) *Double Truncated Cone Model*

It is assumed, as above, that the compressive strength of the PVC rigid foam is reached as a result of instability failure of the load-bearing cell structure elements. The structure of PVC rigid foam can be described reasonably well[4] by the pentagonal dodecahedron model.[14] The mathematical analysis of such a complicated, closed, multi-faced framework is difficult. Compact solutions giving the stiffness and critical compressive force, which are of interest, do not yet exist. However, solutions do exist for bodies of rotation, such as cylinders and cylindrical and spherical shells.

To a good approximation, a pentagonal dodecahedron can be circumscribed by a double truncated cone.[4] Buckling equations and critical loads have been considered previously.[16,17] The dimensions of the pentagonal dodecahedron and the corresponding truncated cone are shown in Fig. 2A.14. Calculation of the critical load of a truncated cone leads to a complex, high order, differential equation. Approximate solutions are obtained using an energy method. The precise, rather complicated derivation will not be given here. It has been re-worked previously[4,18] in accordance with Schnell et al.,[16,17] and can be described as follows.

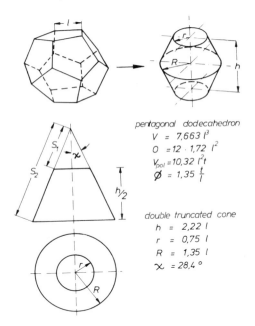

pentagonal dodecahedron
$V = 7.663\ l^3$
$O = 12 \cdot 1.72\ l^2$
$V_{pol} = 10.32\ l^2 t$
$\emptyset = 1.35\ \frac{t}{l}$

double truncated cone
$h = 2.22\ l$
$r = 0.75\ l$
$R = 1.35\ l$
$\chi = 28.4°$

FIG. 2A.14. The geometric relationship between the pentagonal dodecahedron and the circumscribed double truncated cone.

An energy analysis is made on the deformed element immediately after buckling, the deformations that occur being related to defined shear forces. Equilibrium and boundary conditions are obtained for the stress function which is in the form of a fourth-order differential equation. Its solution is obtained using a sine and a cosine function respectively. From this the critical compressive stress of a double truncated cone may be expressed as

$$\sigma_k = \left\{\frac{E_p}{3(1-v_p^2)^{1/2}}\right\} \frac{t}{r} \cos x \left\{\frac{1}{2} \ln \frac{s_2}{s_1} \left[\frac{(1+s_1/s_2)}{(1-s_1/s_2)}\right]\right\}^{1/2}$$

where v_p is the Poisson's ratio of the base material. Substituting in the equation the geometric parameters from Fig. 2A.14, the critical buckling load of a cell is

$$F_k \approx 0.905\, E_p \phi^2\, rt/3(1-v_p^2)^{1/2}$$

Since each cell wall is shared by two cells, only one-half of the number of cells are taken as load bearing in the calculation of the compressive strength. Multiplying the critical buckling load of one cell by half the number of cells per unit area one finally arrives at the compressive strength of the foam as

$$\sigma_{CB} \approx 0.2285\, E_p \phi^2\, a/(1-v_p^2)^{1/2}$$

where a takes account of the influence of the specimen preparation, and is assumed to be $a \approx 0.8$.

(b) *Comparison of Predicted and Measured Values*
The compressive strength of the PVC rigid foam can be calculated using the characteristic data of the base material; $E_p \approx 3 \times 10^4$ kgf cm^{-2}, $v_p \approx 0.4$. Predicted and measured values of the compressive strength are compared in Fig. 2A.15. In addition, values for the compressive strength calculated from the pentagonal dodecahedron model[14] and the tetrahedral decahedron[15] are plotted as a function of the density. It can be seen that with the double truncated cone model a considerable improvement is achieved over the other models, that is, a better approximation to the actual behaviour. This confirms the assumption that the deformation and the failure mechanisms of a closed load-bearing system, such as is represented by closed-cell PVC rigid foam, cannot be analysed using excised single elements.

As with PUR rigid foam, the tensile strength over the reduced cross-section has been estimated here. It is seen in Fig. 2A.15 that there is good agreement between the calculated and experimental values.

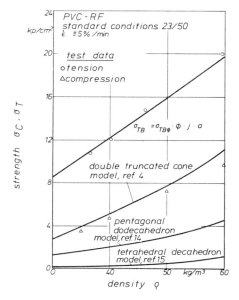

FIG. 2A.15. The strength of PVC rigid foam as a function of density.

2A.2.3 Comments

Since the materials being considered are plastics, whose mechanical properties are time and temperature dependent, it is of interest to examine if these dependencies can also be explained using the model representation. In order to do this it is necessary to know the time and temperature dependence of the properties of the base polymer. If this is so then it would be possible, for example, to investigate if the elastic behaviour of PRF could be represented by an equation of the form

$$E_{(t,\theta)} = C_1 E_{p(t,\theta)} \phi^2/(C_2 + \phi)$$

Since, for the investigations described here, the time and temperature dependence of the properties of the base polymer was not known this could not be done in detail.

Other ways of improving the mathematical prediction of the mechanical properties have been studied.[18] These have included (i) the use of combined models, (ii) the production of statistically valid empirical equations from a large number of measured data and (iii) an optimum model calculation. Previous discussions indicate that the last of these methods, which involves the application of finite element techniques, might prove to be successful.[18]

2A.3 MEASUREMENT OF THE MECHANICAL PROPERTIES OF RIGID FOAMS

2A.3.1 General Considerations

The determination of the characteristic data necessary for design with structural materials presupposes the existence of suitable test procedures. Strength data, with associated strains, and the stress at certain deformations can be determined according to the standard specifications and proposals for the testing of PRF which are valid at this time. Apart from the fact that in the standards no consideration is given to the determination of the elastic properties other characteristic data for dimensional design can only be determined with qualification.

Reproducible results, suitable for dimensional design, can only be obtained if disturbing factors are either excluded or taken into account. Disturbing effects may arise because of: location from which specimen is taken; the geometry of the specimen; the experimental arrangement and the test procedure. The following discussion indicates what difficulties are to be expected in the mechanical testing of PRF and under what conditions useful test results may be obtained.

2A.3.2 Taking of Samples

In addition to having correct test specifications, with PRF the preparation of specimens has to be given special consideration. The dependence of the mechanical properties on density and cell structure has already been described in detail. Because of flow phenomena and the variation of process parameters in the foam blocks or plates, density variations and differences in cell structure, due to flow orientation, occur depending on the foaming process. Differences in material properties are to be expected according to the position from which the specimen is taken relative to the subsequent direction of loading because of differences in cell structure and density. The density distribution and the shape and alignment of the cells has been investigated in blocks and plates of PUR rigid foam and plates of PVC rigid foam along the width and the height directions.[4,6]

The PUR rigid foam was usually in blocks of dimensions 20 × 1·0 × 0·3 m³ and plates of different thickness. A qualitative description of flow orientation, and the cell forms resulting from it in a PUR rigid-foam block is shown in Fig. 2A.16. There are four regions of different cell structure: a homogeneous region with round, model, cells in the centre of the block; elongated cells with varying orientation, in the

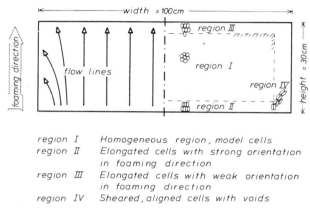

region I   Homogeneous region, model cells
region II   Elongated cells with strong orientation in foaming direction
region III Elongated cells with weak orientation in foaming direction
region IV   Sheared, aligned cells with voids

FIG. 2A.16. Flow orientation and the alignment of cells in a PUR rigid-foam block.

direction of foaming, in the vicinity of the upper and lower boundaries; elongated cells, with many irregularities and voids, deformed as a result of differing flow velocities. Fig. 2A.17 shows the density distribution in

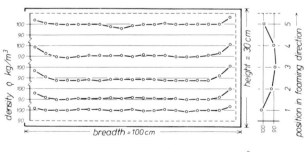

average density $\varrho \approx 100\,kg/m^3$

FIG. 2A.17. The density distribution in a PUR rigid-foam block.

the block studied. To determine the density distribution the compact outer skin of about 1 cm was cut off and the block divided into cubes of side length 5 cm. An increase in density of up to 10% towards the edges is observed across the width and the height of the block.

Plates of PVC of average dimensions $1 \cdot 50 \times 1 \cdot 20 \times 0 \cdot 10$ m^3 with different density were investigated. In contrast to PUR rigid foam no definite statement concerning the distribution of density in the plane of the plates could be made. The density fluctuated by approximately 5% about a mean value.

However, a description of the density distribution, cell size and cell shape in the direction perpendicular to the plane of the plate can be given. The density, cell diameter and cell-orientation factor, which describes the cell shape, are shown in Figs. 2A.18, 2A.19 and 2A.20. In contrast to PUR rigid foam, a decrease in the density towards the edges can be observed. The higher the density the smaller the decrease and the larger the region of uniform density in the central region (Fig. 2A.18). The variation of the mean cell diameter and the degree of orientation as a

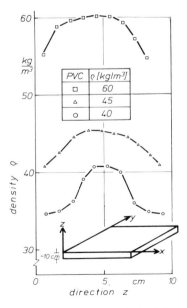

FIG. 2A.18. The density distribution for PVC rigid-foam plates perpendicular to the plane of the plate.

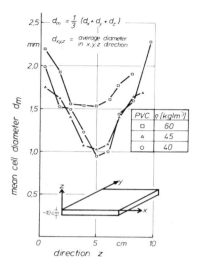

FIG. 2A.19. The distribution of cell diameter for PVC rigid-foam plates perpendicular to the plane of the plate.

FIG. 2A.20. The distribution of the degree of orientation for PVC rigid-foam plates perpendicular to the plane of the plate.

function of the position in the thickness direction, shown in Figs. 2A.19 and 2A.20, are miror images of the density variation. For lower densities, the cell diameter in the outer regions is about twice that in the central region. The highest density foam has a higher orientation which is approximately uniform over the plate thickness whereas with the lower density foam the orientation varies more strongly from a low degree of orientation in the centre towards the edges.

The degree of orientation, OD, of a cell is the ratio of the maximum cell dimension perpendicular to the plane of the plate to the maximum cell dimension in the plane of the plate.

$$\mathrm{OD} = (d_z)_{\max}/(d_{xy})_{\max}$$

The mean diameter, d_m, is calculated as

$$d_m = (k/3)(2 + \mathrm{OD})\sqrt{A/n}$$

where n is the number of cells in a surface which has area A and k is a statistical coefficient which takes account of the random way of sectioning. For approximately spherical cells this was determined[4] as $k = 4/\pi \approx 1\cdot27$.

With PVC rigid foam, irregularities in the cell structure lead one to expect differences in the mechanical properties for different directions of loading. Furthermore, the PRF in the boundary regions will have different mechanical properties from that in the central region. These studies show the scatter in the mechanical data that is likely to occur because of the variation in the cell structure and the density as well as the direction of loading and the size of the specimen. In principle, the material should be stressed during testing in the same way as it will be when it is used as a load-bearing structure.

2A.3.3 Test Specification

From the observations described in the preceding section it is concluded that the size and shape of the specimen must be considered when determining the mechanical properties since the changes in structure, and the density variation within a test piece, will affect the measured values. It is also known that there are general relationships between specimen geometry and material parameters. Consequently with PRF both aspects have to be taken into account. This means that on one hand the specimen should have a homogeneous material structure and on the other, a geometry that will enable an accurate recording of the measured values. The complex problem of an optimum specimen geometry for compression, tension, shear and bending deformations is considered below.

(a) *Compression*
According to DIN 53 421, the compressive strength, and associated fracture strain, for a PRF should be determined on specimens having a square base with a side length of 5 cm. The height of the specimen should be 0·5 to 1·0 times the base dimension.

From work done on the testing of concrete[19] there is a known relationship between the compressive strength of the concrete and the ratio of sample height to sample thickness. This ratio will be referred to as the specimen slenderness ratio. The compressive strength decreases with increasing specimen slenderness ratio. This behaviour is explained as follows.[19]

In the region of the applied force, the transverse strain is restricted because of friction between the pressure plate and the specimen. For small specimen heights, where the regions of restricted transverse strains overlap, this results in an apparently greater material strength. Definitive material data will only be obtained from tests on larger specimen heights where there is a state of stress without restricted transverse strain in the central one third of the specimen. For this range of slenderness ratio, the sample height might be considered as an optimum. If, as is the case with the testing of concrete, the ratio of the cube strength to prism strength of PRF is known, the compressive strength can be determined using cubes and this value reduced by a geometric factor.

(i) *Specimen geometry*

An optimum specimen geometry has been established for PRF[6,20] where base dimension as well as the slenderness ratio were varied. The compressive strength, yield strain and elastic modulus were measured. In the determination of the tangent modulus the cross-head movement, i.e. the total deformation, during buckling was used as well as the measured deformation in the central region. The latter method however could only be employed for specimens with height, h, larger than a certain value.

The base side length, d, of the samples was varied from 4 to 10 cm and the ratio h/d from 0·5 to 4·0. The results are shown in Fig. 2A.21, from which it can be seen that the compressive strength is independent of the height and the base area of the specimen. However scatter of the test results decreases with increasing base area because for larger cross-sections, random inhomogeneities have less effect. Also the compressive strain at buckling, ϵ_B, measured in terms of cross-head movement decreases with specimen height. On the other hand, the elastic modulus determined from these measurements increases with specimen height.

The dependence of the elastic properties on the specimen slenderness ratio can be attributed to the following disturbing effects in the region of the applied force: (a) cut cells in the specimen surface; (b) non plane-parallel cut specimens or non-parallel pressure plates; (c) an unknown multi-axial state of stress and strain due to friction between the pressure plates and the specimen surface; (d) mechanical play in the test equipment. Added to this is the density variations over the height of the specimen.

The elastic behaviour of PUR specimens can be explained with the aid of the model shown in Fig. 2A.22. The regions adjacent to the applied force are equivalent to springs of low stiffness, and the central region, where there is a defined state of stress, corresponds to a spring of higher stiffness. Since the effect of the disturbed regions on the total deformation decreases with increasing specimen height, smaller characteristic

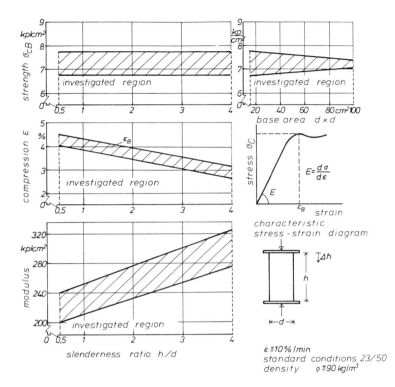

FIG. 2A.21. The mechanical properties of PUR rigid foam as a function of specimen geometry for compressive loading.

compressive strain and hence higher modulus of elasticity, will be obtained for larger specimen heights. If the slenderness ratio is too large there is also the possiblity of the transverse displacement of the entire sample. The size of the disturbed region depends upon, amongst other things, the cell diameter.

According to the results shown, to obtain the highest possible accuracy in the measurement of the characteristic data the largest possible base area should be used so as to reduce the scatter in the compressive strength. With a slenderness ratio of about 4, or a height of about 20 cm, the elastic properties can also be determined with reasonable accuracy by measuring the total deformation. Since such large PRF specimens can rarely be made, and in any case they would be uneconomical, it is

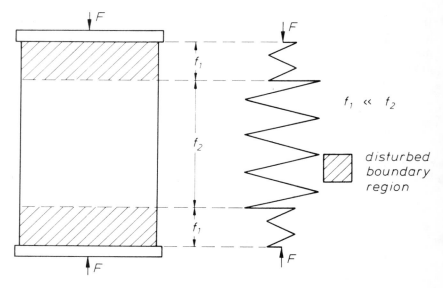

FIG. 2A.22. The two-spring model for the compression test.

necessary to consider other possibilities, particularly with regard to the measurement of deformation.

(ii) *Measurement of deformation*

An improvement of about 10% can be achieved by a pre-treatment of surfaces at which the forces are applied. This was done by applying a tough resin to both faces and grinding the faces plane-parallel after hardening.

However, it makes more sense, especially for the determination of the elastic modulus, if the deformation in the central, undisturbed region of the specimen is measured. These measurements were made using inductive displacement transducers, which were attached to the specimen, as well as a contact-free electro-optical measuring device. In order to compare these measurements with those of the cross-head movement, all measurements were made simultaneously on each specimen, having $h/d = 2$ and $d = 5$ cm. The elastic modulus in the undisturbed region measured in this way was about 25% larger than that obtained using the total deformation. This means that for a specimen height $h = 2d$ the modulus obtained by measuring the central region is about the same as that given by measurement of the cross-head movement on a specimen of height $h = 4d$. The moduli of elasticity, for a particular specimen geometry, determined according to the different measurement techniques are shown in Fig. 2A.23.

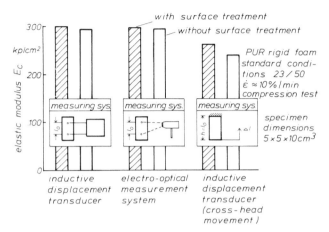

FIG. 2A.23. Comparison of the elastic moduli given by the different strain-measuring techniques.

The influence of gauge length is shown in Fig. 2A.24. According to these results the elastic properties of a specimen of height $h = 10$ cm are independent of the gauge length up to a gauge length of approximately 8 cm. The disturbing boundary effects decrease to zero within a boundary zone of 1·0 to 1·5 cm. The scatter of the measured values decreases with increasing gauge length. In the range $1·0 \leq \dot{\epsilon} \leq 10\%$ min^{-1} the elastic modulus was independent of the strain rate. A slightly smaller value was obtained with strain rates below $1·0\%$ min^{-1}.

(iii) *Test procedures*

As specified in DIN 53 421, cube-like specimens of side length 5 cm are suitable for the determination of compressive strength.

For determining the elastic modulus, a prismatic bar specimen with square base of side length 5 cm should be used. When the deformation is measured in the central region, the specimen height should be 10 cm. If the modulus of elasticity is to be determined from the total specimen deformation the specimens would have to be about 20 cm in order that the measured value of the modulus is not too low. With cube shaped speci-

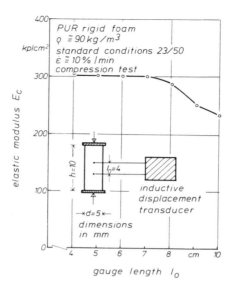

Fig. 2A.24. The influence of the gauge length on the elastic modulus in the compression test.

mens of PUR rigid foam the modulus of elasticity in compression determined using this last method was too low by about 30 to 50%.

(b) *Tension*

According to DIN 53 430 the tensile strength of a PRF, together with the corresponding fracture strain, is determined on shoulder specimens. In one case, they are clamped and in another they are suspended in the shoulder region on two threaded rods. In the second case, shaped metal plates should be attached to both sides in the region of the application of the force. Here again, as with the compression test, no indication as to the determination of the elastic properties are given.

The problems associated with specimen size, strain measuring method and speed of deformation are the same as those in the compression test so they will not be considered further. However, in tension tests special consideration must be given to the application of the force. In order to obtain an unambiguous state of stress in the gauge length, the specimen must be loaded along the central axis so that it is free of any moment. Furthermore, for an accurate tensile test, failure should occur within the gauge length.

A specimen geometry different to that proposed in DIN 56 430 was used in this work because if the DIN proposal is followed high contraction ratios can only be achieved with difficulty, since the specimens have a shoulder in one plane only. For the tension test a prismatic bar with square base and a side ratio $h/d = 2$ was chosen as the basic body. A tension specimen, having shoulders on all sides, is milled from this prismatic bar. Suitable tests can be carried out provided there is a large enough contraction ratio and a uniform transition in the shoulder region. By definition, the contraction ratio is the ratio of the cross-section area A_R in the region of the applied force and the cross-section area A_M in the measurement region.

For the PUR rigid foam studied, a contraction ratio of $A_R/A_M \approx 2 \cdot 5$ was adequate, whereas for the PVC rigid foam it was increased to about $6 \cdot 0$. These differences were due to the varying density distribution over the specimen height and the differing cell diameters. A circular transition in the shoulder region with a radius of between 1 and 2 cm proved to be useful. The force is applied using end-wise steel T profiles. These profiles are stuck on in such a way that an articulated suspension is effective in both directions. The experimental arrangement is shown in Fig. 2A.25.

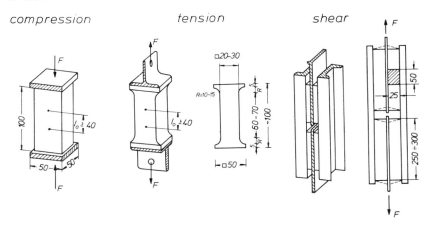

FIG. 2A.25. Specimen shape and test equipment for the short-term tests.

(c) *Shear*
The shear strength of PRF can be determined according to DIN 53 422 as well as DIN 53 427. Whereas DIN 53 422 deals with the punch-through test on circular specimens in DIN 53 427 the shear strength between two

steel plates is determined. When PRF is used as a core material in sandwich construction the determination of structural design parameters makes sense only according to the last mentioned standard. Hence only these experimental arrangements will be considered. Apart from the specimen geometry, the experimental arrangement itself plays a more important rôle than in compression and tension tests.

The shear specimen specified in DIN 53 427, with length (l) $25 \cdot 0 \pm 0 \cdot 05$ cm, width (b) $5 \cdot 0 \pm 0 \cdot 05$ cm and thickness (h) $2 \cdot 5 \pm 0 \cdot 05$ cm, represents a very short bar clamped on both sides and stressed in shear and bending.[21] In the test the clamped sides are moved relative to each other. The contribution due to bending is neglected in the calculation of the breaking stress, shear deformation and shear modulus. The error resulting from this simplification will be neglected for test specimens with dimensions as specified in DIN 53 427.

(d) *Bending*

According to DIN 53 423, the bending strength, ultimate bending stress and the deflection at failure is determined for specimens having dimensions $l = 12 \cdot 0$ cm, $b = 2 \cdot 5$ cm and $h = 2 \cdot 0$ cm. The results obtained from DIN 53 423 are of limited use even for comparative purposes, more so than the previously discussed specifications. With the suggested ratio of 5 for the support width to specimen thickness, the influence of shear on bending is neglected, although it is of the order of 10%. Moreover, in determining the bending strength, a linear stress distribution and equal elastic behaviour for compressive and tensile stress is assumed. Furthermore, for small support width relative to specimen thickness, local deformation of the foam is to be expected because of the strong forces.

For these reasons the bending test according to DIN is in no way suitable for the determination of material parameters and has only limited usefulness for comparative studies.

Comparative bending tests have been carried out on PVC rigid foam with different densities. Figure 2A.26 shows a comparison, as a function of the density, for different modes of loading. The modulus of elasticity obtained from the three-point bending test is less than the compressive modulus, although it should actually be between the compression and tensile moduli, because the influence of shear has been neglected. Also the bending strengths are not compatible with the compression and tension tests.

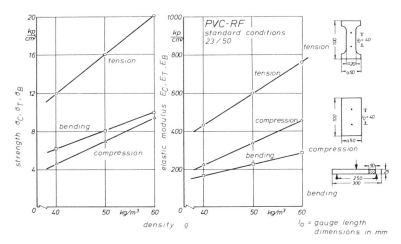

Fig. 2A.26. The mechanical properties of PVC rigid foam as a function of density under compressive, tensile and bending loading.

2A.4 MECHANICAL PROPERTIES

A systematic determination of the mechanical parameters necessary for the structural design of PRF parts has not been carried out. Although the behaviour under compression has been described many times in the literature, there has been little discussion of the behaviour in tension and shear. Direct comparison of data is difficult because of insufficient information concerning cell structure, chemical composition of the base material, the manufacturing process, specimen geometry and test method. Furthermore, fairly large differences in the parameters occur because of the different test conditions and foaming conditions. Even with results that are comparable in some respects, there are differences in the dependence of the mechanical properties on the density because of scatter. Although parameters such as the tensile and shear strengths and the elastic properties for all modes of loading are required comparatively reliable data are only available for the compressive strength. Tests aimed at determining systematically the characteristic mechanical properties as a function of density, time and temperature have recently been reported.[22] These are summarised below. Further details of these tests and the experimental results can be obtained from the original publication.

2A.4.1 Short-Term Investigations

As a result of the investigations described in Section 2A.3 for optimising specimen geometry and test procedures, the specimen shapes and test equipment shown in Fig. 2A.25 were employed for the short-term tests. The tests were run on a universal test machine in line with DIN 51 221 (Class 1) according to DIN 51 220. An electrical strain sensor was used as a displacement transducer. In tensile and compressive tests the knife edges of the sensor were pressed onto both sides of the specimen. In the shear tests they were fastened to the central steel plates. All tests were run strain controlled, and in order to complete the test within about one minute the rate of deformation was in the range $5\% \leq \dot{\varepsilon} = \dot{\gamma} \leq 20\%$ min^{-1}. In the tensile and compressive tests the gauge length was usually 40 mm. In the shear test the relative displacement of the force applying elements was measured.

In order to determine accurately the effect of temperature on the mechanical behaviour the temperature must be uniform throughout the test specimen. The heating of the specimens was determined according to the Bender–Schmidt method.[23] For compression specimens of the two rigid foams studied the maximum heating times were found to be; PUR-RF, 20–80°C ≈ 80 min, PVC-RF, 20–80°C ≈ 40 min.

(a) *Stress–Strain Diagrams and Elastic Behaviour under Tensile and Compressive Loading*

Characteristic stress–strain diagrams for PVC-RF and PUR-RF, having a density of about 60 kg m^{-3}, in a standard climate (23/50) are shown in Fig. 2A.27 for the three types of loading. It can be seen that for both foams, the strength and the elastic properties decreased in the order tension, compression, shear. For the same density the PVC-RF exhibits the better properties. Although both PRF show approximately the same characteristic behaviour in compression and tension, differences are apparent for shear loading. The shear stress–strain diagram of PVC-RF can be divided into two parts exhibiting significantly different stress–strain behaviours. The first region, having a relatively high modulus, is followed by a region where there is a large increase in strain for a small increase in stress. The transition between the two regions can be specified quite accurately. For the PUR-RF the behaviour was different, a small quasi-linear range followed by gradual transition to an increasingly smaller modulus being observed. These differences were detected with all of the densities and temperatures investigated.

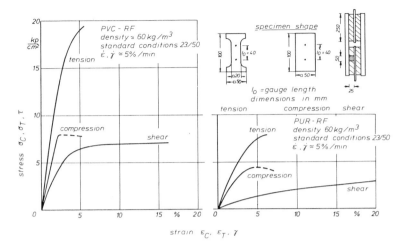

FIG. 2A.27. Characteristic stress–strain diagrams for PVC and PUR rigid foams under compressive, tensile and shear loading.

(b) *Elastic Behaviour Under Compressive and Tensile Loading*
Under normal test conditions it has been observed that the mechanical behaviour of PRF under compressive loading is different to that under tensile loading. Although the differences in strength are easily explained, in terms of material failure in tension and instability failure of the cell walls in compression, the elastic behaviour at small deformations should be the same.

In a special series of tests with PUR-RF an attempt was made to find the characteristic limits of strain for compressive and tensile loading up to which there is a quasi-linear relationship between σ and ϵ, and as a result determine a modulus, and to find, if possible an explanation for the differences in elastic behaviour in tension and compression. To do this, limits of linearity were sought in terms of the first-time permanent material change using the energy-absorption method.[24]

In addition, alternating compression–tension tests were used to show if discontinuities occur in the σ–ϵ behaviour on transition from the compression to the tension mode. In these cyclic tests the compressive/tensile specimen was of the same shape as the tensile specimen shown in Fig. 2A.25. These were fitted into a special reversing jig which gave a positive transmission of force in both deformation modes. The same arrangement was used for the work input tests.

The stress–strain diagrams in Fig. 2A.28 are original curves from a tension–compression test. It can be seen that at the origin, the compressive and tensile moduli have approximately the same value. However, if the initial modulus is determined graphically from such a test record by laying a tangent to the line, differences between tensile and compressive moduli would be observed.

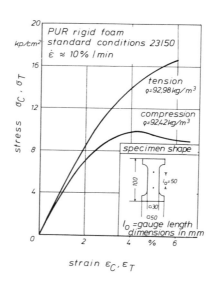

FIG. 2A.28 Stress–strain diagrams (original curves) for PUR rigid foam under compressive and tensile loading.

Because of greater measuring accuracy, a better description of the actual stress–strain behaviour can be obtained from the σ–ϵ curves provided by the energy-input tests (Figs. 2A.29 and 2A.30). It can be seen from Fig. 2A.29 that in compression two regions of quasi-linear behaviour can be defined. The actual initial modulus, E_0, is about 10% higher than the modulus E_1 in the second linear region. In the normal test the modulus E_1 is usually measured because the first linear region cannot be observed.

In tensile testing, a similar error in measurement is made, as can be seen from Fig. 2A.30, which is for a PUR rigid foam of lower density. The quasi-linear region, close to the origin, is followed by a region of higher

STIFFNESS AND STRENGTH—RIGID PLASTIC FOAMS

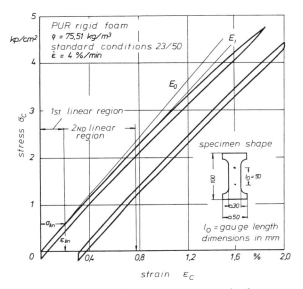

FIG. 2A.29. The determination of the damage energy in the compression test.

σ–ϵ gradient so that here also there are two possibilities of constructing a tangent. However, in this case E_0 is smaller than E_1. Here, the limit of the second linear region is defined as being where the tangent E_0 intersects the curve. This region disappears with increasing density. It was not possible to detect it with a density of approximately 90 kg m^{-3} using the measuring technique available.

The limits of the first permanent material change were determined using the energy-absorption test. These are the limits up to which the strain energy expended during stressing is recovered, except for the damping energy which is converted into heat. In cyclical loading and unloading tests at constant strain rate, $\dot{\epsilon}$, it may be assumed that the specimen remains undamaged in its internal structure as long as the area enclosed by the σ–$f(\epsilon)$ curve remains constant for a number of sequential deformation cycles. If, during the first cycle, damage occurs, resulting in fracture surfaces, irreversible stretching of the cell structure and collapse of the cell rods, then the area will be larger than in subsequent cycles in proportion to the amount of energy expended on the damage. Thus the damage limit, defined as the maximum strain at which a specimen just shows no damage, is obtained by cyclically loading and unloading specimens with different maximum strains.

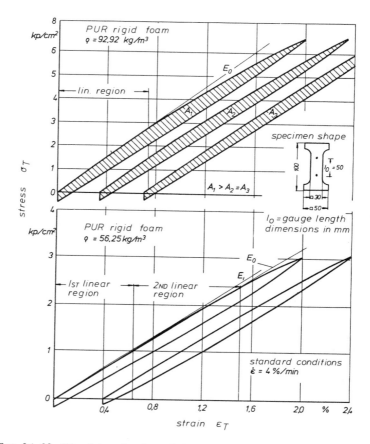

FIG. 2A.30. The determination of the damage energy in the tensile test.

As an example, it is shown in Fig. 2A.30 for a density of 92 kg m^{-3} that the area enclosed by loading and unloading in the first cycle is larger than subsequent cycles; that is the damage limit has been exceeded. The limits for the first linear region determined in this way are shown in Fig. 2A.31. In tension they are independent of density, whereas in compression they increase with density. The limits for the second linear region on the other hand decrease in both cases.

The moduli in the first region are shown as a function of density in Fig. 2A.32. It can be seen that within the scatter band the tensile and compressive moduli agree quite well. If this modulus, which is the same

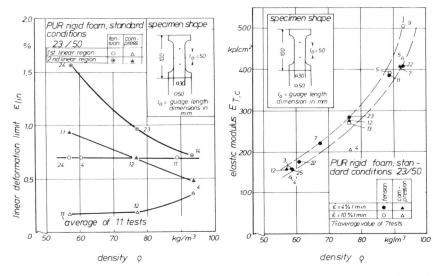

FIG. 2A.31. Linear deformation limits of PUR rigid foam for compressive and tensile loading as a function of foam density.

FIG. 2A.32. The elastic modulus of PUR rigid foam in the first linear region for compressive and tensile loading as a function of foam density.

in compression and tension, is compared with that given in Fig. 2A.33, which is different in tension and compression, it is seen that the actual initial modulus lies between the tensile and compressive moduli given by the less accurate test. In practice differing moduli should be expected because of the difficulties with measuring an accurate value and because the first linear region in compression is so small that it is of no interest for the application to design work.

The differing deformation behaviour in tension and compression can be interpreted qualitatively as follows: basically the total deformation of the foam is composed of two parts, one part due to the normal stresses and the other due to bending. The contribution from bending decreases with increasing density because the stiffness of the rods increases. In tension, particularly with low densities, the first linear region is followed by a region of higher modulus, which can be attributed to the alignment of the load carrying cell elements in the direction of loading. This alignment is assisted by cracks and damage at the nodal points where, particularly with low densities, an increased notch effect is to be expected

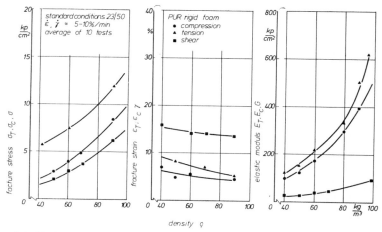

Fig. 2A.33. The mechanical properties of PUR rigid foam as a function of density under compressive, tensile and shear loading.

because of the smaller transition radii between rod and node. The modulus decreases again only when, after further stressing of the rods, the specific material limits of quasi-linear stress–strain behaviour is reached. The fact that in tension these two characteristic deformation regions merge with increasing density is qualitative proof for this assumption.

In compressive loading, on the other hand, alignment of the load-carrying cell elements in the direction of loading does not occur. Instead, because of eccentricities and consequently higher bending stresses, the compressively stressed rods begin to deform in a direction transverse to the direction of loading. This occurs earlier the lower the density and the more slender the rods. These pre-strained rods, in contrast to the ideal Euler rods, continue to show quasi-linear behaviour, although with a smaller modulus. This manifests itself, especially with low densities, in a distinct second linear region. Here too, both regions merge as the density is increased and, above a certain density, pure material failure is to be expected in compression. However, at low densities the load bearing limit of PUR-RF will be determined by the unstable failure of the rods.

(c) *Mechanical Properties as a Function of Density and Temperature*
The influence of density on the fracture stress, fracture strain and moduli for compressive, tensile and shear loading is shown for PUR-RF

in Fig. 2A.33. Although the strength and moduli increase with density, to varying degrees, a small but measurable decrease in the fracture strain is observed. Similar behaviour is observed for PVC-RF, but in the same density range the increase in strength and modulus is larger than that of PUR-RF.

It is clear that there are differences between uniaxial and multiaxial stress states as with solid materials. However, as described above the measured differences between the compressive and tensile moduli are, strictly speaking, due to the accuracy of measurement. The modulus given here is for the second linear region since the initial compression modulus holds only over a very limited range of loading.

Short-term tests at four different temperatures have been carried out for three densities of both of the rigid foams. It was found that the characteristics of the individual stress–strain curves remained essentially the same, as described above, and that in all cases there was a decrease in the mechanical properties with increasing temperature. However, in another series of tests[23,25] an increase in the compression modulus of PVC-RF and PUR-RF at elevated ambient temperature (up to $\theta = 50°C$) was observed and occasionally for shear loading also. For the PUR-RF the cause of this may be due to post-curing if the foam is tested relatively soon after manufacture. It has been shown[20] that when PUR-RF is stored under normal conditions, its mechanical properties increase measurably for a period of up to about ten weeks. With PVC-RF it is possible that annealing occurs within the cell structure which had been highly stretched during the sudden expansion. As a result, the load-bearing cell elements can take-up a somewhat more stable position, and contingent on this, exhibit a better mechanical behaviour. Such behaviour was observed only occasionally in these investigations so that generally speaking a fairly definite dependence of the mechanical properties emerged. The variation of the mechanical parameters of PUR-RF is shown as a function of temperature in Fig. 2A.34. In all cases the strength and the modulus decreased with increasing temperature. This was also found for PVC-RF. In general the scatter in the value of the elastic modulus and the fracture strain was of the order of ± 10 to 15% whereas the fracture stresses deviated from the average by about ± 5%.

For a particular rigid foam, the decrease in the strength and the elastic properties is of the same order of magnitude for the three modes of loading, the temperature dependence of thermoplastic PVC-RF being larger than that of PUR-RF. Although the fracture strain for PUR-RF is approximately constant, the embrittlement being of the same order of

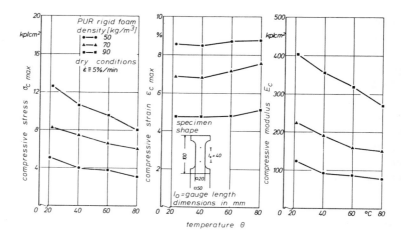

FIG. 2A.34. The mechanical properties of PUR rigid foam as a function of density and temperature under compressive loading.

magnitude as the decrease in the modulus, the breaking strain for PVC-RF increased measurably over the entire temperature range.

2A.4.2 Long-Term Investigations

In the long-term investigations the deformation behaviour under compressive, tensile and shear loading has been measured for both PRF as a function of time at different ambient temperatures and for different densities. The specimen shapes corresponded to those in the short term tests. The test arrangement is shown in Fig. 2A.35; for tension it was

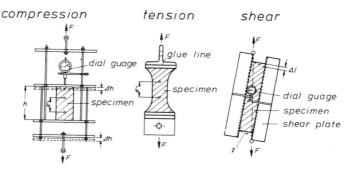

FIG. 2A.35. Test equipment for the long-term tests.

the same as that for the short-term tests, and for compression a reversing mounting was used. In contrast to the short-term tests, the shear measurements were carried out in accordance with ASTM C 273–61. The specimens were stored in heating cabinets at the appropriate elevated temperature for about 12 hours before testing. In all compression and shear tests, the deformation measurements were carried out by means of dial gauges attached to the test rig, as shown in Fig. 2A.35. In some cases the deformation in the compression test was also measured optically using a cathetometer over a gauge length l_0 marked by two points. This was done so that the two methods of measurement could be compared and the dial gauge reading related to the actual deformation in the central region of the compression specimen. Deformations in the tensile tests were usually measured by means of the cathetometer.

A series of test programmes designed to determine the creep behaviour of PVC-RF and PUR-RF for different densities, temperatures and selected modes of deformation have been carried out.[26] Some of the results of these tests are discussed briefly below.

(a) *Deformation–Time Curves*

Deformation–time curves for PVC-RF under compressive loading are shown in Fig. 2A.36. Although the shape of the curves were different for the different modes of loading, these results are reasonably representative

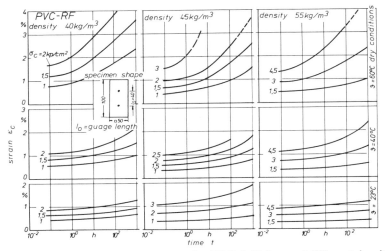

FIG. 2A.36. Deformation–time curves for PVC rigid foam of different densities under compressive loading at different temperatures.

of the behaviour of the two foam systems studied. At comparable stresses the instantaneous deformation and the tendency to creep increase with decreasing density and increasing temperature for all modes of loading.

Little can be said about the nature of the creep from these results because, as mentioned above, the deformation of cellular materials is made up of pre-buckling deformation, the deformation of the cell elements due to normal stresses and the bending deformation of the cell structure skeleton. This makes the characterisation of the long-time deformation behaviour even more difficult than with unfoamed materials. One, so far unexplained, phenomenon observed with PUR-RF under tensile loading should be mentioned. For two low-density foams at a temperature of 60°C an initial period where the strain increased by only a small amount was followed by a region of progressively increasing strain. It will be necessary to carry out structural investigations on loaded specimens in order to explain this behaviour.

(b) *Creep Modulus Curves*

In order to establish the time-dependent elastic behaviour, the initial modulus was obtained from the isochronal at the origin of co-ordinates. The creep moduli for PVC-RF are plotted on a semi-logarithmic scale in Fig. 2A.37. In general it can be stated that, irrespective of the density

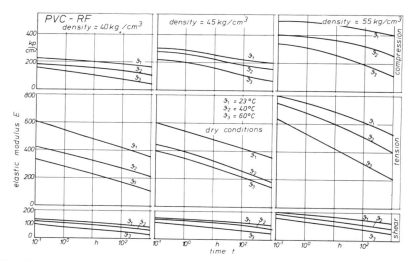

FIG. 2A.37. Creep modulus curves for PVC rigid foam for different densities and temperatures under compression, tension and shear loading.

and the mode of loading, the modulus decreases very much more rapidly with time at higher temperatures. Even in the cases where the curves are running parallel, such as with the foam of density 45 kg m^{-3} for example, the relative decrease in the modulus is larger at the higher temperatures. With specific reference to PVC-RF of densities 40 and 45 kg m^{-3} it can be observed that the room-temperature tensile moduli are about the same. However differences arise with increasing temperature such that the modulus for the higher density decreases more rapidly than that for the lower density at the intermediate temperature. Such peculiarities show, by way of example, what scatter can be expected in the determination of the mechanical properties of commercially manufactured foams.

REFERENCES

1. MULLER, D. (1970) Zum Deformationsverhalten harter Kunststoffschäume Dissertation Technische Universität Hannover, Archivnr. B. 3.45.
2. STASTNY, F. (1969) Polystyrol-Schaumstoffe, Bedeutung des Gebietes, pp. 668–673; Herstellung von Schaumstoffen aus blähfähigen PS-Partikeln, S. 697–737; Eigenschaften von Polystyrolschaumstoffen, pp, 738–757; Vieweg/Daumiller, Kunststoffhandbuch Bd. V.
3. MULLER, D. (1970) Das Verhalten von Kunststoff-Hartschäumen bei kurzzeitiger Druckund Biegebelastung Gummi, *Asbest, Kunst.*, **23**, 740–744.
4. ZIMMERMANN, H. (1974) Diplomarbeit am IKV, Betreuer: F. Knipschild, Archivnr. D 74/38.
5. KREFT, H. and WAGNER, D. (1969) Mechanische Eigenschaften von Schaumstoffen im Temperaturbereich von 300° K bis 20° K. *Kältetechnik-Klimatisierung*, **21**, 258–265.
6. EVERS, K. (1973) Diplomarbeit am IKV, Betreuer: F. Knipschild, Archivnr. D 73/18.
7. MOOS, J. and WEHMEYER, H. P. (1973) Studienarbeit am IKV, Betreuer: F. Knipschild, Archivnr. W 73/46.
8. MENGES, G. and KNIPSCHILD, F. (1974) Vorschlag zur Abschätzung der mechanischen Eigenschaften von Kunststoffhartschäumen auf der Basis einer Modellvorstellung, *Plasticonstruction*, **4**, 205–211; (1974), Berichte zum 7. Kunststofftechnischen Koloquium Aachen, pp. 52–57; (1974). Proposition to the Estimation of Mechanical Properties of Rigid Plastic Foams Based on a Model-Conception. Society of Plastics Engineers 32nd Annual Technical Conference, San Francisco, California, pp. 321–325; (1975). *Polymer Engineering and Science*, **15**, 623–627.
9. DEMENTEV, A. G. and TARAKANOV, O. G. Model Analysis of the Cellular Structure of Plastic Foams of the Polyurethane Type. UDC 678.5.539.32, pp. 744–749.

10. GROSSKOPF, P. and WINKLER, Th. (1973) Auslegung von GFK-Hartschaum-Verbundwerkstoffen, *Kunststoffe*, 881–888.
11. MATONIS, V. A. (1964) Elastic Behaviour of Low Density Rigid Foams in Structural Applications, *SPEJ*, 1024–1030.
12. GENT, A. N. and THOMAS, A. G. (1973) Mechanics of Foamed Elastic Materials: Society of Plastic Industry, Cellular Plastics Division Annual Technical Conference Proceedings, Section 7.
13. DEMENTEV, A. G. and TARAKANOV, O. G. (1972) Die Besonderheiten des Einflusses der Zellstruktur auf die mechanischen Eigenschaften der Schaumstoffe, *Mechanika Polimerov*, **8**, 976–981.
14. CHAN, R. and NAKAMURA, M. (1969) Mechanical Properties of Plastic Foams. The Dependence of Yield Stress and Modulus on the Structural Variables of Closed Cell and Open Cell Foams, *Jour. of Cell. Plast.*, 112–118.
15. DEMENTEV, A. G. and TARAKANOV, O. G. Effect of Cellular Structure on the Mechanical Properties of Plastic Foams. UDC 678.405.8, pp. 519–525.
16. SCHNELL, W. and SCHIFFNER, K. (1963) Experimentelle Untersuchungen des Stabilitätsverhaltens von dünnwandigen Kegelschalen unter Axiallast und Innendruck, Bericht Nr. 243 (243 Za 2496) Deut. Versuchsamt f. Luft-und Raumfahrttechnik.
17. SCHNELL, W. (1962) Die dünnwandige Kegelschale unter Axial-und Innendruck, *Z. Flugwiss.*, **10** 154–159, 314–321.
18. STEINHARD, U. Diplomarbeit am IKV, Betreuer: F. Knipschild, Archivnr. D 75/40.
19. LEWANDOWSKI, R. Beurteilung von Bauwerksfestgkeiten an Hand von Betongütewürfeln und Bohrproben, Dissertation TH Braunschweig.
20. ROTH, E. and KNIPSCHILD, F. (1973) Unveröffentlichte Arbeiten am IKV.
21. VÖLKER, A. (1972) Quasistatische und dynamische Verformungsuntersuchungen am Hartschaumstoff aus Polyurethan u. Polyvinylchlorid, Dissertation an der Universität Karlsruhe.
22. KNIPSCHILD, F. (1975) Ein Beitrag Zur Abschätzung and Ermittlung Mechanischer Eigenshaften von Kunststoffhartschaumen, Institut für Kunststoffverarbeitung (IKV), Aachen.
23. BÄCKERS, R. (1975) Studienarbeit am IKV, Betreuer: F. Knipschild, Archivnr. SA 75/05.
24. VERREHT, R. Diplomarbeit am IKV, Betreuer: F. Knipschild, Archivnr. D 74/69.
25. SCHLEICHER, F. (1955) Taschenbuck für Bauingenieure Zweite Auflage, Band 1, S. 217, Springer - Verlag.
26. BRUMMER, H. (1975) Studienarbeit am IKV.

Chapter 2B

STIFFNESS AND STRENGTH— FLEXIBLE POLYMER FOAMS

N. C. HILYARD
Department of Applied Physics, Sheffield City Polytechnic, Sheffield, UK

NOTATION

A	Cross-section area of cell rod
D	Diameter of dead volume
E	Elastic modulus
F	Force
$F(\epsilon)$	Shape function (Rusch)
I	Area moment of inertia of cell rod
N	Number of cell rods per unit volume
d	Cell dimension
$f(\epsilon)$	Shape function (Gent–Thomas)
l	Free length of cell rod
l_0	Free length of undeformed cell rod
p_u	Probability distribution function
r	Distance between dead volumes
r_0	Initial distance between dead volumes
t	Side length of cell rods (cubic model)
β	D/l_0, t/l_0
ε	Tensile, compressive strain
ε'	Fractional deformation of a cell rod
ζ	Volume fraction of open cells
Θ	Angle
λ	Rod extension ratio
ν	Poisson's ratio
σ	Stress
ϕ	Volume fraction of polymer, angle

Subscripts

1, 2, 3	Principal directions in foam
25	25% deformation
f	Foam
g	Gas
p	Matrix polymer

2B.1 INTRODUCTION

At small strains the deformation mechanisms in flexible polymer foams are essentially the same as those in rigid plastic foams. Consequently much of the discussion given above for rigid foams regarding the influence of the mechanical properties of the base polymer, foam density, cellular structure, cell size and structural anisotropy is relevant to flexible systems. However, with flexible foams other factors must be considered. Two of the most important are that these materials are often subjected to large strains and their use performance is sometimes strongly related to the shape of the stress–strain diagram. This is particularly relevant to deformation in the compression mode.

Relatively few systematic studies of the stress-strain behaviour and the elastic properties of flexible foams have been reported in the literature, and our understanding of these materials is still incomplete. However, the theoretical analyses that are available enable us to give a reasonably accurate interpretation of the observed mechanical properties. Most of the reported data are concerned with the behaviour in the tensile and compressive deformation modes, little information being available for deformation in shear.

As an introduction to this discussion of the mechanics of flexible foams it is useful to compare the stress–strain behaviours in the three deformation modes. Stress–strain diagrams for a low-density reticulated PUR flexible foam with volume fraction of polymer $\phi \approx 0.03$ are shown in Fig. 2B.1. Although these allow a general description of the behaviour they should not be considered as being typical of all flexible foams. This is particularly true of the compressive stress–strain diagram the shape of which is highly dependent on the details of the cell structure geometry.

With most flexible foams under tensile loading, the stress initially increases approximately in proportion to the strain so that it is possible to

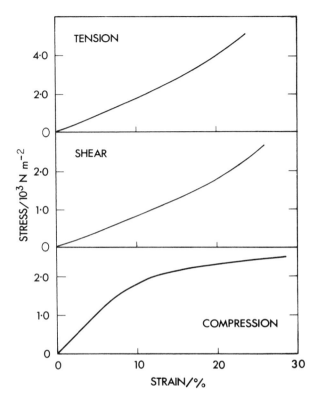

FIG. 2B.1. The stress–strain behaviour of a low density PUR flexible foam in tension, shear and compression.

determine a tensile modulus with reasonable accuracy. The value of the proportional limit varies according to the type of foam. For PUR flexible foams Rusch[1] has reported a tensile proportional limit of about 10%, which is the case for the data of Fig. 2B.1. Gent and Thomas[2] found for latex rubber foams that the tensile stress–strain diagram was rectilinear for strains exceeding 10%. The shape of the stress–strain diagram in shear is similar to that in tension but for the data of Fig. 2B.1, the proportional limit was larger than that for tensile loading. In compression the stress initially increases approximately in proportion to the strain. The proportional limit depends upon the detailed cell structure geometry and usually lies in the range 4–10%. This, almost Hookean region, is followed by a region in which the strain increases rapidly with little or no increase in the stress. The slope of the stress–strain diagram in this plateau region is strongly dependent on the cell structure. Of particular interest, from a fundamental point of view, is the relationship between the initial tensile and compressive moduli. This has been discussed in some detail by Menges and Knipschild (Chapter 2A) for rigid foams. Such a detailed study for flexible foams has not been reported. However, Rusch[1] has investigated this relationship in flexible foams and his results indicate that the two moduli have the same value.

In the second part of this Chapter we consider the relationship between the stress–strain behaviour and elastic properties of flexible foams and their cellular structure and the mechanical properties of the matrix polymer. We are concerned primarily with low-density foams. The geometry of the cell structures and the models used to represent these structures have been described in Chapter 1. It is assumed in the following discussion that the deformation is quasi-static so that rate effects and phenomena associated with the flow of fluid through the cellular matrix can be ignored. These effects are considered in Chapters 4 and 5. It is often observed in compression tests that at small strains the stress–strain diagram is curvilinear because of the incorrect alignment of the pressure plates with respect to the surfaces of the specimen. Following this region is a rectilinear response, the slope of which can be taken as the tangent modulus of the foam. Unless otherwise stated, the compressive moduli discussed below are the tangent moduli.

2B.2 TENSILE ELASTIC MODULUS

2B.2.1 Structural Theories

One of the earliest, and probably the most widely employed, structural theories for the mechanical behaviour of open-cell flexible foams is that due to Gent and Thomas.[2] Although they were concerned specifically with latex rubber foams which have an irregular cell structure (see Fig. 1.10), the equations derived using their random-rod structural model are sufficiently general to apply to most open-cell polymer foams. The model, which is shown in Fig. 1.19, consists of thin rods of elastomer with undeformed length l_0 and cross-section area A, orientated randomly in space and connected at their ends by undeformable spheres of polymer having diameter D. These spheres were considered as dead volume and because of their presence the strain in the cell rods is larger than the macroscopic strain of the foam specimen. For small strains it is assumed that the rods are in simple tension or compression.

It can be shown that the strain energy w in a single deformed rod is

$$w = (1/2) l_0 E_p A (1 + D/l_0)^2 (r/r_0 - 1)^2$$

where r is the distance between adjacent connecting spheres in the deformed state and r_0 the value in the undeformed state. By assuming the spheres move affinely with the bulk of the material and ignoring second and higher order terms in the strain ε, the strain energy can be related to the normal strain components $\varepsilon_i (i = 1, 2, 3)$. For spherical co-ordinates (r, Θ, ϕ) the expression is

$$w = (1/2) l_0 E_p A (1 + D/l_0)^2 (r_0/r)^2 \{\varepsilon_1 \sin^2\phi \cos^2\Theta +$$

$$\varepsilon_2 \sin^2\phi \sin^2\Theta + \varepsilon_3 \cos^2\phi\}^2$$

If the rods are considered as randomly orientated vectors with origin at the centres of co-ordinates the number dN_{Θ_o} in a sphere of radius r_0 having directions between $\Theta + d\Theta$ and $\phi + d\phi$ is

$$dN_{\Theta_o} = (N/4\pi r_0^2) r_0^2 \sin\phi \, d\phi \, d\Theta$$

where N is the number of rods per unit volume. The strain energy per unit volume, W, in the deformed system is obtained by summing the contributions $w dN_{\Theta_o}$ for all possible rod orientations,

$$W = (N/2\pi) \int_0^{\pi/2} \int_0^{\pi/2} w \sin\phi \, d\phi \, d\Theta$$

This gives

$$W = (Nl_0 E_p A/30)(1 + D/l_0)^2 \{3(\varepsilon_1^2 + \varepsilon_2^2 + \varepsilon_3^2) + 2(\varepsilon_1\varepsilon_2 + \varepsilon_2\varepsilon_3 + \varepsilon_3\varepsilon_1)\}$$

Comparing this equation with that for the stored-energy density in a material of modulus E_f and Poisson's ratio v_f subject to normal stress $\sigma_i (i = 1, 2, 3)$ gives for the modulus of the foam

$$E_f = E_p (NAl_0/6)(1 + D/l_0)^2 \tag{2B.1}$$

and the Poisson's ratio $v_f = 0.25$. The term NAl_0 is the volume fraction of foam which is in the form of cell rods. For a structural model in which the entire surface of the spherical dead volume is covered by rods, $NAl_0 = 3\beta^2/(1 + \beta)^3$ where β is the ratio of the sphere diameter to the undeformed rod length, $\beta = D/l_0$. This parameter is directly related to the density ratio of the foam as discussed in Section 1.4.3. Substituting into eqn (2B.1) gives for the elastic modulus of the foam

$$E_f/E_p = \beta^2/\{2(1 + \beta)\} \qquad (2B.2)$$

According to Lederman[3] the Gent–Thomas theory does not take full account of variations in the structure which result from differences in the processing conditions. He established more generally applicable equations which not only allow the prediction of the initial elastic modulus but also the influence of the initial cell orientation on the modulus. The structural model employed by Lederman is essentially the same as the random-rod model of Gent and Thomas. However, in his treatment he used a stress-analysis approach rather than a strain-energy method. By considering the number of spheres intersecting an elemental plane, area dA, in the foam, the number of rods n connected to each sphere and the stress in each rod he derived an equation for the traction across this area. The stress acting on each rod is assumed to be some function $f(\lambda_1)$ of the rod extension ratio λ_1. By expressing the overall deformation of the foam specimen in terms of λ_1 he related the stress applied to the foam to the resulting strain.

The effect of rod orientation was introduced by means of a probability distribution function p_u which he defined as the probability of finding a rod within a given solid angle. For a random distribution of rod orientations $p_u = 1/(2\pi)$. At small strains, for which it is assumed that the behaviour of the rod elements is Hookean, these equations have simple solutions. The relationship for the initial elastic modulus of a foam with random rod orientation was shown to be

$$E_f/E_p = (nA/\pi D^2)\left[\beta^2/\{2(1 + \beta)\}\right] \qquad (2B.3)$$

This differs from the Gent–Thomas equation only in terms of the cell structure parameter $nA/\pi D^2$ which is the amount of surface area of the connecting spheres covered by the cell rods. This parameter might be considered as a fitting constant since there appears to have been no work done on establishing its value by independent measurement. However, Lederman has shown that different values are obtained for latex rubber foams prepared in different ways, and he concluded that this parameter is related, in some way, to the processing variables. The Gent–Thomas model corresponds to a cell structure parameter of unity. It was also shown that for a given volume fraction of polymer the foam/polymer modulus ratio, E_f/E_p, has a maximum value when $nA/\pi D^2 = 0.5$. This condition was considered to be the optimum cell structure to achieve maximum stiffness for a given foam density.

Structural models more closely related to the highly ordered cellular structure of PUR flexible foams have been proposed by Ko[4] (see for example Section 1.3.3). By considering the normal and bending deformations of the cell struts in a free-standing unit cell, he related the modulus ratio to the slenderness ratio l_0^2/A of the strut elements. For the model generated by the hcp packing of spherical voids (Fig. 1.16) the expression is

$$E_f/E_p = (\sqrt{3}/4)(A/l_0^2)\{1 + 9/[4(17 + 2\sqrt{3}l_0^2/A)]\}$$

where the slenderness ratio is related to the volume fraction of polymer by $A/l_0^2 = 2\sqrt{3}\phi/9$. Little work has been done on establishing the validity of the relationships derived using this type of model.

In the Gent–Thomas and Lederman theories, the parameter $\beta = D/l_0$ is related to the volume fraction of polymer ϕ as described in Section 1.4.3. If it is assumed that $nA/\pi D^2 = 1 \cdot 0$ these expressions simplify for low-density foams $\phi \to 0$ to $\beta^2 \approx \phi/3$. All of the structural theories cited above predict that for foams of low density the modulus ratio E_f/E_p increases in proportion to ϕ and that for $\phi \to 0$, $E_f/E_p \approx \phi/6$.

Gent and Thomas have also investigated the tensile elastic behaviour using the less complicated cubic-rod model[5] shown in Fig. 1.17. A detailed analysis of this model is given in Sections 2B.3 and 2B.5 below. The relationship for E_f/E_p predicted using the model differs from eqn (2B.2) only in terms of a numerical coefficient. These workers concluded that the form of the relationship for the modulus ratio is not critically dependent on the detailed structure of the foam.[2]

2B.2.2 Empirical Relationships

From the preceding discussion it is clear that the most important structural parameter governing the elastic modulus of a flexible foam is the ratio D/l_0 where l_0 is the free length of the undeformed cell rod and D is related to the cross-sectional area of the rod. It has also been shown that D/l_0 is directly related to the volume fraction of polymer in the foam ϕ. Consequently the modulus should be independent of the cell size provided the ratio D/l_0 remains constant.

By analysing the stress–strain behaviour of several series of PUR flexible foams of different density, cell size and cell-size distribution,

Rusch[1] showed that this is essentially true. He found for a wide range of densities and cell sizes that the modulus ratio E_f/E_p is related approximately to the volume fraction of polymer by

$$E_f/E_p = (\phi/12)(2 + 7\phi + 3\phi^2) \qquad (2B.4)$$

This equation predicts that as $\phi \to 0$ the modulus ratio $E_f/E_p \approx \phi/6$ as given by the structural theories, but unlike the Gent–Thomas theory, eqn (2B.2), it also predicts that $E_f/E_p \to 1 \cdot 0$ as $\phi \to 1 \cdot 0$. Although this expression was established for deformation in compression it is assumed that it is also valid for tensile loading.

2B.2.3 Comparison between Predicted and Measured Values

Experimental data reported by Gent and Thomas[2] for a series of natural rubber latex foams and by Lederman[3] for SBR latex foams are given in Fig. 2B.2. This shows the relationship between E_f/E_p and ϕ. The data of Lederman have been shifted upwards by an order of magnitude for the sake of clarity. It is seen in both cases that over the density range investigated the experimental results agreed quite well with the behaviour predicted by the respective theories. The empirical relationship of Rusch, eqn (2B.4), which was established for PUR flexible foams, has been applied to the data of Gent and Thomas. This is shown as curve (b), and it can again be seen that there is good agreement between the predicted and measured values.

From this it is concluded that the structural theories and the empirical relationship provide adequate mathematical models for the tensile elastic modulus of flexible foams. However, these data could equally well be fitted over a wide range of density ratios by an expression of the form $E_f/E_p = \phi^n$ as discussed in Chapter 1. The value of the exponent n is 1·68 for the Gent–Thomas data and 1·39 for the Lederman data.

2B.3 STRESS–STRAIN BEHAVIOUR AND ELASTIC MODULUS IN COMPRESSION

2B.3.1 Stress–Strain Behaviour

The compressive stress–strain behaviour of representative types of flexible foam have been compared by Rusch.[1] His results are shown in Fig. 2B.3. In order to make this comparison he reduced the stress–

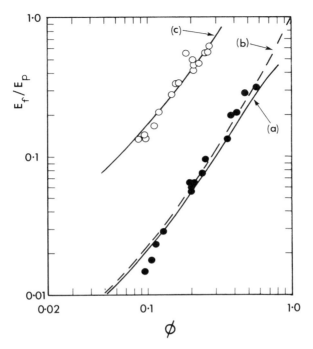

FIG. 2B.2. Variation of E_f/E_p in tension as a function of the volume fraction of polymer ϕ for NR latex foams[2] (closed circles) and SBR latex foams[3] (open circles). The data for SBR have been shifted upwards by one decade. Curve (a) eqn (2B.2), (b) eqn (2B.4), (c) eqn (2B.3) with $nA/\pi D^2 = 0.15$.

strain behaviours to a common initial compression modulus, $E_f = 10$ psi. Of particular importance in the following discussion are the details of the cell-structure geometries. Foam E was a latex rubber foam which had open cells and a wide distribution in the size of the cell elements. A photomicrograph of a latex rubber foam is given in Fig. 1.10. Foam F was a non-reticulated PUR flexible foam with irregular cell dimensions and some cell membranes. Foams G and H were reticulated PUR foams with highly regular cell structures (i.e. uniform cell size) and many cell membranes.

It is seen from Fig. 2B.3 that the compressive stress–strain diagram of a flexible foam can be divided into three regions; an initial Hookean region in which the stress increases in proportion to the strain, a plateau region in which the slope is much less than the initial part of the stress–

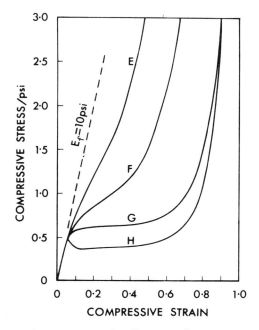

FIG. 2B.3. Compressive stress–strain diagrams for representative types of flexible foam.[1]

strain diagram, its value depending strongly on the cellular structure, and a region in which the stress increases rapidly with increasing strain. The deformation mechanisms in the first two of these regions are essentially the same as those for rigid foams described by Menges and Knipschild (Chapter 2A). These are the simple compression of the cell rods under the action of normal forces followed by the buckling of these cell elements.

When a flexible foam is subjected to cyclical compressive deformation, it is found that the initial elastic modulus and the load-bearing capacity decrease with increasing number of deformation cycles. The change in the mechanical properties is largest for the first and second compressions and approximately equilibrium values of the properties are obtained after a small number of cycles.[1] It has also been observed that a significant proportion of the loss in load-bearing capacity caused by prolonged compression at high strains is recovered after the removal of the stress.[6,7] The amount of recovery increases with increasing time. These observations indicate that the plateau region in the compressive stress–strain diagram of a flexible foam is due to the combined action of permanent material damage[1] and a retarded elastic buckling instability.

Some structures, such as foams E and F in Fig. 2B.3, do not exhibit the abrupt decrease in the slope of the stress–strain diagram that would be expected of a simple rod at the point of unstable buckling. Since the critical buckling load for a rod depends upon the size and the geometry of the rod,[8] this behaviour can be attributed partially to a wide distribution in the dimensions of the cell elements in these polymer foams.[2] Foams G and H which had a more uniform cell structure geometry exhibited stress–strain behaviours closely resembling the ideal buckling form.

The third region in the compression stress–strain diagram, in which the stiffness increases rapidly with increasing strain, is due to the interaction between the cell elements as the cellular structure collapses to form a compact material system.

The important features of the stress–strain behaviour in simple compression are the gradient of the curve in the first region, i.e. the initial elastic modulus, the strain at the proportional limit, the gradient of the curve in the plateau region and the strain at the upper limit of the plateau.

2B.3.2 Empirical Relationships

Because of the complexity of the deformation mechanisms and their dependence on the cell-structure geometry, it has not been possible to establish equations relating the stress–strain behaviour in compression to the structure of the foam and the mechanical properties of the base polymer. However, some understanding of this relationship has been obtained by factoring the compressive stress into two parts; an elastic modulus, and some function $F(\varepsilon)$ of the strain. This function reflects the influence of the buckling of the cell elements and the subsequent compaction of the cellular structure on the shape of the compressive stress–strain diagram. Empirical and analytic expressions for $F(\varepsilon)$ have been described in several publications.[1,2,9,10]

One of the most useful applications of this approach is that due to Rusch[1] who expressed the relationship between compressive stress and strain as

$$\sigma = E_f \varepsilon F(\varepsilon) \tag{2B.5}$$

where E_f is the initial compressive modulus of the foam. This is related to the volume fraction of polymer by eqn (2B.4). In principle, the shape function $F(\varepsilon)$ should be independent of ϕ and should depend only on the cell-structure geometry. An alternative way of expressing the stress–

strain relationship, in terms of the stress at 25% compression and a strain function, is discussed in Chapter 5.

The variation of the shape functions $F(\varepsilon)$ with compressive strain for the data of Fig. 2B.3 is shown in Fig. 2B.4. At small strains, where the stress increases in proportion to the strain, $F(\varepsilon) = 1.0$. At the onset of buckling the slope of the $F(\varepsilon) - \varepsilon$ curve becomes negative. With a further

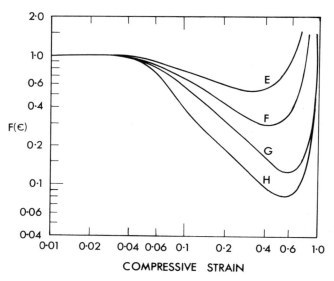

FIG. 2B.4. The variation of the Rusch[1] shape function $F(\varepsilon)$ with strain for the stress–strain diagrams of Fig. 2B.3.

increase in strain the function passes through a minimum value $F(\varepsilon)_{\min}$, at which no further collapse of the cell structure is possible, and subsequently increases. In the high-strain region the condition where $F(\varepsilon) = 1.0$ corresponds to the strain at which all of the voids in the cellular system have been compressed. The function $F(\varepsilon)$ has been related to the strain by the analytic expression[9]

$$F(\varepsilon) = a\varepsilon^{-p} + b\varepsilon^q \qquad (2B.6)$$

where a, b, p and q are empirically determined curve-fitting constants. The significance of these constants is shown in Fig. 2B.5.

According to Rusch, $F(\varepsilon)$ and hence the shape of the compressive stress–strain diagram, can be characterised in terms of four quantities; the critical buckling strain ε_b, which is defined as the strain at which

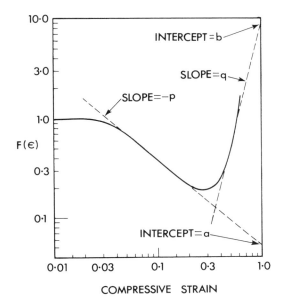

FIG. 2B.5. An idealised shape function $F(\varepsilon)$ showing the origin of the coefficients in equation (2B.6).[9]

$F(\varepsilon) = 0.95$, the average value of the slope post yield, the value of $F(\varepsilon)_{min}$, and the strain ε_{min} at the minimum in the curve. He found for PUR flexible foams that the value of ε_b was essentially independent of the average cell size, but increased with increasing density ratio. This density dependence was most marked in systems with a wide distribution in cell size and was small in foams with a uniform cell structure. For specimens having stress–strain diagrams similar to curve F in Fig. 2B.3 the value of ε_b varied from about 4% to 10% for foams having volume fractions of polymer in the range $0.037 \leq \phi \leq 0.34$.

The most important factor governing the shape of the $F(\varepsilon) - \varepsilon$ curve and the value of $F(\varepsilon)_{min}$ is the geometry of the cellular structure; the average cell size and the density ratio having only a small effect. In addition, $F(\varepsilon)$ is independent of the rate of deformation and the temperature, except below the glass transition temperature of the matrix polymer. Above T_g the temperature influences the stiffness of the foam through its effect on E_p. It is seen in Fig. 2B.4 that the value of ε_{min} and hence the upper limit of the plateau region in the stress–strain diagram depends upon the uniformity of the cellular structure, ε_{min} decreasing as the width of the cell-size distribution increases.

These results indicate that the load-bearing capacity of a flexible foam, taken for example as the stress at 25% compression σ_{25}, is governed by the initial compression modulus and the cell-structure geometry. From eqn (2B.5), $\sigma_{25}/E_f = 0.25F(0.25)$. Foams with similar, and uniform, cell structures should have approximately the same value for $F(\varepsilon)$, so that the ratio σ_{25}/E_f should be approximately independent of cell size. Data from another investigation of the mechanical behaviour of PUR flexible foams having uniform cell structures[10] support this observation. The shapes of the stress–strain curves for these foams were intermediate between curves G and H in Fig. 2B.3, $\phi \approx 0.03$, and the average cell size varied between 0.28 and 2.25 mm. The value of the σ_{25}/E_f ratio for these foams was in the range 5.5×10^{-2} to 6.9×10^{-2} with an average value of 6.7×10^{-2}. (These data are for foam samples B–J in the original publication.[10]) This range and average value for σ_{25}/E_f is in close agreement with that obtained from Rusch's experimental results for PUR foams with uniform cell structure.

2B.3.3 Structural Models

Further insight into the factors responsible for the mechanical behaviour under compressive loading can be obtained by considering the earlier work of Gent and Thomas.[5] For this analysis they assumed a model consisting of a cubical array of square-section rods of undeformed length l_0 and side t as shown in Fig. 1.17. The rod intersections, which form cubes of side t, are assumed to be substantially undeformable. The deformation of the foam specimen is made up of two parts; the simple compression of the rods aligned in the direction of the applied stress, giving a strain component σ/E_f, and the strain associated with the buckling of the rods. At small strains these components are assumed to be additive. However, for foams of low density the strain associated with the normal forces was taken to be negligibly small so that in the first instance only the buckling strain was taken into account.

By considering the fractional approach ε' of the two ends of a cell rod orientated in the direction of the applied load they were able to express the force F supported by this rod in terms of its area moment of inertia I as

$$F = E_p I f(\varepsilon')/l_0^2$$

where $f(\varepsilon')$ is an unknown function of ε' independent of E_p and ϕ. This function is determined empirically from the stress–strain behaviour of representative samples of the foam. For cell rods of square cross-section the parameter I is proportional to t^4. Introducing geometric detail from the model and absorbing the constant of proportionality into $f(\varepsilon')$ gives the compressive stress as

$$\sigma = E_f f(\varepsilon')\{\beta^2/(1+\beta)\} \tag{2B.7}$$

where, for the cubic-rod model, $E_f = \beta^2/(1+\beta)$ and $\beta = t/l_0$. The strain ε' associated with the buckling of the rod elements is related to the total strain ε of the foam specimen by $\varepsilon' = \varepsilon/(1+\beta)$. For foams of low density β is small so that $\varepsilon' \approx \varepsilon$, $\beta^2 \approx \phi/3$ and

$$\sigma \approx E_f \phi f(\varepsilon)/3 \tag{2B.8}$$

This relationship differs from the empirical Rusch expression, eqn. (2B.5), in that the term $\varepsilon F(\varepsilon)$ is replaced by $(\phi/3)f(\varepsilon)$. Equation (2B.7), and in its simpler form eqn. (2B.8), only takes account of the buckling deformation of the cell rods. The influence of the simple compressive deformation of the rods on the shape of the stress–strain curve is discussed in the original publication.[5] The function $f(\varepsilon)$ can be determined from measured data using eqn. (2B.7). The variation of $f(\varepsilon)$ as a function of ε for a low-density ($\phi \approx 0.03$) reticulated PUR flexible foam and low-density natural rubber foams is shown in Fig. 2B.6. The curve for the rubber foams is the average behaviour of a series of specimens having different densities.[5]

Equations (2B.7) and (2B.8) can be expressed in terms of the modulus E_p of the base polymer. When this is done it is seen that for a given base polymer and shape function $f(\varepsilon)$ the load-bearing capacity of low-density foams should increase in proportion to ϕ^2. This can be compared with the tensile behaviour predicted by the Gent–Thomas analysis which indicates for foams with $\phi \to 0$ that the elastic modulus should increase in proportion to ϕ.

2B.4 CLOSED–CELL STRUCTURES

The inclusion of closed cells within the cellular matrix of a flexible foam can have a profound effect on its mechanical behaviour. Skochdopole and Rubens[11] found for a semi-rigid polyethylene foam that not only did the compression modulus and the stress at 25% compression increase with

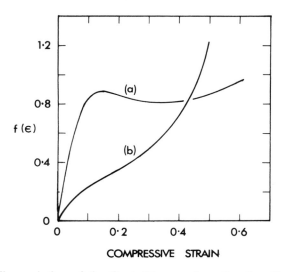

FIG. 2B.6. The variation of the Gent–Thomas shape function $f(\varepsilon)$ with strain for (a) a low-density reticulated PUR flexible foam and (b) NR latex foams.[2]

increasing volume fraction of closed cells, but the shape of the stress–strain diagram was changed also. Some understanding of this behaviour can be obtained using the simple model put forward by these workers. This is shown in Fig. 2B.7(a) and consists of an enclosure filled with a gas arranged in parallel with the polymer matrix. When the foam is deformed the cell elements deform and buckle as described above and the gas is compressed. The stress required to deform the foam is the sum of the stresses resulting from the two processes. Since the stress–strain relationships for these processes are very different, as shown in Fig. 2B.7(b), the overall stress–strain behaviour of a foam containing closed cells is strongly dependent on the contributions which each mechanism makes.

If it is assumed that the walls of the enclosure are undeformed during compression, the stress needed to deform the gaseous phase can be calculated approximately using the ideal gas law. The volume V_g of gas enclosed within the system is related to the fractional compression ε by $V_g = V_{g0}(1 - \phi - \varepsilon)/(1 - \phi)$ where V_{g0} is the volume of the gas in the undeformed foam. The corresponding pressure is $P_g = P_{g0}(1 - \phi)/(1 - \phi - \varepsilon)$ where P_{g0} is the initial pressure. Assuming that P_{g0} is equal to atmospheric pressure, P_A, the stress required to compress the gas so as to produce a strain ε in the foam is $\sigma_g = P_g - P_A = P_A \varepsilon/(1 - \phi - \varepsilon)$. For a

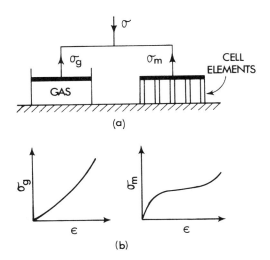

FIG. 2B.7. (a) A model for a closed-cell foam.[11] (b) The stress–strain behaviour of the two mechanisms; σ_g, the stress associated with the gas phase, σ_m, the stress associated with the polymer matrix.

partially-closed-cell system the stress associated with the enclosed gas phase will be proportional to the volume fraction of closed cells ζ provided there is continuity between the cells such that the strain is uniformly distributed throughout the foam. If this is the case, $\sigma_g = P_A \zeta \varepsilon / (1 - \phi - \varepsilon)$. Combining this with the stress needed to deform the polymer matrix gives the overall stress as a function of strain. For example, using the Rusch expression, eqn. (2B.5), gives

$$\sigma = E_f \varepsilon F(\varepsilon) + P_A \zeta \varepsilon / (1 - \phi - \varepsilon) \qquad (2B.9)$$

This equation was first proposed by Rusch[12] who commented that the second term is negligible if (i) $E_f \gg P_A$, (ii) the foam is open cell, $\zeta \to 0$, or (iii) the matrix is very brittle, such that the closed cells are ruptured during compression. However, when $E_f \ll P_A$ it is predicted that the second term will dominate the response so that the stress–strain behaviour of the polymer matrix will have only a small effect on the stress–strain behaviour of the closed-cell foam.

The relative importance of the two mechanisms on the overall mechanical behaviour in compression can be demonstrated by considering a specific example. The influence of the volume fraction of closed cells on the stress–strain behaviour of a cellular polymer with $E_f = 1.0 \times 10^6 \mathrm{Nm}^{-2}$, $\phi = 0.03$ and shape function $F(\varepsilon)$ as in curve G, Fig. 2B.4, is shown in Fig. 2B.8. It is seen that the load-bearing capacity is predicted to increase with increasing volume fraction of closed cells, as found by Skochdopole and Rubens, and that the influence of the enclosed gas phase is most marked at

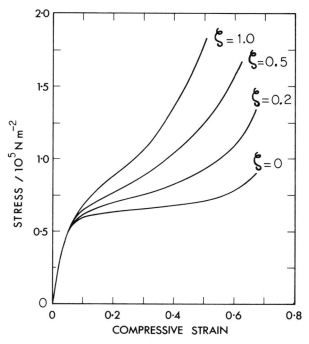

FIG. 2B.8. The influence of the volume fraction of closed cells ζ on the compressive stress-strain diagram of a flexible foam predicted by eqn. (2B.9).

large strains. For this particular system the initial compression modulus is essentially independent of ζ. The influence of closed cells on the modulus of a foam will increase as the stiffness of the polymer matrix decreases.

The two important deficiencies of the model just described are the assumptions that the polymer matrix is undeformed by the elevated gas pressure, and, for small values of ζ, that there is continuity between the closed cells. Gent and Thomas[5] have established equations relating the elastic modulus of closed-cell flexible foams to the stiffness of the matrix and the volume fraction of polymer using a strain-energy method. Assuming that the initial pressure of the gas in the closed cells is equal to atmospheric pressure they showed for a foam deformed under isothermal conditions that the tensile elastic modulus E'_f is

$$E'_f = \frac{2E_f}{5} \left\{ \frac{2E_f(1-\phi) + 3P_A}{(4/5)E_f(1-\phi) + P_A} \right\}$$

where E_f is the tensile modulus of the corresponding open-cell foam. When $E_f \gg P_A$ the elastic modulus of the closed-cell foam is approximately the same as that of the open-cell foam. For soft closed-cell foams, $E_f \ll P_A$, the modulus E'_f is predicted to be at most 20% larger than E_f,

2B.5 MECHANICAL ANISOTROPY

2B.5.1 General Considerations

One of the main difficulties associated with investigations of the relationship between structural and mechanical anisotropy is the measurement of the basic cell-structure parameters and the description of their orientation distribution. In early studies of foams exhibiting two dimensional anisotropy,[13,14] the ratio of the maximum dimension of a complete structural cell to the minimum dimension, $d_{\max}/d_{\min}$, was used to describe the degree of orientation. The nature of the cell orientation and the value of the degree of orientation is strongly dependent on the processing conditions as described in Chapter 1. The maximum cell dimension lies in the direction of material movement during foaming, so that for free-rising foams measurements have been made in directions parallel and perpendicular to the direction of foam expansion. Experimental studies have shown that in general the elastic modulus and strength are a maximum in the direction of cell elongation.[13-16] The data of Harding[14] show that the compressive strength of PUR rigid foams increases approximately in proportion to $d_{\max}/d_{\min}$ for values of the degree of orientation in the range $1\cdot0 \ll d_{\max}/d_{\min} \ll 1\cdot6$, the constant of proportionality increasing with increasing foam density.

An alternative way of describing and measuring structural anisotropy in thermoplastic cellular polymers has been proposed by Mehta and Colombo.[17] Their work was concerned with mechanical anisotropy in extruded cellular polystyrene boards, and they expressed the degree of orientation in terms of the dimensional changes that occur when the extrudate is relaxed at an elevated temperature. Again they were interested in anisotropy in two dimensions; the machine and transverse directions. They were able to relate the structural parameters employed in the Halpin–Tsai equations[18] to the empirically determined dimensional changes.

In order to obtain a complete description of the cell-structure geometry, it is necessary to take account of the three-dimensional anisotropy that may exist in the system. From a practical point of view, this is most conveniently done in terms of the average undeformed cell size in three mutually perpendicular directions (1,2,3) orientated in the directions of maximum and minimum cell dimensions. The discussion given in Sections 2B.3.1 and 2B.3.2 above indicates that in order to understand the relationship between structural and mechanical anisotropy it is also necessary to take into consideration the directional dependence of the cell size distribution.

2B.5.2 Experimental Results

Some indication of the behaviour to be expected with anisotropic cellular systems can be obtained from the experimental data given in Fig. 2B.9 which is for PUR flexible foams (A,B,C and D) with approximately the same density ratio but different commercial origins.[19] This shows the ratios of the tensile elastic moduli, E_T, the compression moduli, E_C, and the load-bearing capacity, σ_{25}, in two mutually perpendicular directions. Direction 3 is the direction of maximum cell dimension and direction 1 the direction of minimum cell dimension. The tensile measurements were made on square-section bars of length 100 mm and side 20 mm and the compression measurements were made on cubes of side 50 mm. The strain rate was 10% per minute and each data point is the average value for five specimens. The directional dependence of the average cell size and the cell-size distributions for foams A,B and C are given above in Figs. 1.12, 1.13 and 1.14, respectively. The cell-size distribution in foam D was similar to that of foam C.

It is seen that in most cases the mechanical properties were enhanced in the direction of maximum cell dimension and that, as expected, the mechanical anisotropy increased with the degree of structural anistropy (expressed in terms of the ratio d_3/d_1). It is also seen that, except for two data points, the enhancement in the properties for tensile loading was larger than that for compressive loading. The dependence of the degree of property enhancement on the deformation mode has been reported previously by Kanakkanatt.[16] The inconsistency between the compres-

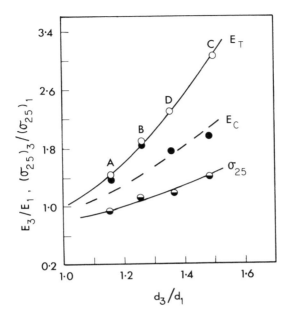

FIG. 2B.9. The variation of the mechanical anisotropy in structurally anisotropic PUR flexible foams as a function of the ratio of the maximum to minimum cell dimensions.[19] E_T, tensile moduli ratio; E_C, compressive moduli ratio; σ_{25}, the ratio of the stresses at 25% compression.

sive modulus data for foams A and B and foams C and D can be attributed in part to differences in the cell-size distributions and the directional dependence of the distribution of cell membranes. The density of cell membranes in foams C and D was considerably larger than that in foams A and B. In addition the number of cell membranes was different in each direction. Anisotropy in the distribution of cell membranes in PUR flexible foams has been reported previously by Jones and Fesman.[20] The scatter in these data illustrates the difficulties associated with the comparison of the mechanical behaviour of foams with different commercial origin.

2B.5.3 Structural Models

The influence of structural anisotropy on the elastic modulus of flexible foams has been studied theoretically by Lederman.[3] This was done by modifying the probability distribution function p_u (Section 2B.2) which describes the orientation of the cell rods. He considered two extremes of structure; (i) where all of the rods are orientated in the direction of loading ($\Theta = 0$) for which $p_u \to \infty$ as $\Theta \to 0$ and $p_u = 0$ for $\Theta \neq 0$ and (ii) where all of the cell rods are orientated perpendicular to the direction of loading. Assuming that the parameter $\beta = D/l_0$ remains constant, the theory predicts that in the first case the initial elastic modulus is six times that of a structurally isotropic foam and in the second the modulus is zero. He assumed that this would apply to both tensile and compressive loading. Although this approach is useful, in that it allows us to predict the upper and lower bounds to the modulus, it has not been applied in practice because of the difficulty in establishing empirically the orientation distribution function.

An understanding of the relationship between structural and mechanical anisotropy can be obtained more easily by considering the simple, but less rigorous, cubic rod model of Gent and Thomas[5] described in Section 2B.3.3. This has been used previously by Kanakkanatt.[16] A section from a cell represented by this model is shown in Fig. 2B.10. The three orthogonal directions (1,2,3) are assumed to be parallel to the cell-rod elements which have undeformed length l_i ($i = 1,2,3$) and side t_i. For simple tension or compression in the 3 direction, the number of rods N_{33} per unit area of foam, orientated perpendicular to the direction of loading, supporting the applied stress σ_3 is $N_{33} = 1/(l_1 + t_1)(l_2 + t_2)$. The load on each rod in the 3 direction is $F_3 = \sigma_3/N_{33}$. Assuming the behaviour of

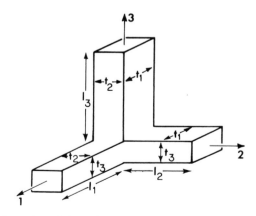

FIG. 2B.10. A section of the unit cell for the anisotropic cubic-rod model.

the cell elements is Hookean, the deformation, Δl_3, of each cell rod is $\Delta l_3 = (F_3/t_1 t_2)(l_3/E_p)$. This is equal to the total deformation of the unit cell so the overall strain in the foam specimen is $\varepsilon_3 = \Delta l_3/(l_3 + t_3)$. Combining these terms gives the initial elastic modulus in the 3 direction as

$$E_3 = \frac{E_p t_1 t_2 (1 + \beta_3)}{l_1 l_2 (1 + \beta_1)(1 + \beta_2)}$$

where $\beta_i = t_i/l_i$. For structurally isotropic foams this equation reduces to the expression given by Gent and Thomas.[5] Carrying out the same procedure in the 1 direction and taking the ratio of the moduli it is easily shown that

$$E_3/E_1 = (t_1/t_3)(l_3/l_1)\{(1 + \beta_3)^2/(1 + \beta_1)^2\} \quad (2B.10)$$

For low-density foams, $\beta_i \ll 1 \cdot 0$ so that the ratio of the initial elastic moduli for simple tensile and compressive loading is predicted to increase in proportion to the structural anisotropy factor l_3/l_1, the constant of proportionality being t_1/t_3.

With low-density systems, $t_i \ll l_i$ so the ratio of the free-rod lengths l_3/l_1 is approximately equal to the ratio of the cell dimensions d_{max}/d_{min}. Assuming that these conditions prevail it is possible to relate the data of Fig. 2B.9 to the mathematical model in eqn. (2B.10). The tensile modulus data indicate that for these foams the ratio of the side dimensions of the cell struts, t_1/t_3, was greater than unity and, because of the upwards curvature of the line, its value increased with increasing l_3/l_1.

The influence of structural anisotropy on the load-bearing capacity can be determined using the Gent–Thomas analysis for the buckling of the rod elements (Section 2B.3.3). For an anisotropic system the stress σ_3 is related to the shape function $f(\varepsilon_3')$ by

$$\sigma_3 = \frac{E_p I_{21} f(\varepsilon_3')}{l_3^2 l_1 l_2 (1 + \beta_1)(1 + \beta_2)}$$

where ε_3' is the fractional approach of the ends of the cell rods in the direction of loading and I_{21} is the area moment of inertia for buckling deformation perpendicular to the direction of loading. For rectangular section rods $I_{ij} = t_i t_j^3/12$ where t_j is the smallest dimension. Using the above equation, and that corresponding to compressive deformation in the 1 direction, the ratio of the stresses is shown to be

$$\sigma_3/\sigma_1 = (I_{21}/I_{32})\{f(\varepsilon_3')/f(\varepsilon_1')\}(l_1/l_3)\{(1 + \beta_3)/(1 + \beta_1)\}$$

where the macroscopic strain of the material ε_i is related to the fractional approach of the ends of a rod ε_i' by $\varepsilon_i = \varepsilon_i'/(1 + \beta_i)$. If it is assumed that the shape function $f(\varepsilon')$ is independent of direction, this analysis predicts for low-density foams ($\beta_i \ll 1$) that the stress at 25% compression should decrease in proportion to the structural anisotropy factor l_3/l_1 provided I_{21}/I_{32} remains constant. That is, the load-bearing capacity should be a minimum in the direction of maximum cell dimensions, which is opposite to the behaviour predicted for the initial tensile modulus. Although the data of Fig. 2B.9 shows that σ_{25} is much less dependent upon the structural anisotropy than E_T a small increase in the σ_{25} ratio was observed for three of the foams. This behaviour might be attributed to two factors; (i) the increase of the t_1/t_3 ratio with increasing l_3/l_1, which was indicated by the tensile data; (ii) anisotropy in the shape function $f(\varepsilon')$ resulting from anisotropy in the width of the cell-size distribution. Rusch[1] has shown that the wider the distribution the larger the slope in the plateau region of the compression stress–strain diagram. For foam A, Fig. 2B.9, which had small structural anisotropy and a narrow cell-size distribution, the width of which was approximately the same in each direction, the σ_{25} ratio was less than unity.

ACKNOWLEDGEMENT

The author wishes to thank the editors of the *Journal of Applied Polymer Science* and the *Journal of Cellular Plastics* for permission to use material contained in their publications.

REFERENCES

1. RUSCH, K. C. (1969) *J. Appl. Polym. Sci.*, **13**, 2297.
2. GENT, A. N. and THOMAS, A. G. (1959) *J. Appl. Polym. Sci.*, **1**, 107.
3. LEDERMAN, J. M. (1971) *J. Appl. Polym. Sci.*, **15**, 693.
4. KO, W. L. (1965) *J. Cell. Plast.*, **1**, 45.
5. GENT, A. N. and THOMAS, A. G. (1963) *Rubber Chem. Tech.*, **36**, 597.
6. DWYER, F. J. (1976) *J. Cell. Plast.*, **12**, 104.
7. KANE, R. P. (1965) *J. Cell. Plast.*, **1**, 217.
8. TIMOSHENKO, S. P. and GERE, J. M. (1961) *Theory of Elastic Stability*, McGraw-Hill, Chap. 2.
9. RUSCH, K. C. (1970) *J. Appl. Polym. Sci.*, **14**, 1433.
10. HILYARD, N. C. and DJIAUW, L. K. (1971) *J. Cell. Plast.*, **7**, 33.

11. SKOCHDOPOLE, R. E. and RUBENS, L. C. (1965) *J. Cell. Plast.*, **1**, 91.
12. RUSCH, K. C. (1970) *J. Appl. Polym. Sci.*, **14**, 1263.
13. HARDING, R. H. (1965) *J. Cell. Plast.*, **1**, 385.
14. HARDING, R. H. (1967) *Morphologies of Cellular Materials. Resinography of Cellular Plastics*, ASTM STP 414, Am. Soc. Testing Mats., 1967, p3.
15. HERMANSON, R. (1968) *J. Cell. Plast.*, **4**, 466.
16. KANAKKANATT, S. V. (1973) *J. Cell. Plast.*, **9**, 50.
17. MEHTA, B. S. and COLOMBO, E. A. (1976) *J. Cell. Plast.*, **12**, 59.
18. ASHTON, J. E., HALPIN, J. C. and PETIT, P. H. (1969) *Primer on Composite Materials: Analysis*, Technomic Publishing Company Inc.
19. HILYARD, N. C. Unpublished data.
20. JONES, R. E. and FESMAN, G. (1965) *J. Cell. Plast.*, **1**, 200.

Chapter 3

CUSHIONING AND FATIGUE

H. W. WOLFE

E. I. du Pont de Nemours and Company, Wilmington, Delaware, USA

3.1 INTRODUCTION

Up to the beginning of the World War II post-war era, most cushioning designed for comfort and durability was generally based on a composite of coiled steel springs or elastic designed steel supports and some other materials such as burlap, cotton, curled hair, etc. However, by the early 1950s a new and different type of cushion, cellular polymeric foams, had gained acceptance in a number of different types of applications and was showing tremendous potential for growth. These cellular polymeric foams obtained their cushioning characteristic solely from the cellular nature of the material and not from any internal spring support. Table 3.1 lists the three cellular polymer cushions that pioneered this development and growth.

Cellular polymer cushions[1] owes its real beginning to natural rubber latex foam, which was initially developed by Untiedt in the late 1920s. These foams were, and currently are, made by stabilising the natural rubber 40% solids latex and then mechanically beating air into it followed by drying to eliminate the water. Initially these foams were high in density, but as improvements were made to this technology, foam densities were reduced to acceptable levels. Several years later, in the early 1930s, similar

TABLE 3.1
Cellular Polymeric Foams

Date	Polymer	Cell structure	Developed by
Late 1920s	Natural rubber latex	Mechanical	Untiedt
Early 1930s	Polychloroprene latex	Mechanical	Du Pont
1937	Polyurethane	Chemical	Otto Bayer

latex foams were developed by the Du Pont Company. These foams were based on a synthetically produced polymer polychloroprene (neoprene).

Later, in 1937, Otto Bayer and his co-workers discovered a new chemical method for synthesising high molecular weight polyurethane polymers by reacting glycols with di-isocyanates. These polymers were rapidly recognised as having an excellent basis for still another type of foam cushioning, and by the mid-point of this century, along with the latex foam cushions, they had made their moderate but firm penetration into numerous cushion applications. Concurrent with this penetration, numerous other potential cellular polymeric foams and low-density sponges were being developed from a variety of different polymers, such as shown in Table 3.2.

TABLE 3.2
Polymers Evaluated in Cellular Polymer

Chlorosulphonated polyethylene	Polyisoprene
Ethylene/propylene	Polypropylene
Fluorocarbon	Polysilicone
Nitrile	Polystyrene
Polyacrylate	Polysulphide
Polybutadiene	Polyvinyl
Polyethylene	Styrene–butadiene rubber

Ironically these foams and sponges, with the exception of those based on styrene–butadiene rubber, could not meet the demanding requirements (comfort, cost, density, durability, processability) of most cushion applications as well as the cellular polymeric foams developed in the late 1920s and 1930s.

As the requirements for cushioning applications increased during the decade of the 1950s, the technological competition between the survivors, polyurethane and latex foams, became more intense. By the beginning of the 1960s, polyurethane foams had clearly demonstrated their superiority over latex foams in more fully meeting the demanding requirements mentioned above. As a result of this superiority, polyurethane at the beginning of the 1980s had captured about 90% of all cushioning applications.

3.2 COMFORT CUSHIONING

3.2.1 Force–Deflection Behaviour

Simply stated, any material that provides a force of fight back when subjected to indentation by a load can qualify as a cushion. Figure 3.1(a)

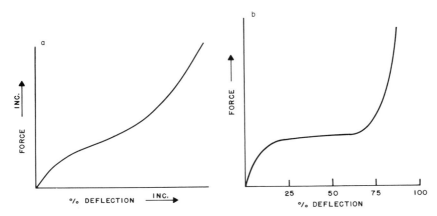

FIG. 3.1. Indentation force deflection (IFD) of (a) typical cushion and (b) a cellular polymer with uncomfortable plateau. Reproduced from ref. 2 with permission.

shows this relationship of force (load) to indentation (deflection).

As the force or load on a cushion increases, the indentation increases as result of the load pressing further into the cushion. The shape of this curve in essence also defines the comfort of the cushion. Most of the cushion materials listed in Table 3.2 will provide some degree of fight back, but will not provide the high degree of comfort of either polyurethane foams or latex foams.

Comfort in any cushion is a subjective property that can vary from individual to individual. It can be partially illustrated by means of the IFD (Indentation Force Deflection) curve. Figure 3.1(b) shows an IFD curve that contains a significant plateau, which would be indicative of a foam cushion that could be rated low in comfort. As a force is applied to this foam cushion, a 'fight back' is present up to about 15% deflection, and then the force or load factor remains fairly constant. At about 75% there is a sharp rise in the curve, or a sharp increase in fight back. In terms of human comfort, an individual sitting on this foam cushion would find an initial 'fight back' but then would settle very rapidly into the cushion. If the density of this foam was low, i.e. 0·04 kg m^{-3} or lower, the individual would 'bottom out'. This phenomena is undesirable in any foam cushion, and can be eliminated or minimised by the chemistry of the foam polymer and the foam density. The plateau can also be minimised, and in some foams, eliminated, by properly designing the polymer structure. When these conditions are obtained the force–indentation relationship would approach that of the ideal curve shown in Fig. 3.1(a).

3.2.2 Material Requirements

Comfort or lack of it can further be demonstrated and defined by means of measuring the hysteresis, sag factor, breathability and resilience of a foam cushion.

Hysteresis
This is a measurement of energy lost or absorbed by a foam when subjected to a deformation. Foams with a high level of hysteresis show poor comfort, since the energy lost through cycling of loading and unloading is not able to do work. The net result of this may be either a deformation of the original shape of the foam, or the foam may be slow in regaining its original configuration. A ball or some other light object when dropped into the foam surface will show a low rebound. Contrasted to this type of foam is one with a low hysteresis which would have a high rebound of the same object dropped from an identical height. These differences in hysteresis are shown in Fig. 3.2.

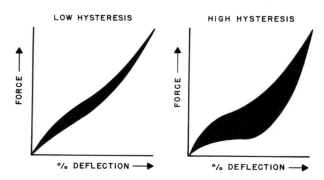

FIG. 3.2. Cellular polymers with low to high hysteresis.

Both of these examples of hysteresis loss are measured by applying a force to a foam over a deflection of 75% and then allowing the foam to relax back to zero deflection by removing the deforming force. The areas under each curve reflect the loss of energy.

Sag factor
This is simply the ratio of the force required to compress a specimen to 65% and 25% deflection.

$$\text{Sag factor} = \frac{65\% \text{ IFD}}{25\% \text{ IFD}}$$

Foams having sag factors of about 2·8 or greater are generally credited with having an extra comfort rating, provided that the initial 25% deflection force is lower, but this can be misleading. As this lower value increases due to the foam polymer design and construction, and the ratio remains constant due also to an increase in the 65% deflection value, the comfort factor will change to the point where the foam cushion could no longer be considered as comfortable. Table 3.3 illustrates this change or shift in comfort even though the sag factor remains constant.

TABLE 3.3
Variations of IFD or 'Fight Back'

Foam	25% Deflection (N)	65% Deflection (N)	Sag factor
A	110	308	2·8
B	130	364	2·8
C	150	420	2·8
D	170	476	2·8

As is clearly shown in Table 3.3, all four foams have a sag factor of 2·8 which raises the question of which one is the more comfortable. The answer, of course, is both subjective and relative and is dependent upon the application. Some individuals have a preference for firmness in all cushion applications of automotive, furniture and bedding, while others prefer softer cushions that will not 'bottom out'. In the softer grades of cushioning having high sag factors, the initial low 25% IFD gives a luxurious feel while the correspondingly higher IFD at 65% deflection provides necessary firmness for support.

Breathability
This is the rate at which air can be expelled from the matrix structure of a foam pad when a force is applied. Normally in the fabrication of a cellular polymer, such as polyurethane foams, the cellular matrix that is formed initially contains a large number of closed cells. As the foam continues to expand, many of these closed cells will rupture, and the remaining ones generally are ruptured mechanically by a post treatment to prevent shrinkage of the foam matrix. The degree to which all of the closed cells are ruptured and size of the cell rupture will determine the ease of breathing air in and out of each individual cell.

When a force is applied to a cellular polymer cushion, the air contained in each cell must be expelled in equilibrium with the force applied. Otherwise, if the expulsion of air is inhibited because of cell orientation and cell-rupture opening size is too small, the cushion will resist the force applied which in turn adversely affects the comfort. This phenomenon is illustrated in Fig. 3.3, which shows the compressive force vs. deflection for three foams which are chemically identical, but different mechanically. Foam A has a small amount of closed cells. Foam B has all of the closed cells ruptured, but the rupture holes in the individual foam cells do not permit a rapid exchange of air, in and out of the cell. The cells in Foam C have been ruptured, i.e. opened, and each cell can breathe very freely as a force is applied. This behaviour is in general agreement with that predicted by the phenomenological model described in Section 2B.4.

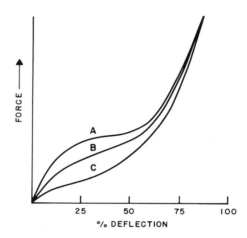

FIG. 3.3 Comfort comparison by IFD.

Foam C, as previously shown, would be the most comfortable of the three cushions. Foam A would initially give the feeling of comfort cushioning for about the first 5–10% deflection, but then the feeling would be similar to sitting on a wooden board or block of concrete. Obviously, the complete openness of cells and uninhibited breathability is desirable in any cellular polymer foam if comfort is to be achieved.

One means of measuring the breathability of most foam cushions is the use of a Gurley densometer. Basically this instrument is placed against the surface of the foam, and the time required to pass a given volume of air through the foam is observed.

Resilience

To some degree, resilience is a function of the properties of compressive force deflection, sag factor and breathability, but it also obtains its resilient or bouncy character from the chemical design of the cellular polymer, the cellular matrix and the mechanical fabrication of the foam (see Chapter 5). There are numerous methods of measuring resilience, but probably the most impressive is the ball-rebound test. Hartings and Hagan[2] have shown that this test was correlatable to comfort as rated by a panel of judges. These researchers repeatedly dropped $2\frac{1}{4}$ inch diameter steel balls, weighing 764 g on a series of foams rated for comfort by the judges. The percentage rebound (resilience) of the steel ball was observed and is plotted against comfort as shown in Fig. 3.4.

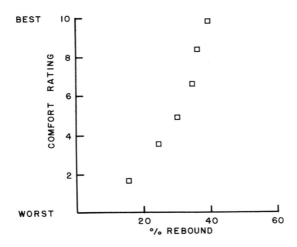

FIG. 3.4. Comfort rating by ball-drop resilience. Reproduced from ref. 5 with permission.

Figure 3.4 clearly shows that as the resilience increases, the comfort rating of the foam cushion also increases. According to the average opinion of the panel of judges, 40% rebound was considered to be the critical value for initial comfort. Foam cushions having rebound values greater than 40% begin to approach a rating of ultimate in cushion luxury.

3.3 MECHANICS OF CUSHIONING

3.3.1 General Considerations

Cushioning as initially defined in Section 3.2 referred to any material that provides a 'fight back' with some degree of comfort. Some materials as also shown previously can qualify for cushioning applications according to this definition. However, polyurethane (henceforth to be referred to as urethane) foam, currently because of its relatively low cost, comfort, ease of fabrication (processability), density and durability is the material of choice in the greatest majority of cushion applications.

Urethane foam is a cellular polymer that derives its cushioning character and properties in part from the cell matrix formed during fabrication. This formation in situ of a cell matrix makes urethane foams unique amongst all other cushioning material, and provides a significant amount of flexibility in designing the cell both physically and chemically.

Each cell of the foam matrix is, to a degree, very similar to the action of a steel spring. When a force or load is applied, the spring is compressed, and as it is compressed because of its design, metallurgy, etc., develops a given amount of potential energy. This energy is constantly available for 'fighting back' against the applied force or load.

The 'fight back' ability of any urethane foam has been shown[3] to be a function of: (i) shape of the cells; (ii) cell-wall thickness; (iii) tackiness of the cells; and (iv) closed-cell content.

3.3.2 Cell Geometry

Many foam cells when formed during the fabrication of urethane foam generally are conical in shape rather than spherical. This is due to the elongation of the cell formed by gas bubbles and their direction of blow (see Section 1.2). When a foam contains a goodly percentage of these elongated or conically shaped cells, an increase in the 'fight back' of the foam is obtained provided the force is applied in the same direction in which the foam blew during formation. Table 3.4 illustrates the difference between the compression force at various deflections for foam measured with the direction of foam blow and perpendicular to the direction of blow. Obviously the orientation of the foam in its end use application is very important for maximising 'fight back'.

TABLE 3.4
Effect of Blow Direction on IFD

		Force (N)	
		25% Deflection	50% Deflection
(With)	↓○	138	414
(Perpendicular)	→○	112	325

3.3.3 Cell-Wall Thickness

This is an important variable as it affects the compressive force of 'fight back' properties of a foam. If two chemically identical foams are prepared at a density of 40 kg m^{-3}, but one having a cell count of 20 cells cm^{-1} while the other has a cell count of 5 cells cm^{-1} the foam with the fewer cells per linear cm will also have the thicker cell walls. Generally this type of thicker-celled foam will have the greater compression–deflection or 'fight back' properties.

3.3.4 Tackiness

Tackiness in the cell walls of any foam is an undesirable property. When compressed by a force, a 'tacky' foam will be slower in recovering from the deformation because of many of the cell walls collapsed by the compressive force have an adhesive attraction for each other. This condition when present in a foam is mostly due to the chemical character of the polymer, which can be either eliminated or minimised by redesigning the polymer.

3.3.5 Closed-Cell Content

Closed cells in a foam are desirable if high compressive properties or 'fight back' is the only property desired. As the closed-cell content increases in a foam, the 'fight back' properties of the foam will also increase. In many urethane cushions, however judged to have closed cells, the cells are not actually closed completely, but have small rupture holes that do permit the passage of air in and out of the individual cells. This type of foam cushion is pneumatic in character and will have a very undesirable comfort index. When a compressive force is applied to the surface of the foam, a large 'fight back' is observed, but this is slowly followed by a settling into the foam by the compressive force as the air is slowly squeezed out of the 'so called' closed cells. The influence of foam permeability on the behaviour under dynamic loading is discussed in detail in the following chapters.

3.4 URETHANE FOAM PREPARATION

3.4.1 Chemistry

An understanding of the formation of urethane foams and all of the related ramifications first involves a consideration of the organic chemistry of the reaction leading to gas formation and polymer growth, bubble nucleation and growth and rheology.

Basically the reactive ingredients of a foams system are normally a hydroxylated-terminated resin, known as a polyol, a di-isocyanate and water. The reaction[4] of the di-isocyanate with the polyol produces a urethane polymer:

URETHANE

$$OCN-R-NCO \;\; + \;\; HO\!\sim\!\sim\!\sim\!OH$$
$$\text{diisocyanate} \qquad\qquad \text{polyol}$$

$$\sim\!\sim\!OCN-R-\underset{H}{\overset{O}{N}}-\overset{\parallel}{C}-O\!\sim\!\sim\!\sim\!O-\overset{O}{\overset{\parallel}{C}}-\underset{H}{N}-R-NCO\!\sim\!\sim\quad (1)$$

The reaction of the di-isocynate with water produces a urea molecule plus carbon dioxide via an amine intermediate.

UREA

$$OCN-R-NCO + HOH \longrightarrow OCN-R-\underset{H}{\overset{O}{\overset{\parallel}{N}}}-C-OH$$

$$OCN-R-\overset{O}{\overset{\parallel}{C}}-OH \longrightarrow OCN-R-N\!\!\begin{array}{c}{\scriptstyle H}\\{\scriptstyle H}\end{array} + CO_2\uparrow$$

$$OCN-R-N\!\!\begin{array}{c}{\scriptstyle H}\\{\scriptstyle H}\end{array} + OCN-R-NCO \longrightarrow OCN-R-\underset{H}{N}-\overset{O}{\overset{\parallel}{C}}-\underset{H}{N}-R-NCO \quad (2)$$

Common to these two reactions is an active hydrogen atom of the polyol hydroxyl group, the hydrogen on the water and amines, and is the source of chemical bond formation.

Other reactions of the di-isocyanate, which basically are side reactions, that can occur during foam preparation are:

BIURET

$$\sim\sim\sim\overset{O}{\underset{}{C}}-\underset{H}{N}-R-\underset{H}{N}-\overset{O}{\underset{}{C}}-\underset{H}{N}-R-\underset{H}{N}-\overset{H}{\underset{}{C}}\sim\sim\sim + OCN-R-NCO$$

$$\downarrow$$

$$\sim\sim\sim\overset{O}{\underset{}{C}}-\underset{H}{N}-R-\underset{H}{N}-\overset{O}{\underset{}{C}}-\underset{\underset{H-N-R-NCO}{C=O}}{N}-R-\underset{H}{N}-\overset{O}{\underset{}{C}}\sim\sim\sim \quad (3)$$

ALLOPHONATE

$$\sim\sim\sim\overset{O}{\underset{}{C}}-\underset{H}{N}-R-\underset{H}{N}-\overset{O}{\underset{}{C}}-O\sim\sim\sim O-\overset{O}{\underset{}{C}}-\underset{H}{N}-R-\underset{H}{N}-\overset{O}{\underset{}{C}}\sim\sim\sim + OCN-R-NCO$$

$$\downarrow$$

$$\sim\sim\sim\overset{O}{\underset{}{C}}-\underset{H}{N}-R-\underset{H}{N}-\overset{O}{\underset{}{C}}-O\sim\sim\sim O-\overset{O}{\underset{}{C}}-\underset{\underset{H-N-R-NCO}{C=O}}{N}-R-\underset{H}{N}-\overset{O}{\underset{}{C}}\sim\sim\sim \quad (4)$$

DIMER

$$OCN-R-NCO \longrightarrow OCN-R-N\overset{\overset{O}{\underset{}{C}}}{\underset{\underset{O}{\overset{}{C}}}{}}N-R-NCO \quad (5)$$

TRIMER

$$OCN-R-NCO \longrightarrow OCN-R-N\overset{O=C}{\underset{\underset{O}{\overset{}{C}}}{}}\overset{\overset{R-NCO}{\underset{}{N}}}{\underset{}{}}\overset{C=O}{}N-R-NCO \quad (6)$$

Equations (1) and (2) are the two key chemical reactions involved in the preparation of most urethane foam cushions, since they contribute to the polymer growth and formation and its cellular character. The other reactions, (3)–(6), in some foam systems can contribute in the mechanics of foaming and in the chemical modification of the final polymer structure.

3.4.2 Structural Consideration of Reactants

The di-isocyanate used in preparing most foam cushions is usually an 80:20 mixture of TDI isomers (2,4- and 2,6-tolylene di-isocyanate).

2,4   2,6   TDI

MDI

To a lesser extent, another di-isocyanate, MDI (4,4'-diphenylmethane di-isocyanate) is used to prepare foam cushions, usually as a blend with the 80:20 TDI.

MDI is a symmetrical di-isocyanate having the isocyanate groups in the *para* position of each aromatic ring and separated by a methylene bridge. The isocyanate groups thus have equal activity, and the reaction of one of these groups during the polymerisation phase of foaming, does not affect the activity of the other NCO group. TDI is an unsymmetrical molecule, and the isocyanate groups do not have equal reactivity. The isocyanate group in the *para*, or 4, position is the most reactive, at least 20 times faster than either the 2- or 6-positioned isocyanate group. This difference is due to the electronic structure of the molecule as well as steric hindrance effect on the 2 or 6 groups after the reaction of the *para* group with a bulky molecule, such as a polyol resin, has occurred.

These unsymmetrical and reactivity properties of TDI give it significant advantages over MDI when used in urethane foam cushions: processing is safer, more controllable; foams have less tendency to shrink; foam physical properties generally are somewhat better; and durability of the foam is improved. Most high-resilience urethane foam cushions currently being made however do require the use of about 20% MDI as part of the isocyanate requirements.

Polyol resins available for use in foam polymerisation also vary both in chemical composition and reactivity. Basically there are two types: a polyester and a polyether, and both types are available as diols or triols.

Diol   Triol

Polyester polyols by the very nature of their chemistry of preparation, i.e. the reaction of a diol with a dibasic acid, enjoy the flexibility of producing a wide variety of different structures designed to meet the requirements of specialised applications. Processability of most of these polyesters is usually more difficult than the processability of a polyether. For this reason, and others such as basic costs, comfort and basic foam properties, polyether polyols are predominately used in the production of urethane foam cushions for automotive, furniture and bedding applications.

Polyether diol polyols are synthesised by the polymerisation of propylene oxide to yield linear diols while the polyol triol is produced by using a monomeric triol as a nucleus initiator for the propylene oxide polymerisation.

```
C-O-C-C-O-(C-C-O)_x-C-C-OH
    |        |        |
    C        C        C
C-O-C-C-O-(C-C-O)_x-C-C-OH   TRIOL
    |        |        |
    C        C        C
C-O-C-C-O-(C-C-O)_x-C-C-OH
    |        |        |
    C        C        C
```

The degree of this polymerisation determines the flexibility and rubber-like character of the urethane polymer because of its glass transition T_g phenomenon.[5]

Figure 3.5 shows the effect of the molecular weight of a triol on the T_g of the urethane polymer. As expected, the T_g decreases from 125°C for a urethane polymer based on a 500 molecular weight triol polyol to $-50°C$ for a polymer based on a 3000 molecular weight triol polyol. This urethane polymer glass transition response appears to result from motions of the

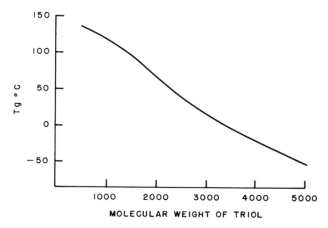

FIG. 3.5. Glass transition (T_g) versus molecular weight of polymer.

oxypropylene segments positioned between fairly rigid urethane and urea bonds in the polymer chain. Decreasing the oxypropylene chain length results in a restriction to chain flexibility which produces an increase in the T_g. The net result of this shifting of the T_g in a polyol triol will be translated into the cellular polymer as stiffness or lack of it.

When low molecular weight polyols are used to prepare the cellular polymer, the T_g will be well above room temperature and the polymer will be very stiff or even rigid. As the molecular weight of the polyol triol is increased, the T_g decreases below room temperature, and the stiffness of the cellular polymer decreases sharply due to the shift to the rubber state from that of the glass state.

Stiffness is an important property of urethane cellular polymers, since it contributes to the load support or 'fight back' characteristics of a cushion. Cellular polymers having a T_g considerably below room temperature derive their stiffness from the structure of the urethane polymer, the disubstituted urea bonds and hydrogen bonding.

3.4.3 Catalyst

The preparation of a urethane cellular polymer cushion involves the initial stoichiometric mixing of the liquid ingredients of polyol triol, di-isocyanate and water to form a master homogeneous liquid mix. Carbon dioxide gas which starts to form within a few seconds, is initially soluble in the liquid reactant mix. This gas becomes insoluble in about 10 seconds and forms a tremendous number of tiny bubbles, and the liquid reaction mix takes on the appearance of a creamy fluid. At this point, the bubbles must be stabilised while the viscosity of the liquid medium increases very rapidly due to the urethane and urea reactions. The way in which the bubbles expand and coalesce to form a low-density cellular structure is described in Section 1.2.

The selection of catalyst to promote the combined polymerisations is very important to the success of the foaming phase. The carbon dioxide must be generated at a rate that is co-ordinated with the urethane reaction which contributes to the viscosity increase of the liquid medium, and the strength and stability of the bubble. When this strength and/or stability is insufficient at the critical stage of the foam rise, i.e. 75–100% completion, coalescence of the bubbles/cells will begin, resulting in a partial or complete collapse of the cellular urethane polymer.

Organic and inorganic bases are used as catalysts to initiate and sustain the water reaction with di-isocyanates, and organo-metallic compounds are used to catalyse the urethane reaction. When these various catalysts are combined in the foaming medium, various combinations of the organic bases and metal organics show a synergystic increase in the reaction rates for the individual reactions.

3.4.4 Surfactant

Formation of a bubble in the liquid phase of a urethane foaming system is called nucleation. It is assisted by a second phase or nucleating agent, by ensuring the stabilisation of the bubble as it forms and grows; without this nucleating agent, foam collapse will result. Nucleating agents (surfactants) as used in cellular urethane polymer based on polyether polyols are generally silicone oils. These silicone oils, by means of variations in the typical molecular structure of dimethylsiloxane-polyoxyethylene.

$$CH_3\ SiO(Me_2\ SiO)_x - \underset{\underset{CH_3}{|}}{\overset{\overset{CH_3}{|}}{Si}} - O\ CH_2 CH_2 (OCH_2 CH_2)_Y\ OR$$

dimethyl siloxane polyoxyethylene

can be made to control the cell size, amount of cells, the openness of the cells and foam height. The overall cell characteristics of a cellular urethane polymer will have a measure of influence on the ultimate properties of the foam such as density, tensile elongation, compression set, breathability, comfort and durability. Selection of the right silicone surfactant is therefore as critical as the choice of the polyol, di-isocyanate and catalysts in designing a cellular urethane polymer to meet the requirements of specific applications.

3.5 DESIGNING A POLYMER FOR COMFORT CUSHIONING

3.5.1 General Considerations

During the period of growth after final acceptance as a cushioning material, cellular urethane polymers evolved[6] through a series of foam systems as listed below:

(1) Polyester prepolymer-slab
(2) Polyester one-shot-slab
(3) Polyether prepolymer-slab or moulded
(4) Polyether one-shot-slab
(5) Polyether hot-cure moulded
(6) Polyether cold-cure-slab or moulded

The first two systems of this series were based on a polyester and differed slightly in the chemistry of preparation. In system (1), the prepolymer or two-step reaction utilised a prereaction of part of the urethane polymer before the foaming phase:

$$2\ UCN-R-NCO + HO\!\sim\!\sim\!\sim\!OH$$
$$\text{Di-isocyanate} \quad \text{Polyester}$$

$$OCN-R-N-\overset{O}{\underset{}{C}}-O\!\sim\!\sim\!\sim\!O-\overset{O}{\underset{}{C}}-N-R-NCO$$
$$\text{Prepolymer}$$

Normally an excess of di-isocyanate is used in the reaction in order to:

1. Ensure a simple capping of the polyol molecule. In theory a 2:1 ratio should be adequate, but usually if this ratio is used a condensation reaction, producing higher molecular weight fractions, results.

$$\text{ISO} - \text{POLYOL} - \text{ISO} - \text{POLYOL} - \text{ISO}$$

2. Keep the viscosity as low as possible. The excess isocyanate acts as solvent diluent and the polyol capped with the di-isocyanates is inherently lower in viscosity than condensed or partially polymerised reaction depicted in 1 above.

The polyester prepolymer containing excess di-isocyanate is reacted with water.

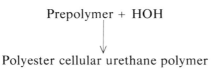

Prepolymer + HOH
↓
Polyester cellular urethane polymer

System (2), a one-shot, which quickly replaced the two-step prepolymer system, is a process in which the di-ioscyanate, polyester polyol and water along with the catalyst and surfactant are all mixed simultaneously.

Although polyester foams were excellent for many applications, they were not readily adaptable to seating applications, due to their initial stiffness at 25% deflection which adversely affected the comfort properties of the foam. Currently, polyester foam has found wide use in speciality applications such as filters, gasketing sponges, novelties and as a fabric foam-laminating material.

As polyether polyol diols became available commercially and were used to prepare urethane foams, it readily became apparent that these cellular polymers potentially offered more comfort than a polyester based foam. Initially the foams (system (3)) were made with diols via the prepolymer technique due to a lack of a proper catalyst system. These foams had improved comfort, but marginal processability, especially for applications such as automotive seating that required a moulding operation.

In order to prepare a satisfactory foam from a polyether diol, it was necessary to build into the prepolymer[7] a small amount of crosslinking sites by means of a biuret or allophonate reaction, as shown in Table 3.5. These prepolymers had very high viscosity at room temperature and were extremely difficult to process in a mixing head with the small amount of water, catalyst and surfactant required for foaming, as the following formulation indicates: prepolymer (9·5% NCO), 100; silicone oil, 0·5; N-methylmorpholine, 1·0; triethylamine, 0·3; water, 2·25.

Many of these polyether prepolymer systems required high-pressure piston-type pumps to transfer the reactants to the mixing head, as did the

TABLE 3.5
Preparation of Biuret and Allophonate Prepolymers

	Biuret	Allophonate
Polypropylene ether glycol, mol. wt. 2000	1·00	100
Water	0·4	0·15
80/20 TDI	15	9
Reaction temp. (°C)	120	80
Reaction time (min)	90	120
80/20 TDI	25	30
Reaction temp (°C)	—	140
Reaction time (min)	—	120
Viscosity at 250°C (Pa)	20	17
% NCO	9·5	9·5

polyester systems. However, with discovery of organotin compunds as being excellent urethane catalysts, and the availability of functional high molecular weight polyols, a simplified low-pressure one-shot system became practical to produce polyether foam. Table 3.6 illustrates a typical one-shot polyether foam formulation divided into components as mixed in processing mixers.

TABLE 3.6
Typical One-shot Polyether One-shot Foam

Compound A—Polyol masterbatch		
	Polyether triol, 3000 mol.wt.	100
	Water	2·6
	Silicone oil	1·0
	Dibutyltin dilaurate	0·03
	Triethylenediamine	0·05
Compound B—Isocyanate		
	80/20 toluenedi-isocyanate	36

Most commercial foams currently produced for use in automotive, furniture or bedding applications use the basic one-shot polyether foam technology for either slabstock or moulded foam. Slabstock is produced by discharging the mixed components into an open mould and allowing it to expand to its designed density. Because of the size of commercial slabstock foam as made, a significant amount of exothermic heat is generated, which is usually sufficient to cure the slabstock foam to its desired properties.

Moulded foam also prepared by using slabstock one-shot polyether foam technology was known as hot-cure foam, because it required the use of a post-cure at high temperatures in order to achieve good foam properties. Comfort in these foams was satisfactory, but the desire by some applicators for the 'ultimate' in seating comfort led to the development of a new generation of foams identified as cold-cure or HR (high-resilience) foam.[8-10] This latest generation of foam developed good foam properties at ambient cure conditions. A comparison of the evolution of comfort in the various urethane foam systems is shown in Table 3.7.

TABLE 3.7
Comfort Evolution in Urethane Foams

	IFD ($N/323\ cm^3$)		Sag factor
	25% Deflection	65% Deflection	
Polyester	148	267	1·8
Polyether prepolymer	134	267	2·0
Polyether 1-shot slab (conventional)	121	267	2·2
Polyether (hot cure)	121	267	2·2
Polyether (HR)	89	267	3·0

The HR foam of Table 3.7 represents the sag factor of moulded foams, whereas the HR slabstock would be slightly lower, i.e. about 2·8. Nearly all automotive cushioning currently uses HR foams, while most of the furniture and bedding applications continue to use conventional one-shot polyether foam.

In addition to the difference in comfort as measured by sag factor, hand feel and ball-drop resilience, other differences exist, as shown in Table 3.8.

TABLE 3.8
Property Comparison between Conventional and HR Foams

	Conventional	HR
Processability	Good	Excellent
Density ($kg\ m^{-3}$)	38	38
Tensile (kPa)	240	200
Elongation, %	280	180
Sag factor	2·2	3·0
Ball-drop resilience, %	52	65
Compression set, %	5	5

At comparable densities, by an averaging of properties, conventional one-shot foams usually have higher stress–strain, comparable compression sets and lower comfort than the HR foams. Processability, while rated as adequate for the conventional one-shot foam, is poorer by comparison to HR foams. Each foam type possesses inherent properties that can be either maximised or minimised in order to adjust and vary its total cushioning performance.

3.5.2 Processability

Fabrication of a successful one-shot polyether foam, with a regular and fine cell structure that is free of splits, voids, closed cells and shrinkage is dependent upon both the techniques of mixing as well as the 'fine tuning' of the chemical components. As shown in Table 3.9 the conventional one-shot foams are more difficult to produce than the HR foams, which is due to basic differences of the chemical components used in each system.

TABLE 3.9
One-shot Conventional and HR Foams

	Conventional	HR
Polypropylene polyol, 3000 mol.wt.	100	—
Polypropylene polyol, 4800 mol.wt.	—	25–75
Polymer polyol, 6000 mol.wt.	—	25–75
Water	2·5	2·7
Stannous octoate	0·4	—
Dibutyltin dilaurate	—	0·01
Triethylenediamine	0·15	0·2
Silicone oil	0·5	1·0
80/20 TDI	35	—
Polymeric MDI 80/20	—	34

Polyols normally used in conventional foams have molecular weights of 3000–4000 and contain three secondary hydroxyl groups. These hydroxyl groups are very slow to react during the foaming reaction, but when properly catalysed, the trifunctional character contributes to a rapid development of the urethane polymer molecular weight and cross-linking density. Concurrently with this reaction, urea bonds are inserted into the developing urethane polymer as carbon dioxide is being generated by the water–isocyanate reaction. When the urethane reaction vs. the water reaction proceeds too rapidly, unacceptable foams possessing one or combinations of splits, voids, shrinkage or high density can result. If the water reaction proceeds too rapidly, foam cells do not form properly, the cell walls are weak, and foam settling or foam collapse will usually result. Many years of both laboratory and production experience have shown the balance of catalytic activity in the conventional foams for most processing applications to be very critical and the latitude for variation very small. HR foams, by contrast, have a significantly larger latitude for controlling the urethane and water reactions through catalyst adjustments. The term 'catalytic window' (depicted in Fig. 3.6) clearly shows this difference.

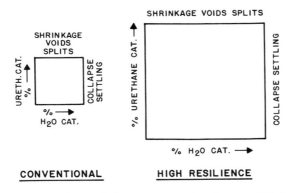

FIG. 3.6. Catalytic windows—conventional versus HR foam.

HR foams enjoy the greater latitude of catalytic variations because of the type of polyols and isocyanates used. Table 3.9 shows a typical HR foam that is based on a 4800 molecular weight polypropylene ether glycol and a 'grafted' polypropylene ether glycol having a 6000 molecular weight. Both of these trifunctional polyols contain primary hydroxyl groups and are about 15 times more reactive than the polyols used in conventional foams. The isocyanate used to react with these polyols, a blend of 80/20 TDI and polymeric MDI, is more reactive than that used for the conventional foam and requires significantly lower amounts of the urethane tin catalyst. The combined effect of the increased reactivity of the polyols and isocyanate blend is one of the basic contributing factors for HR foams being easier to process by enjoying the larger latitude of catalyst variations or 'catalytic window'.

During the initial phase of the foaming reaction, immediately after the homogeneous mixing of the chemical ingredients, the polymeric MDI reacts rapidly with the highly reactive polyols, initiating the beginning of polymerisation and a rapid increase in viscosity. The slower-reacting TDI at this point becomes very active due to the combined effect of the tin catalyst and the exothermic heat of the initial polymerisation.

This step-wise technique of the foam polymerisation permits the use of significantly lower concentration of tin catalyst, which partly contributes to making HR foam considerably easier than conventional foams to process.

3.5.3 Hardness

In either of the two foam systems, hardness can be varied by numerous techniques, all of which will affect the foam comfort to varying degrees.

Polyol Molecular Weight
This influences hardness and, probably more than any other factor, the feel and comfort of a urethane foam. By holding the water isocyanate within the foaming system constant and thus controlling stoichiometrically the generation of carbon dioxide available for blowing, it is possible to achieve a softer foam as the molecular weight of the polyol is increased[11] as shown in Fig. 3.7.

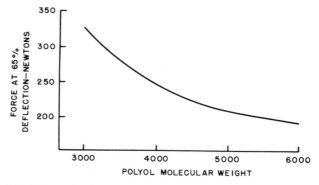

FIG. 3.7. Effect of polyol molecular weight on IFD at 65% deflection.

This trend occurs basically because the cross-linking density decreases, and with this softening of the foam the hand feel and 'creature' comfort increases to a finite point. Beyond this comfort begins to decrease due to loss of 'fight back'.

Graft Polyols
Graft polyols or polymer polyols, a recent innovation,[12,13] are produced by the in situ polymerisation of vinyl monomers in the presence of a conventional polyol. The monomers are usually styrene and or acrylonitrile, and the conventional polyol is a 5000 molecular weight or higher polypropylene ether triol capped to a primary hydroxyl content of 65–90%. These polyols complement the conventional polyol by increasing the hardness at any given molecular weight as shown in Fig. 3.8 and help to retain firmness in the foam as additional comfort is achieved by increasing the polyol molecular weight.

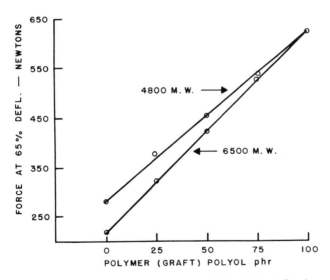

FIG. 3.8. Effect of polymer polyol on IFD at 65% deflection.

Comfort, according to Fig. 3.8, can be maximised without a loss of hardness (IFD) or 'fight back' by increasing the amount of polymer polyol in the foam polymer. A hardness of 270 N for the 4800 molecular weight conventional polyol using 0 phr of polymer polyol is retained when using a 6500 molecular weight conventional polyol along with about 25 phr of the polymer polyol.

Water

As the water reacts with the di-isocyanate during the foaming phase, it not only generates carbon dioxide which blows the foam mass to low densities, but also produces disubstituted urea molecules and polymers which immediately react into the forming urethane foam polymer. As the water concentration is increased in a conventional foam, the density decreases as shown in Table 3.10, but the hardness or 'fight back' of the foam at 65% deflection remains fairly constant.

TABLE 3.10
Effect of Water Concentration on Foam Comfort

Water concentration, phr	1·5	1·65	1·92	2·75
Density (kg m^{-3})	57	52	40	31
IFD (323 cm^3) 25% deflection (N)	145	157	165	175
IFD (323 cm^3) 65% deflection (N)	333	345	329	333
Sag factor	2·3	2·2	2·0	1·9
Yerzley resilience (%)	54	50	36	28

However, the sag factor is shown to decrease with the decreasing density as a result of the increase in the IFD at 25% deflection. The net result of this phenomena, is a trend toward less comfort in the foam as also indicated by the decreasing resilience of the foam. Figure 3.9 shows the change in comfort as the water is increased.

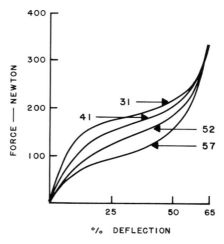

FIG. 3.9 Effect of foam density on IFD.

Concurrently with the loss of comfort as a result of the increase of IFD at 25% deflection, there is also a loss of hand feel or harshness of the foam. These changes of properties are due primarily to the progressive increases of the urea bond as a function of increased water content.

Lower densities, with certain limitations can be achieved without losing as much comfort by holding the water concentration constant and using auxiliary blowing agents, such as trichlorofluoromethane (b.p., 23°C).

As the water is held constant and the auxiliary blowing agent is added in increasing amounts, the density decreases and the IFD both at 25% and 65% deflection decrease at proportional rates, causing the sag factor to remain constant. The decreased IFD in the lower density foam as shown in Table 3.11 probably would fall outside the acceptable specifications for comfort in many cushion applications.

TABLE 3.11
Effect of Auxiliary Blowing Agent on Comfort

Trichlorofluoromethane	0	10	13	16
Density (kg m^{-3})	57	40	38	31
IFD (323 cm^3), 25% deflection (N)	145	108	79	70
IFD (323 cm^3), 65% deflection (N)	333	238	167	146
Sag factor	2·3	2·2	2·1	2·1
Yerzley resilience (%)	54	55	55	56

Dry Mineral Fillers

When added as an ingredient to a foaming mixture, dry mineral fillers can improve foam comfort. These fillers are chemically inert, but do have a modifying influence on the formation of the foam polymer and reinforcing action of the cellular polymer. Table 3.12 shows these two possible responses.

TABLE 3.12
Effect of Mineral Fillers on Comfort

Stannous octoate	0·3	0·55	0·6
Talc 42R	—	6	—
BaSO$_4$	—	—	10
Density (kg m^{-3})	29	29	30
IFD (323 cm^3) 25% deflection (N)	151	105	129
IFD (323 cm^3) 65% deflection (N)	257	220	270
Sag factor	1·7	2·1	2·1

Other properties not shown in the table, such as tension and elongation, are adversely affected, but the level of comfort achieved was significantly increased due to the decrease of the IFD values at 25% deflection plus the increase in IFD ratios.

Isocyanate Index

The isocyanate index is the stoichiometric ratio of isocyanate groups to the combined hydroxyl groups and water. If the ratio is less than 1, a stoichiometric deficiency of the isocyanate exists, whereas a ratio greater than 1 indicates an excess of isocyanate groups available for reacting into the urethane polymer. The highest theoretical polymer molecular weight is achieved by a 1·0 index. Comfort, as measured by sag factor and resilience is optimised, within the limits of this variable, while also achieving other properties such as tension, elongation and compression sets that are satisfactory by using an index of between 1·0 and 1·5 (see Table 3.13). As the index is increased to 1·1 and above, the foam becomes stiffer, sag factor decreases slightly and the foam begins to lose some of its rubbery feel characteristic.

TABLE 3.13
Effect of Isocyanate Index on Foam Comfort

Isocyanate index	0·95	1·0	1·5	1·1
Density (kg m^{-3})	37	35	35	38
IFD (323 cm^3) 25% deflection (N)	93	97	95	93
IFD (323 cm^3) 65% deflection (N)	285	271	265	250
Sag factor	3·0	2·8	2·8	2·7

This trend is due to the formation of many undesirable by-products and side reactions such as biuret and allophonate branches and poly urea molecules which increase the cross-linking density, and low molecular weight polymers which has a net effect of decreasing the number average molecular weight of the cellular urethane polymer.

Temperature

The temperature of the foaming ingredients influences the comfort of most conventional and HR foams and also the reaction rates of the urethane and water reaction. As the temperature increases, the reaction rates increase, necessitating a reduction in the amount of catalyst, primarily the urethane catalyst, in order to retain good control of the foaming mass. Significant improvements in the comfort of two classes of cellular urethane polymers can be achieved by this means, as shown by the sag factor in Table 3.14.

TABLE 3.14
Effect of Ingredient Temperature on Comfort

Foam A			
Ingredient Temp (°C)	22	32	38
Density (kg m^{-3})	29	29	30
IFD (323 cm^3) 25% deflection (N)	138	131	117
IFD (323 cm^3) 65% deflection (N)	249	249	236
Sag factor	1·8	1·9	2·0
Foam B			
Ingredient Temp (°C)	27	41	
Density (kg m^{-3})	45	43	
IFD (323 cm^3) 25% deflection (N)	110	73	
IFD (323 cm^3) 53% deflection (N)	306	222	
Sag factor	2·8	3·1	

Openness of Cells
The openness of cells varies as a function of the foam type and how it is processed. Most of the HR foams as prepared have closed cells which must be broken in order to maximise comfort of the foam. If these cells can not be broken as shown previously, either foam shrinkage will occur or the foam will have a pneumatic feel which will greatly minimise much of the comfort that is designed into the foam. Of all the techniques available for minimising the development of closed cells in a foam as it is forming and 'setting' after completion of foam rise, the correct choice of surfactant to stabilise the foaming mass has generally been the most important one. Many surfactants for stabilising polyether foams are available from numerous suppliers and their performance must be matched to the specific requirements of each foam system. More recently,[13] it has been shown that polymer (graft) polyols are excellent in HR foams for not only building hardness or 'fight back' in a foam, but are excellent for controlling the closed-cell content and reducing the force required to crush the foams to eliminate residual closed cells.

Hydrogen Bonding
Hydrogen bonding is a phenomenon that occurs in any chemical system that possesses labile hydrogen atoms and oxygen atoms of carbonyl groups. In a urethane foam, a hydrogen bond exists between the hydro-

gen of either a urea or urethane bond and the oxygen of a carbonyl group of either one of these two bonds in another chain (see Fig. 3.10.)

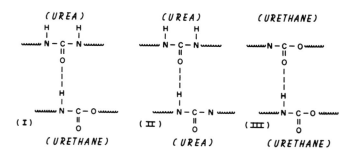

FIG. 3.10. Hydrogen bonding in urethane foam.

It is this newly formed but weak bond that forms automatically as a cellular urethane polymer develops and grows in molecular weight, that contributes much of the IFD character of 'fight back' in foams designed for cushioning applications. A foam cushion based on a 6500 molecular weight triol has crosslinking sites spaced geometrically in the polymer by about 2000 molecular weight units. If hydrogen bonds did not form in this polymer, the cushion would have very little load-carrying capacity or 'fight back'. A foam cushion based on a 300 molecular weight triol with crosslinking sites spaced by about 1000 molecular weight units also must rely on the formation of hydrogen bonds for load support. Because the 3000 molecular weight foam cushion polymer has a higher crosslinking density it has a greater load-carrying capacity than the 6500 molecular weight based cushion at comparable foam densities.

By comparison, rigid foams which are not suitable for any type of cushioning application have an extremely high crosslinking density and do not need the phenomenon of hydrogen bonding for developing sufficient IFD or hardness.

3.5.4 Ultimate Design

Property requirements for cellular-foam cushions can vary considerably as a function of the application. By virtue of its definition, cushioning does require some degree of comfort. Some applications, such as automotive, require the best comfort possible. This requirement can be achieved by combining the principles previously discussed. Other properties, such as compression set, durability, stress–strain, humid ageing, etc., should be considered when combining these principles. It may be necessary to compromise some of the ultimate in comfort in order to achieve the desired combination of foam properties. Such an optimised foam cushion is illustrated in Table 3.15.

TABLE 3.15
Urethane Foam Cushion Optimised for Comfort

Formulation	
Polypropylene ether triol mol.wt. 6500	50
Polymer (graft) polyol mol.wt. 6000[a]	50
Water	2.5
Silicone surfactant[b]	1.5
Dibutyl tin dilaurate	0.02
Triethylenediamine	0.2
Bis(2-dimethylaminoethyl) ether[c]	0.15
80/20 TDI/Polymeric MDI (1·05 index)[d]	32
Properties	
Density (kg m^{-3})	43
Sag factor	3.0
IFD (323 cm^3) 25% deflection (N)	138
IFD (323 cm^3) 65% deflection (N)	413
Ball-drop resilience (%)	75
Compression set, 90% deflection (%)	8
Tensile (kPa)	150
Elongation (%)	180

[a] Union Carbide Niax Polyol 34–28.
[b] Union Carbide silicone L5303.
[c] Union Carbide Niax A-1.
[d] 80/20 Blend 2, 4 and 2, 6–TDI.

For those applications that do not require the high comfort level obtained in the HR foam of Table 3.15, less comfortable conventional foams can be manufactured at lower costs with properties acceptable for those applications (see Table 3.16).

The design of these three foams (Tables 3.15 and 3.16) considered basically various comfort levels along with associated foam properties of compression sets, etc. No consideration has been given to the durability, commonly known in foams as fatigue, when used in both dynamic as well as static applications. In-use experience is now showing that the resistance to fatigue is as important a foam property as comfort in many of the current applications.

TABLE 3.16
Conventional Urethane Foams–Two Comfort Levels

	A	B
Formulation		
Polypropylene ether triol, mol.wt. 3000	100	100
Water	2·3	3·6
Silicone surfactant[a]	1·0	1·0
Stannous octoate	0·4	0·3
Triethylenediamine	0·15	0·1
80/20 blend of 2, 4 and 2, 6–TDI (1·05 index)	33	46
Properties		
Density (kg m^{-3})	39	25
Sag factor	2·14	1·75
IFD (323 cm^3) 25% deflection (N)	151	160
IFD (323 cm^3) 65% deflection (N)	322	280
Ball-drop resilience (%)	55	42
Compression set, 90% deflection (%)	6	5
Tensile (kPa)	96	130
Elongation (%)	200	250

[a] Union Carbide silicone L-520.

3.6 FATIGUE IN URETHANE CUSHIONS

3.6.1 General Considerations

Fatigue in urethane foam cushions may be defined as a loss of 'fight back' or load-bearing and/or thickness incurred under a constant deflection or at a constant load in applications including static or dynamic conditions. It is a problem that is found in each of the major cushioning applications of automotive, furniture and bedding.

Fatigue in automotive cushioning results from dynamic conditions and manifests itself with a changing of the eye level as a function of the usage of the car. Eye level for the driver of the car is defined in reference to the top dead centre of the steering wheel. Initially, the eye level might be an inch or more above the steering wheel, but after several years of driving the car, the eye level can fall below the steering wheel. In furniture, fatigue is a result of static conditioning and is manifested in the development of soft spots with time in the areas of the cushion receiving the most usage accompanied by a possible 'bottoming out' if the fatigue is severe. Bedding also suffers from static fatigue by a softening of the foam in the area of the mattress receiving the most usage. The net result is the formation of valleys or 'bellies'.

Since the adoption of polyether urethane foams as a material for cushioning, significant advances have been made in the techniques of minimising *fatigue*. However, it is doubtful that foam fatigue can be completely eliminated.

3.6.2 Mechanism of Fatigue

Theoretical considerations suggests the mechanism of fatigue[3,14–16] in urethane foam cushions is a physical and chemical phenomenon involving the hydrogen bonds. As discussed in Section 3.5.4 hydrogen bonds in an unstressed condition exist between the hydrogen of either a urea or urethane bond and the oxygen of a carbonyl group of either one of these two bonds in another chain. The examples of Fig. 3.11 illustrate the stressed and unstressed conditions of the two chains. Fig. 3.11A shows the original unstressed conditions. In Fig. 3.11B the same two chains are under stress, the bottom chain is pictured to have shifted forming new hydrogen bonds more favourable to the stressed position.

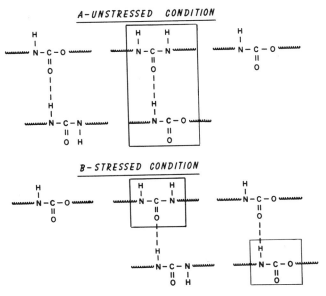

FIG. 3.11. Hydrogen bonding shift—unstressed to stressed.

Upon removal of stress, the hydrogen bonding slowly tends to return to its preferred position in the unstressed condition, but does not completely attain this position. The difference between the original unstressed condition and the final recovered condition represents the phenomenon of fatigue. Since hydrogen bonding contributes to part of the total hardness or load-bearing capacity of the foam, the fatigue is translated into a loss of hardness, or softening of the foam accompanied by a loss in foam height.

The fatigue phenomenon can be demonstrated by the use of the constant deflection static fatigue test ASTM D3574-77.

In the method, a foam specimen of known IFD is compressed to 75% deflection between two platens and held in this position for a period of 22 h at $23 \pm 2°$ C and $50 \pm 2\%$ relative humidity. After release of the specimen the IFD is measured again after 30 min. The difference between the two IFD or load measurements is the loss in IFD and represents the fatigue for the first cycle. The major loss of IFD or fatigue occurs during this first cycle. If the IFD is measured 24 h or longer after release of the specimen from the stress condition, the foam will recover a substantial portion of the IFD measured at 30 min. This recovery of the foam when it is not compressed gives the illusion of a cushion that has not fatigued. However, all foams have a memory by exhibiting a cumulative fatigue (see Fig. 3.12).

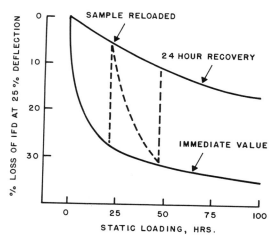

FIG. 3.12. Static loading and recovery.

As this foam sample in Fig. 3.12 is subjected to additional static compression cycles, the IFD at the recovery curve rapidly drops down to the immediate value curve and continues to lose IFD through the completion of cycle 2. After a long recovery time, the hydrogen bonds return to the position of the unstressed condition and the foam again enjoys the illusion of having fatigued but very little when the IFD returns to the recovery curve level. This trend continues through the life of the foam during each cycle of applying a stress and then removing the stress.

The fatigue memory of a foam cushion can further be clarified by the rear-mirror technique of an automobile. Shortly after a driver settles into the car seat when commencing a trip, an adjustment is made for clear viewing in the rear mirror. About ten or more minutes after making this initial adjustment, another adjustment is necessary because the driver has settled a little more into the seat cushion due to the rapid return to the previous IFD fatigue point of the cusion.

If hydrogen bonding is the primary cause of fatigue in urethane foam cushions, then according to the hypothesis, fatigue can be minimised by reducing by any means available, the carbonyl and hydrogen sites in the foam polymer.

3.6.3 Test Methods

During the growth period of the urethane foam-cushion market numerous laboratory techniques for measuring fatigue in a foam under static or dynamic conditions have been used. Currently there are two standard ASTM methods[17] used by researchers to measure accelerated fatigue properties of slabstock and moulded foam cushions.

1. ASTM D-3574-80
Static Fatigue, Constant Deflection—22 h
2. ASTM D-3574-80 Procedure A
Dynamic Fatigue, Constant Load—8000 cycles
ASTM D-3574-80 Procedure B
Dynamic Fatigue, Constant Load—20 000 cycles

These two methods are based on previous standard ASTM methods and are basically similar with minor modification. For example, the static method currently requires 22 h of constant deflection at a temperature of 23°C. Previously the method required 17 h at the same temperature, and was used primarily by most researchers.[14,15,18]

Other methods used by these early researchers include:

3. Low Frequency, Constant Deflection—250 000 cycles
4. Static Fatigue, Constant Load—17 h
5. Linear Shear, Constant Deflection—20 000 cycles
6. Linear Shear, Constant Load—20 000 cycles
7. Rotary Shear, Constant Load—50 000 cycles

In addition to these methods, equipment such as the jounce test machine has been and is currently being used to test cushions dynamically for automotive applications. Other innovative tests, based on the compressive-deflection method have been used to measure fatigue on small samples.

3.6.4. Prediction of Service Performance

The reliability of using accelerated laboratory fatigue test data in predicting in-use performance has been the objective of all previous and current studies.[2,15,19-21] Investigators,[22] in one of these studies found a correlation in performance trends of different slab-stock foams by comparing static and dynamic fatigue test results with in-use testing in a public building (see Fig. 3.13). A comparison of the slab-stock cushions A, B and C shows a good correlation between the loss in IFD of the control cushions in laboratory testing and the actual in-use performance of identical cushions.

Another investigator[15] also has found good correlation between accelerated laboratory fatigue testing and in-use behaviour of foam mattresses. The data of Fig. 3.14 show that at 25% force deflection, the IFD value obtained for the mattress sample tested in the laboratory is comparable to the IFD of the same mattress sample fatigued by in-use testing.

In both of these independent studies it was generally concluded that the accelerated test for fatigue by the static method at constant deflection was a useful technique for predicting field performance. However, in contrast to these findings, a statistical study[18] of fatigue by static constant loading showed the best correlation to in-use behaviour with respect to thickness loss of the cushion but not the IFD.

The results of these conflicting studies and others[23,24] illustrate the problem of attempting to predict the service performance of cellular urethane polymers in cushioning applications from the data of accelerated fatigue testing. Nevertheless, accelerated fatigue tests do offer a good technique to measure response in chemical modifications of the foam polymer as well as mechanical processing as they affect the fatigue and estimating comparative performances in field applications.

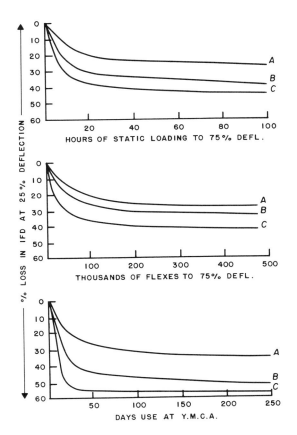

FIG. 3.13. Fatigue test comparison of furniture cushions.

3.7 OPTIMISING FATIGUE RESISTANCE IN CUSHIONS

3.7.1 General Considerations

All urethane foams as shown previously are based on a number of ingredients which contribute to the formation of the molecular structure of the foam polymer. According to the hydrogen-bond hypothesis, fatigue can be minimised in urethane foam cushions by reducing the concentration of labile hydrogens and carbonyl groups. Both of these factors are inserted into the molecular structure by the formation of urethane bond/linkages and urea bond/linkages. Thus a careful selection and concentra-

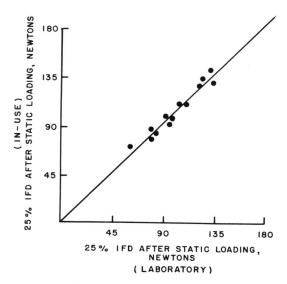

Fig. 3.14. Fatigue test comparison of foam mattresses.

tion of the bond-linkage forming ingredients along with the other foam ingredients that influence and control foam formation is very critical.

3.7.2 Effect of Formulation

Polyol

The polyol is one of the major components in a urethane foam system that contains functional groups through which the urethane polymer is formed. A recent statistical study[25] using an adduct isocyanate in a fixed foam formulation (see Table 3.17) examined the structural variation of molecular weight functionality, primary vs. secondary hydroxyl groups and conventional vs. graft of polyether polyols. Each one of these variables were examined individually and many blends were studied; equivalent weights were held constant within a given series (see Fig. 3.15).

A number of trends were observed from this study (see Fig. 3.16):

  (i) The polyol, 6500 molecular weight, represented by point 1 of Fig. 3.15 gives the best fatigue resistance.
  (ii) Foams based on trifunctional polyols have better fatigue resistance than those based on lower functionality polyols.
  (iii) Dilution of 6500 molecular weight trifunctional polyol with lower

TABLE 3.17
General Formula for Evaluating Polyol Structure and its Effect on Fatigue

Polyol	100
Silicone[a]	0.02
Water	2.7
Dibutyltin dilaurate	0.025
Triethylenediamine	0.2
Niax A-4[b]	0.9
HRL-3895[c]	1.04 index

[a] Dow Corning F11630 silicone.
[b] Union Carbide–amine catalyst.
[c] Du Pont–adduct isocyanate.

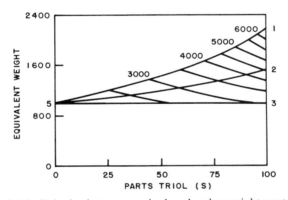

FIG. 3.15. Polyol mixtures nominal molecular weight contours.

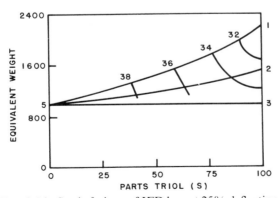

FIG. 3.16. Static fatigue of IFD loss at 25% deflection.

molecular weight diol and triols leads to a reduced fatigue resistance.

These trends support the hydrogen-bonding hypothesis in two ways: (i) by increasing the polyol molecular weight, the concentration of carbonyl groups and labile hydrogens are reduced; and (ii) foams based on triols depend less on hydrogen bonding for load bearing than foams based on polyols or polyol blends with lower functionality. Thus when hydrogen bonding shifts under stress, and does not completely reverse when stress is removed, the loss of IFD is less for the higher molecular weight triol based foam.

Isocyanate

There are at least three basic chemical systems for preparing a HR urethane foam, all of which possess good foam physical properties including excellent processing and comfort (see Table 3.18). System B foams, based on a polyol of 6500 molecular weight and a nominal functionality

TABLE 3.18
HR Foam Chemical Systems

	A	B	C
Polyol	×	×	×
Graft polyol	×	×	
Crosslinker (amine)	×		
Undistilled TDI	×		
TDI/Polymeric MDI		×	
Adduct isocyanate			×

of three, have better resistance to fatigue when statically tested than System A or C foams based on the identical polyols (see Fig. 3.17). This difference in fatigue performance can be explained by the hydrogen-bond hypothesis. Foams made by System A contain an amine crosslinker which is needed to provide good gelation and stabilisation during processing. System C uses an isocyanate adduct which is prepared by reacting TDI with a polyamine to form a biuret. Both these Systems, A and C, contain within the formed foam polymer more carbonyl and labile hydrogen bonds than System B, and thus are more sensitive to fatigue.

Water

Water reacts with the isocyanate to produce disubstituted ureas which are inserted into the urethane polymer as both monomeric and polymeric

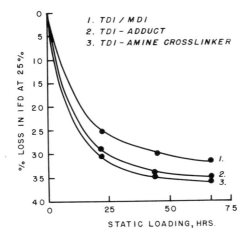

FIG. 3.17. Effect of isocyanate type on IFD loss at 25% deflection.

linkages. Figure 3.18 shows the effect of adding increasing amounts of water on IFD loss at 25%. As the water is increased, the fatigue of the resulting foam also increases.

As shown previously, the water reaction is needed to make a successful

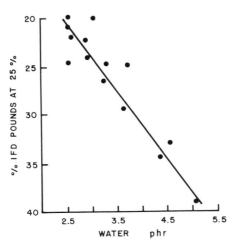

FIG. 3.18. Effect of water on IFD loss at 25% deflection.

foam cushion, but the presence of these urea bonds and linkages in the foam polymer chain are not desirable since they also contain carbonyl groups and labile hydrogen atoms which enhance the adverse effect of fatigue through hydrogen bonding. The increase in foam fatigue is not due to the mechanical effect of reduced foam density.

Density reduction by the use of water has normally been the technique to achieve reduced foam costs in primarily furniture, bedding and recreational vehicle applications. Foam quality as a result of this practice has generally suffered in the properties of fatigue and comfort. A specific example can be cited concerning the foam cushion in a recent vintage recreational vehicle, having fatigued very badly over a period of $1\frac{1}{2}$ years of normal use.

This fatigue was manifested by severe pocketing and 'bottoming out' of the foams. Apparently the trend to reduce foam density will continue due to the continued spiralling cost of foam ingredients. HR foam is currently available only in automotive seating and the top lines of furniture.[26]

Fatigue or loss of IFD can be minimised at lowered foam densities by using an auxiliary blowing agent such as fluorotrichloromethane or methylene chloride. The foam will generally be softer, i.e. the IFD will be lower but the loss of IFD after static loading will show little or no loss.[27]

Isocyanate Index

Increasing the isocyanate index, or the amount of isocyanate available to react with the reactive hydrogens of polyols, water, crosslinkers or similar compounds has a beneficial affect on fatigue resistance of foam cushions. Figure 3.19 illustrates this point by showing an increase in fatigue resistance by a decrease in % IFD loss after static loading at both 25 and 65% deflections. At some higher index, the improvement will plateau and the fatigue improvement is reversed as indicated by an increase in the %IFD loss. Other foam properties would also be affected adversely by using a higher isocyanate index to prepare the foam.

According to the hydrogen-bonding hypothesis, the excess isocyanate groups available above an index of 1·0 will react with the labile hydrogens of both the urethane and urea bond linkages to form allophonate and biuret linkages. The total amount of hydrogen atoms available for bonding is decreased, there is less rearrangement of crosslinking during a stress condition and the foam thus shows less fatigue as measured by the % of IFD loss.

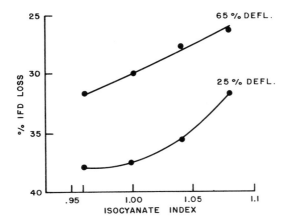

FIG. 3.19. Effect of isocyanate index on fatigue at 25% and 65% deflection.

Catalysts

Both conventional and HR urethane foams are normally prepared using a dual catalyst of metal salt, e.g. organo-tin, and an amine or a combination of amines. Fatigue in conventional foam[18] using a stannous octoate and triethylene catalyst combination was recently found to increase slightly as the concentration of catalyst was increased. In a similar investigation,[25] but based on HR foams, four catalysts were studied using a 6500 molecular weight polyol and an isocyanate adduct. Study was made using statistical techniques of varying the catalysts of dibutyltin dilaurate, triethylenediamine (DABCO), Niax A-1 and Niax A-4. The effect of these catalysts on foam fatigue as measured by % IFD loss under static loading is significant (see Fig. 3.20).

Analysis of the % IFD loss as a function of varying the combinations of catalysts shows that fatigue resistance is optimised:

(i) at low levels (including zero) of DBTDL.
(ii) at moderate to low levels of Niax A-1.
(iii) at high levels of Niax A-4.
(iv) at moderate to high levels of triethylenediamine (DABCO)

The catalytic activity of each of the catalysts (urethane promoting—dibutyltin dilaurate; urethane and urea blowing promoting—triethylenediamine and Niax A-1; urea (blowing)—Niax A-4) confirm these trends towards reducing fatigue. Polyols used in preparing HR foams are very reactive because of the primary hydroxyl groups, and excessive catalysis

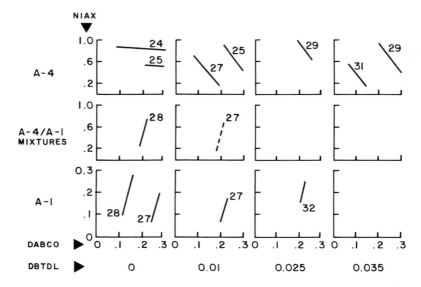

FIG. 3.20. Catalyst study—static fatigue 25% IFD loss.

of the urethane reaction throws it out of balance with the urea reaction needed for minimising fatigue.

Surfactant
Silicone surfactants, as shown previously, are used to stabilise the rising foam mass and control the formation of foam cells. The size and porosity of the foam cells have an effect on foam fatigue. Foams containing fine cell structure and a high degree of porosity generally exhibit better durability or fatigue resistance than foams with coarse cell structure and/or porosity.

3.8 DESIGN FOR HIGH COMFORT AND GOOD FATIGUE RESISTANCE

Both comfort and fatigue resistance have much in common; both are desirable in foam cushions, most of the chemical variables which improve the one, improve the other, and while achieving high comfort and good fatigue resistance, the other foam properties of stress–strain, compression set and others are not compromised. Economics will generally tend to fix the foam density as low as possible while maintaining the optimisation of comfort and fatigue resistance. Table 3.19 shows an ideal formula designed to achieve the maximum in the performance of a cellular polymer cushion.

TABLE 3.19
Urethane Foam Optimised for Comfort and Fatigue

Formulation	
Polypropylene ether glycol (primary OH), 6500 mol.wt.	50
Polymer polyol, 6500 mol.wt.	50
Surfactant[a]	1·5
Water	2·7
Triethylenediamine[b]	0·3
Niax A-4[c]	1·0
80/20 TDI/polymeric MDI[d]	1·08 index
Properties	
Density (kg m^{-3})	42
Sag factor	3
IFD (323 cm^3) 25% deflection (N)	138
IDF (323 cm^3) 65% deflection (N)	414
Ball-drop Resilience, %	74
IFD loss after static loading (%) at 25% deflection (N)	29
IFD loss after static loading (%) at 65% deflection (N)	23
Tensile (kPa)	165
Elongation (%)	170
Compression set at 90% (%)	8

[a] Union Carbide—Silicone L5303.
[b] Air Products—DABCO. Texaco Chemical Co.—Thancat TD33.
[c] Union Carbide—amine catalyst.
[d] 80/20 mixture of 2,4- and 2,6-tolylene di-isocyanate.

3.9 SUMMARY

During the growth of cellular polymers for cushioning applications, a considerable amount of research and development effort was expected adequately to define the quality and requirement needs of potential polymeric materials. Polyurethane polymers were ultimately and universally accepted because of economics, ease of processing and foam properties.

The physical properties of these foams are variable but controllable within the inherent limits of the foam molecular structure. Comfort and durability (fatigue), which are probably the two most important foam physical properties in most cushioning applications, are optimised by carefully selecting the foam ingredients and concentrations of polyol, isocyanate, water catalyst, surfactant and other factors, such as isocyanate index.

Accelerated testing of urethane foams for fatigue by means of both static and dynamic methods are useful for relative comparison of different foam systems. To some extent these results are correlatable to in-use performance.

REFERENCES

1. ROGERS, J. H. and HECKER, K. C. (1973) *Rubb. Technol.*, 459.
2. HARTINGS, J. W. and HAGAN, J. H. (1978) *J. Cell. Plast.*, **14**, 81.
3. SAUNDERS, J. H. (1960) *Rubb. Chem. Technol.*, **33**, 1293.
4. ARNOLD, R. G., NELSON, J. A. and VERBANC, J. S. (1957) *Chem. Rev.*, **57**, 47.
5. WHITMAN, R. D. (1963) *7th Annual Technical Conf. Proceedings—SPI*, Sec. 2C.
6. WOLFE, H. W. (1972) Rubber Division, ACS-Boston. Symposium on Urethanes.
7. WOLFE, H. W. and TUFTS, E. (1958) *Du Pont Co., Foam Bulletin, HR-27*.
8. GRAY, A. H., REGES, B. M. and WOLFE, H. W. (1972) *J. Cell. Plast.*, **8**, 214.
9. PATTEN, W. and PRIEST, D. C. (1972) *J. Cell. Plast.*, **8**, 134.
10. WOLFE, H. W. (1972) *Proceedings, 3rd SPI Int. Cell Plastic Conf.*, p. 451.
11. WOLFE, H. W. (1971) *Du Pont Co. Internal Report*.
12. PATTEN, W., ROSE, C. V. and BENSON, A. S. (1973) *J. Cell. Plast.*, **9**, 82.
13. PATTEN, W., SEEFRIED, C. G. and WHITMAN, R. D. (1974) *J. Cell. Plast.*, **10**, 276.
14. BEALS, B., DWYER, F. J. and KAPLAN, M. (1965) *J. Cell. Plast.*, **1**, 32.
15. KANE, R. P. (1965) *J. Cell. Plast.*, **1**, 217.
16. TERRY, S. M. (1971) *J. Cell. Plast.*, **7**, 229.
17. Annual ASTM Standards. (1980) *Part 38*.
18. DWYER, F. J. (1976) *J. Cell. Plast.*, **12**, 20.
19. BALL, G. W. and DOHERTY, D. J. (1967), *J. Cell. Plast.*, **3**, 223.
20. LLOYD, E. T. (1973) *J. Cell. Plast.*, **9**, 262.
21. DOHERTY, D. J. and BALL, G. W. (1967) *J. Cell. Plast.*, **3**, 291.
22. TOUHEY, W. T. and KNOX, R. E. (1961) *Du Pont Co. Foam Bulletin, July*.
23. STEINGISER, S., DARR, W. C. and SAUNDERS, J. H. (1964) *Rubb. Chem. Technol.*, **37**, 38.
24. MARCHANT, R. P. (1972) *J. Cell. Plast.*, **8**, 85.
25. WOLFE, H. W., BRIZZOLARA, D. F. and BYAM, J. D. (1977) *J. Cell. Plast.*, **13**, 48.
26. Markets and Directory. (1979) *U. S. Foamed Plastics*, 6.
27. WOLFE, H. W. (1971) *Du Pont Co. Internal Report*.

Chapter 4

DYNAMIC MECHANICAL BEHAVIOUR

N. C. Hilyard
Department of Applied Physics, Sheffield City Polytechnic, Sheffield, UK

NOTATION

B	Flow inertia coefficient
E_g	Bulk modulus of compressible fluid
$E(\omega)$	Dynamic modulus of fluid-filled foam
$E'(\omega)$	Storage modulus of fluid-filled foam
$E'_m, E'_m(\omega)$	Storage modulus of polymer matrix
$E_t, E_t^*(\omega)$	Effective tensile modulus for transverse deformation
K	Flow-permeability coefficient
$K(\omega)$	Effective flow-permeability coefficient
L	Length of foam test piece
P	Excess pressure
$T(\omega)$	Transmissibility
T_g	Glass transition temperature of matrix polymer
W	Width of test piece
$d(\omega)$	Loss tangent of fluid-filled foam
$d_m, d_m(\omega)$	Loss tangent of polymer matrix
e	Fractional compression
e_o	Amplitude of oscillatory compression
f	Linear frequency
h	Height of foam test piece
j	$\sqrt{-1}$
t	Time
α	Parameter for gas-filled cellular polymers eqn (4.19)
β	Parameter for liquid-filled cellular polymers eqn. (4.8) and (A2)

γ	Parameter for fluid-filled cellular polymers, eqn. (4.7)
δ	Phase angle between stress and strain
ζ	Volume fraction of open cells
η	Viscosity of fluid
ω	Angular frequency
ω_n	Natural frequency

4.1 INTRODUCTION

Theories relating the stiffness of a flexible cellular polymer to its composition and structure, Chapter 2B, show that at small strains the elastic modulus is directly proportional to the modulus of the matrix polymer. As a result it would be expected that the small-strain dynamic properties of a foam will depend on the frequency of deformation and the temperature, in much the same way as the properties of the matrix polymer depend on these variables. In practice this is not always the case. The differences in the viscoelastic behaviour are due to mechanisms associated with the enclosed fluid phase. At low frequencies, the stress required to deform the foam is made up of two components. These are the stress associated with the deformation of the matrix and the stress associated with the deformation or transport of the fluid within the matrix. The fluid-flow process is essentially an energy-dissipating mechanism, and in open-cell foams it can have a significant effect on the dynamic mechanical behaviour.[1] With some foam-fluid combinations the density to modulus ratio is large so that at high frequencies there are additional stresses related to the motion of the distributed mass of the system.

For material systems possessing stiffness and damping the dynamic stress $\sigma(\omega)$ can be related to the strain $\varepsilon(\omega)$ by the expression $\sigma(\omega) = E^*(\omega)\varepsilon(\omega)$ where the complex modulus given by $E^*(\omega) = E'(\omega) + jE''(\omega)$ is both frequency and temperature dependent. The storage modulus $E'(\omega)$ is that part of the stress–strain ratio associated with the elastic process which is in-phase with the applied strain. The loss modulus $E''(\omega)$ is that part of the stress–strain ratio which is $\pi/2$ radians out of phase with the strain. The complex modulus can also be written in the form $E^*(\omega) = E'(\omega)\{1 + jd(\omega)\}$ where the loss tangent, $d(\omega) = E''(\omega)/E'(\omega)$, is proportional to the ratio of the energy dissipated to the energy stored during one deformation cycle.

In this chapter we will first consider the dynamic mechanical behaviour of the polymer matrix in the absence of fluidic effects. The way in which the enclosed fluid influences the small-strain dynamic behaviour is then described. The design of vibration-isolation systems which exploit the useful characteristics of fluid-filled open-cell foams is discussed in the final section. In the theoretical analysis it is assumed that the strain amplitude is small so that for the polymer matrix the stress increases in proportion to strain. Unless otherwise stated, experimental data presented below for compressive deformation were obtained using strain amplitudes of 1% or less.

4.2 THE POLYMER MATRIX

To study the dynamic mechanical behaviour of the matrix of a cellular polymer it is necessary to ensure that the effects of the enclosed fluid are negligible. This can be done in a variety of ways, such as by performing the tests in a vacuum, by using simple shear (isovolumetric) deformation or by using a sample geometry and frequency range such that the forces associated with the fluid phase are negligible compared with those due to the deformation of the matrix polymer. These conditions can be established using the equations given in Section 4.3.

This last procedure was adopted in the investigation of the effects of frequency and temperature on the dynamic mechanical behaviour of reticulated and non-reticulated PUR flexible foams.[2,3] The test specimens were in the form of rectangular blocks which were subjected to forced vibration in the compression mode. The apparatus used has been described previously.[4] The frequency range was from about 0·9 Hz to 12 Hz. Values of the reduced storage modulus $E'_m(\omega)$ were determined and master curves established by shifting $E'_m(\omega)$ along the frequency axis as described in many texts on the mechanics of polymers, e.g. Ferry.[5] Results for a non-reticulated polyester PUR flexible foam are given in Fig. 4.1. It is seen that the data for $E'_m(\omega)$ superposed reasonably well giving a continuous master curve. However, when the same shift factor a_T was used to superpose the $d_m(\omega)$ data there was considerable scatter about the curve. This was attributed primarily to errors in measurement resulting from the use of small deformation amplitudes with systems having low stiffness. The empirically determined shift factors for this foam are plotted as a function of temperature in Fig. 4.2 together with those for a reticulated PUR flexible foam determined in the same way. The reference temperature $T_s = T_g + 50°C$ in each case. Over most of the temperature

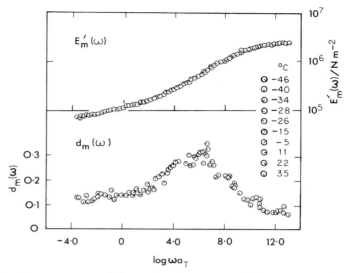

FIG. 4.1. Master curves of the reduced storage modulus $E'_m(\omega)$ and the loss tangent $d_m(\omega)$ for the polymer matrix of a non-reticulated polyester PUR flexible foam with $\phi \approx 0.033$ and $T_g \approx -46°C$.[2]

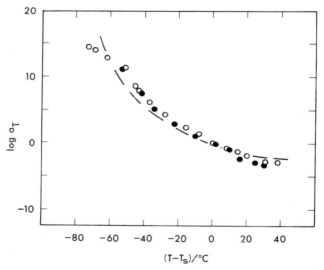

FIG. 4.2. The shift factor a_T as a function of temperature for PUR flexible foams.[2] Closed circles, non-reticulated polyester, $T_s = 6°C$: open circles, reticulated polyester, $T_s = 6°C$; both sets of data from dynamic-compression measurements. The WLF equation with the universal values of C_1 and C_2 is shown as the broken line.

range the data could be fitted quite well by the WLF equation,[5] shown as the broken line, using the universal values for the constants C_1 and C_2.

Deviations from WLF type behaviour can be seen at both high, $T - T_s > 10°C$, and low, $T - T_s < -52°C$, temperatures. This is not surprising since PUR flexible foams are likely to be rheologically complex. If a cellular polymer contains a significant proportion of closed cells, the enclosed gas phase will have a measurable effect on the stiffness of the system, as explained in Chapter 2B. This will be most noticeable when the modulus of the matrix is small, and we would expect for cellular systems containing a proportion of closed cells that the decrease in the modulus with increasing temperature would be less than that for a completely open-cell system. Furthermore, it has been found that many PUR elastomers are essentially block copolymers having a two-phase structure.[6-9] This is discussed briefly below. It has been reported that the size of the rigid domains in these materials is somewhat less than that observed in purely hydrocarbon block copolymers.[9] The shift factor a_T for segmented polyurethanes and other block copolymers does not follow the WLF equation in the rubbery plateau region. This has been discussed, for example, by Cooper and Tobolsky[6] and Fielding-Russel and Fitzhugh.[10]

The relationship between the dynamic mechanical properties of a foam matrix and those of the base polymer is of particular interest. Unfortunately few studies have been reported and this relationship is not well established. This is because of the difficulty of obtaining bulk samples of material suitable for experimental investigations which are representative of the matrix polymer. Results from one study are given in Fig. 4.3, in which $d(\omega)$ for a cellular NR–SBR latex foam is compared with that for a bulk sample of the same polymer cured under similar conditions. The dynamic properties were determined in the compression deformation mode at a frequency of about 2·0 Hz. It is seen that although the two specimens exhibited a maximum in $d(\omega)$ at the same temperature the value of $d(\omega)$ for the cellular rubber was larger than that of the bulk sample over most of the temperature range investigated. Rusch[11] has observed that at room temperature $d(\omega)$ for a series of PUR flexible foams (without fluidic effects) was larger than expected.

The differences between the dynamic properties of the matrix polymer and the bulk polymer may be attributed in part to differences in the processing conditions within the two types of sample, and to structure changes, such as molecular orientation and crystallisation, that result from the expansion of the polymer during foam formation. These problems have been overcome to a large extent by compressing expanded polymer

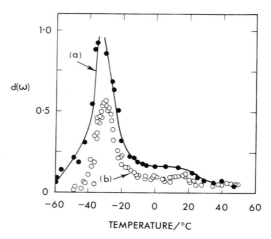

FIG. 4.3. The loss tangent $d(\omega)$ as a function of temperature for an NR–SBR elastomer under dynamic compression at about 2 Hz. (a) Cellular polymer, (b) bulk polymer.

into solid plaques as described by Critchfield et al.[12] In this process urethane systems are reacted and allowed to expand to a semi-gelled state. The cellular gel is then compression moulded and cured under specified conditions of temperature and pressure to give a solid polymer plaque. This procedure has been used by a number of workers investigating the influence of polymer composition and formulation on the properties of the matrix polymer of flexible and high resiliency urethane foams.[12–14] However, it was pointed out[13] that the process conditions experienced by the polymer during the final gelation and cure stages is different to those which exist during foam formation. It was concluded that the mechanical properties measured using solid plaques represent the optimum values that could be attained in the matrix polymer.

PUR elastomers are complex materials and their properties are governed by a wide range of chemical and structural variables.[15] Although it is not possible to enter into a detailed discussion of all the factors that influence their dynamic mechanical behaviour, it is worthwhile to consider briefly some of the more important features relevant to PUR foams. The polymer is made up of relatively high molecular weight flexible segments, either polyether or polyester, and rigid segments which contain urethane, urea and isocyanate groups. There is much evidence to show that molecular aggregation occurs in the solid elastomer with the formation of soft (elastomeric) domains and rigid domains.[6-9,15] Within the rigid domains there may be chemical and hydrogen bonding. Thus, these materials may be considered as block copolymer elastomers in which the properties of the polymer, and hence the foam, are intimately related to the structure, size and composition of the flexible and rigid domains.[16,17]

The first consequence of this two-phase structure is that two glass transitions exist. The primary, or low-temperature glass transition is associated with the soft domains and the high-temperature transition is associated with the rigid domains. The value of the low-temperature T_g is governed primarily by the flexibility of the macromolecule in the soft segments. The high-temperature T_g depends on the softening temperature of the polymer comprising the rigid phase or the temperature at which dissociation of the hydrogen bonds occurs. The mechanical behaviour in the intermediate region, that is the rubbery plateau, is governed to a large extent by the volume fraction of rigid domains in the elastomer. These not only act as pseudo crosslinks but also as reinforcing filler.

The disubstituted urea groups in a PUR macromolecule have strong hydrogen bonding capabilities and the amount of intermolecular hydrogen bonding in the rigid domains can be controlled by varying the number of these groups. Results reported by Seefried $et\ al.$[14] for a series of cellular polymers in solid plaque form are given in Fig. 4.4(a). These polymers

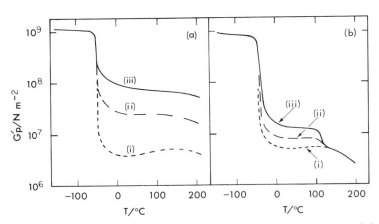

FIG. 4.4. (a) Temperature dependence of the dynamic storage modulus in shear for a flexible PUR elastomer with different amounts of urea structures; increasing in the order (i), (ii), (iii).[14] (b) Temperature dependence of the dynamic storage modulus in shear for a high resiliency PUR elastomer with different amounts of polymer polyol; (i) zero; (ii) 50 pbw; (iii) 100 pbw.[13]

were based on a conventional polyether polyol and the data were obtained using a torsion pendulum at a frequency of about 1 Hz. It is seen that the room temperature modulus of the elastomers increased with increasing urea content but the primary glass transition ($T_g \approx -52°C$) was unaffected. It has also been shown that the substitution of a polymeric isocyanate for the conventional tolylene di-isocyanate[14] and the introduction of a crosslinking agent[8] results in the reduction of the room-temperature stiffness of the PUR polymers. This was attributed in both cases to the disruption in the symmetry of the disubstituted urea sequences which reduced the ability of these chain segments to align and form strong hydrogen bonds.

The room-temperature modulus can also be controlled by incorporating a polymer polyol in the foam formulation.[12-14] These are formed by grafting a styrene–acrylonitrile copolymer onto a conventional polyol. Some of the data reported by Patten et al.[13] for PUR elastomers in solid plaque form are given in Fig. 4.4(b). These specimens were prepared using a blend of tolylene di-isocyanate and a polymeric isocyanate and different concentrations of conventional and polymer polyols. The two glass transitions resulting from the two-phase structure of these materials is clearly seen. The upper transition, at about 110°C, is associated with the glass transition of the styrene–acrylonitrile copolymer. The low-temperature transition, at about $-40°C$, is associated with the flexible polyether segment and is not significantly affected by the amount of rigid styrene–acrylonitrile phase present in the system.

4.3 FLUID-FLOW EFFECTS IN OPEN-CELL FLEXIBLE FOAMS

4.3.1 General Considerations

Under dynamic compression, flexible open-cell polymer foams sometimes exhibit a higher mechanical damping than can be attributed to the polymer matrix alone. This is because of the flow of fluid through the matrix, and it occurs with both compressible and incompressible fluids. In the former case it has been called pneumatic damping. Gent and Rusch[11,18] have shown experimentally that pneumatic damping depends on the mechanical properties of the fluid and the matrix polymer, the cell structure, the geometry of the specimen and the frequency of the mechanical excitation. The variation of the dynamic modulus $E(\omega) = [E'(\omega)^2 +$

$E''(\omega)^2]^{1/2}$ and the loss tangent $d(\omega)$ as a function of frequency for an air-filled open-cell PUR flexible foam at a temperature well above the glass transition of the matrix polymer is shown in Fig. 4.5.[11] It is seen that $d(\omega)$ initially increased with increasing frequency of deformation, passed through a maximum value and then decreased. The value of $E(\omega)$ increased with frequency until an equilibrium value was attained. This be-

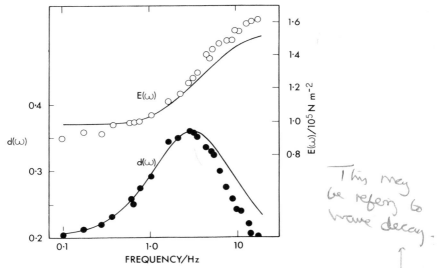

FIG. 4.5. Variation of the dynamic modulus $E(\omega)$ and the loss tangent $d(\omega)$ of an open-cell PUR flexible-foam subject to oscillatory compression showing the mechanical damping resulting from the fluid-flow process.[11]

haviour is similar to that of the polymer matrix at temperatures close to its T_g. However, in this case the transition occurred over a relatively small frequency range and corresponded to the behaviour of a system with a single relaxation time rather than a distribution of relaxation times as exhibited by solid elastomers.

The only theory that has been put forward that allows the prediction of the effects of fluid flow on the mechanical behaviour of a cellular polymer subjected to oscillatory compression from independently measured parameters is that due to Gent and Rusch.[11,18] The most important factor governing fluidic damping is the resistance to the flow of fluid through the polymer matrix. From a detailed investigation of a series of foams having a wide range of cell size, it was shown[19] that the resistance of fluid flow is made up of two parts. These are the frictional interaction between the fluid and the matrix polymer and the kinetic energy lost due to turbulence caused by the irregular nature of the flow path. The pressure gradient required to maintain the flow of a Newtonian fluid, with viscosity η and density ρ, through an open-cell foam was shown to be

$$- dP/dx = (\eta/K)v + (\rho/B)v^2 \qquad (4.1)$$

where v is the average flow velocity of the fluid, K is the permeability of the foam and B is a constant of the foam which reflects the irregular nature of the flow channel. These results are in general agreement with the work of Jones and Fesman.[1]

Gent and Rusch found experimentally that both K and B are independent of the fluid, whether it be compressible or incompressible, and that for a range of open-cell foams with no cell membranes the coefficient $K = 0.012d^2$, where d is the average cell diameter. Using a structural model consisting of a parallel array of tubes of diameter d containing $1/d$ orifices per unit length they showed theoretically that $K = 0.031d^2$ and $B = (D^2 - 1)^2/2d$ where D is the ratio of the diameter of the tube to the diameter of the constriction. They also showed that when the foam specimen is compressed the effective diameter d of the flow channels is $d' = d(1 - e)^{1/2}$ where e is the fractional compression. Consequently both K and B are deformation dependent.

Using eqn (4.1) as the starting point, Gent and Rusch developed theories for the dynamic mechanical behaviour of open-cell flexible foams containing compressible and incompressible fluids. Because of the strain-dependence of the coefficients K and B and the non-linear stress–strain behaviour of the cellular matrix (see Section 2B.3), the resulting equations are only applicable for small fractional compressions. A similar approach was used by Liber and Epstein in their analysis of the effect of air flow on the behaviour of cellular polymers subjected to dynamic compression.[20] Their work is discussed in the following chapter.

4.3.2 Prediction of the Dynamic Mechanical Behaviour

Incompressible Fluids
The model assumed by Gent and Rusch is a rectangular block of foam of length L, height h and width $W \ll L$ with massless plates bonded to the top and bottom surfaces, as shown in Fig. 4.6(a). The origin of coordinates is set at the centre of the lower surface of the block and the

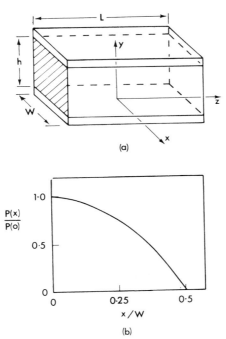

FIG. 4.6. (a) The model of Gent and Rusch.[11,18] (b) The excess pressure $P(x)$ in the test piece as a function of x caused by compression in the y direction.

boundaries at $x = \pm W/2$ are in contact with a reservoir of the fluid filling the matrix. Since the block is assumed to be infinite in length, plane strain conditions prevail. When the upper surface is moved towards the lower surface, the polymer matrix deforms uniformly in the y direction and the fluid is forced to flow in the transverse, $\pm x$, directions. For the flow process, the y–z plane along the centre of the block is a plane of symmetry. Because of the resistance to fluid flow a pressure is developed within the system. From eqn (4.1) it was shown that the pressure gradient varies as a function of x according to

$$-dP/dx = \{(\eta x)/(K\zeta)\}\dot{e} + \{(\rho x^2)/(B\zeta^2)\}\dot{e}^2 \qquad (4.2)$$

where ζ is the volume fraction of open cells.

The fractional compression e is related to the dimensions of the test piece by $de = dh/h$ and is taken as a positive quantity. Since the pressure at $x = W/2$ is the ambient pressure the excess pressure profile $P(x)$ within the foam–fluid system can be determined by integration. If it is assumed that the term containing B is small, which is true for many systems, $P(x) = \{(\eta W^2 \dot{e})/(8K\zeta)\}\{1 - 4(x/W)^2\}$. This profile is shown in Fig. 4.6(b) where $P(0)$ is the excess pressure at the plane of symmetry, $x = 0$. The compressive stress on the upper and lower surfaces necessary to produce this fluid flow is obtained by considering a unit length of block, in the z direction, and integrating $P(x)$ over the width of the block. For small fractional compressions this stress is given by

$$\sigma = [(\eta W^2)/\{12\zeta K(\omega)\}]\, \dot{e} \tag{4.3}$$

where the term in eqn (4.2) containing the coefficient B is incorporated into the effective permeability $K(\omega)$. In general this is dependent on the frequency of deformation as described by Rusch.[11] However, for many foam–fluid combinations $\rho/B \ll \eta/K$ and the flow-inertia term in eqn (4.1) is unimportant at frequencies where the flow process is governing the behaviour of the cellular polymer. In these situations, the resistance to fluid flow can be adequately described by Darcy's law, which is essentially the first term on the right-hand side of eqn (4.1).

For small oscillatory strains of the form $e = e_0 \sin \omega t$ the stress required to deform the polymer matrix is

$$\sigma = E'_m e_0 \sin \omega t + E''_m e_0 \cos \omega t \tag{4.4}$$

The total stress needed to produce the homogeneous deformation of the block and the fluid flow is the sum of the stresses given in eqns (4.3) and (4.4). Since the stress associated with the flow process is out of phase with the applied deformation it results in an energy loss. This increases with increasing frequency.

At relatively high frequencies Gent and Rusch proposed that the viscous interaction between the flowing fluid and the polymer matrix is so large that the matrix is forced to deform in the transverse, $\pm x$, directions, as shown in Fig. 4.7(a). In the extreme case there is no flow of fluid through the matrix so there are no flow energy losses. It is predicted that because of these two conflicting mechanisms the mechanical damping of a fluid-filled foam will pass through a maximum value as the frequency of the compressive deformation is increased.

At high frequencies, the foam block behaves in the same way as a homogeneous block of incompressible elastomer bonded to rigid plates at its upper and lower surfaces. When the upper boundary is moved towards the lower boundary an initially vertical element, of width dx,

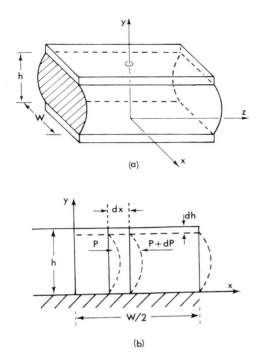

FIG. 4.7. (a) The transverse deformation of the test piece for an infinitely long block of foam. (b) The deformation of the material in the transverse direction.

within the block takes up a parabolic configuration as shown in Fig. 4.7(b). This configuration can be achieved by the action of an excess pressure dP acting on one curved surface. The excess pressure gradient at x is given by[21,22]

$$- dP/dx = (4E_t e/h^2)x$$

where E_t is the effective tensile elastic modulus for deformation in the transverse direction. Integrating across the section of the block gives the excess-pressure profile. From this, the stress acting on the upper and lower surfaces is shown to be

$$\sigma = (E_t/3)(W/h)^2 e_t \qquad (4.5)$$

where e_t is the deformation in the direction of compression associated with the transverse deformation of the material.

To combine the fluid-flow and transverse-deformation mechanisms, they assumed that a compressive stress σ_{ft} applied to the upper and lower surfaces will result in compressive strain components e_t and e_f, where e_f is the component associated with the flow process. The total compressive deformation is

$$e = e_t + e_f \tag{4.6}$$

Equation (4.3) can be written in the form $\dot{e}_f = [\{12\zeta K(\omega)\}/(\eta W^2)]\sigma$, so by substituting in eqn (4.6)

$$\dot{e} = \left[(3/E_t)(h/W)^2\right]\dot{\sigma}_{ft} + \left[\{12\zeta K(\omega)\}/(\eta W^2)\right]\sigma_{ft}$$

For a fractional compression of the form $e = e_0 \sin \omega t$ this has a solution

$$\sigma_{ft} = \frac{E'_m \beta (\gamma/12)^2}{1 + \beta^2(\gamma/12)^2} e_0 \sin \omega t + \frac{E'_m(\gamma/12)}{1 + \beta^2(\gamma/12)^2} e_0 \cos \omega t$$

where

$$\gamma = (\omega \eta W^2)/\{\zeta E'_m K(\omega)\} \tag{4.7}$$

$$\beta = 3(E'_m/E_t)(h/W)^2 \tag{4.8}$$

It is seen that the transverse deformation is an elastic mechanism and contributes to the storage modulus of the system. The total stress needed to produce the homogeneous deformation of the matrix, fluid flow and transverse deformation is the sum of σ_{ft} and the stress given by eqn (4.4). By combining in phase and out of phase components of the stress–strain ratio, σ/e_0, the dynamic storage modulus $E'(\omega)$ and the loss tangent $d(\omega)$ of the fluid-filled foam are predicted to be

$$E'(\omega) = \frac{E'_m\{1 + \beta^2(\gamma/12)^2 + \beta(\gamma/12)^2\}}{1 + \beta^2(\gamma/12)^2} \tag{4.9}$$

$$d(\omega) = \frac{d_m\{1 + \beta^2(\gamma/12)^2\} + \gamma/12}{1 + \beta^2(\gamma/12)^2 + \beta(\gamma/12)^2} \tag{4.10}$$

According to this analysis, the dynamic mechanical behaviour of a liquid-filled foam in compression is governed by three dimensionless parameters; γ, β and d_m. If γ is used as an independent variable, rather than the frequency of deformation ω, families of curves can be generated which will apply to a wide range of situations. Examples, taken from the work of Rusch,[11] are given in Fig. 4.8 in which $d(\omega)$ is plotted as a function of $\gamma/12$ for different values of β, and in Fig. 4.9 in which $d(\omega)$ and $E(\omega)$ are shown for different values of d_m.

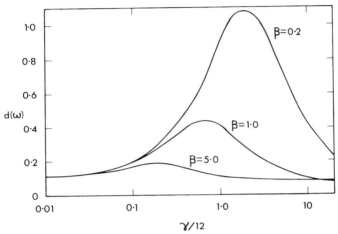

FIG. 4.8. The variation of the loss tangent $d(\omega)$ of a flexible foam filled with an incompressible fluid as a function of $\gamma/12$ predicted by eqn (4.10) for different values of β assuming $d_m = 0 \cdot 10$ (from Rusch[11]).

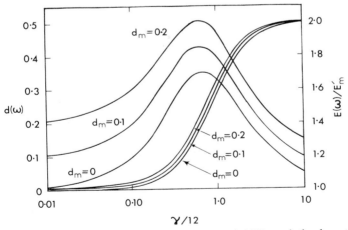

FIG. 4.9. The variation of the modulus ratio $E(\omega)/E'_m$ and the loss tangent $d(\omega)$ of a flexible foam filled with an incompressible fluid as a function of $\gamma/12$ predicted by eqns (4.9) and (4.10) for different values of d_m assuming $\beta = 1 \cdot 0$ (from Rusch[11]).

The value of the parameter γ at maximum fluid-flow damping can be obtained by setting the differential of $d(\omega)$ with respect to γ equal to zero. For small values of d_m this condition is given approximately from eqn (4.10), by

$$(\gamma/12)_{\max} \approx 1/(\beta + \beta^2)^{1/2} - d_m/(1 + \beta) \qquad (4.11)$$

Substituting into eqn (4.10) gives

$$d(\omega)_{\max} \approx \frac{\beta^2 d_m (\gamma/12)_{\max} + 1/2}{(\beta + \beta^2)(\gamma/12)_{\max}} \qquad (4.12)$$

so that $d(\omega)_{\max}$ is a function of β and d_m, and is independent of γ.

At frequencies up to that for the damping maximum, good agreement between the measured behaviour and that predicted by eqns (4.9) and (4.10) is obtained when empirically determined values of γ, β and d_m are used. At higher frequencies the value of $E(\omega)$ tends to be larger and $d(\omega)$ smaller than predicted. This can be attributed to several factors. For example, in the simplified analysis presented here the flow-inertia term containing the coefficient B is not properly accounted for. Also it was assumed that E'_m and d_m are independent of frequency and that the transverse deformation process is elastic. Neither of these assumptions is strictly valid. The frequency dependence of the properties of the matrix can be accounted for by replacing E'_m and d_m by $E'_m(\omega)$ and $d_m(\omega)$ where the latter parameters are measured as described in the previous section. The dissipative nature of the transverse modulus can be introduced by replacing E_t by $E_t^*(\omega) = E'_t(\omega) + jE''_t(\omega)$ and assuming $E''_t(\omega)/E'_t(\omega) = d_m(\omega)$. When this is done the parameter β becomes complex and can be written in the form $\beta^* = \beta' + j\beta''$.

At high frequencies, it is also necessary to take account of additional inertia forces. Two inertia terms can be included.[4] The first is related to the motion of the centre of mass of the system in the direction of compression. This gives rise to a stress component $\sigma = (M_e h/W) \ddot{e}$ or

$$\sigma = -(\omega^2 M_e h/W)e_0 \sin \omega t \qquad (4.13)$$

where M_e is the effective mass for unit length of block, in the z direction. According to the Rayleigh approximation[23] $M_e = \rho Wh/3$. This stress is added directly to the in-phase component of the σ/e_0 ratio associated with the uniform deformation of the polymer matrix, eqn (4.4).

The second inertia term accounts for the effect of the distributed mass on the transverse deformation mechanism at frequencies where the system is behaving like a bonded rubber block. Under these conditions it has been shown[22] that the compressive stress needed to produce the compressive strain component e_t associated with the transverse deformation is

$$\sigma = [\{E_t^*(\omega)/3\}(W/h)^2 - \omega^2 W^2 \rho/15]e_t \quad (4.14)$$

The stress σ_{ft} due to the fluid flow and transverse deformation is obtained by combining eqns. (4.3) and (4.14) using the Gent–Rusch procedure, eqn. (4.6). The resulting expression is given in the Appendix. Adding in-phase and out-of-phase components of the σ/e_0 ratio from equations A(1), (4.4) and (4.13) gives the overall in-phase stress–strain ratio $\sigma'(\omega)/e_0$ for the liquid-filled foam as

$$\sigma'(\omega) =$$

$$\frac{E'_m(\omega)[\{1 - \omega^2 \rho h^2/3E'_m(\omega)\}\{(1 - \beta'\gamma/12)^2 + (\beta'\gamma/12)^2\} + \beta'(\gamma/12)^2]}{(1 - \beta''\gamma/12)^2 + (\beta'\gamma/12)^2}$$

(4.15)

and the phase angle $\delta(\omega)$ between the stress and strain as

$$\tan \delta(\omega) =$$

$$\frac{d_m(\omega)\{(1 - \beta''\gamma/12)^2 + (\beta'\gamma/12)^2\} + (\gamma/12)(1 - \beta''\gamma/12)}{\{1 - \omega^2 \rho h^2/3E'_m(\omega)\}\{(1 - \beta''\gamma/12)^2 + (\beta'\gamma/12)^2\} + \beta'(\gamma/12)^2}$$

(4.16)

The parameters β' and β'' are defined in the Appendix. By making simplifying assumptions, such as $\beta' \approx \beta$, $\beta'' = \beta d_m(\omega)$ and that $\gamma/12$ is small, it can be shown that at low frequencies these equations reduce to those of Gent and Rusch, eqns (4.9) and (4.10), respectively. It should be noted that eqns (4.15) and (4.16) are for the in-phase stress–strain ratio and the phase angle between stress and strain. These parameters are only equal to the storage modulus and the loss tangent of the system when the inertia terms are negligible compared with those for the elastic behaviour.

Compressible Fluids
Although the dynamic mechanical behaviour of gas-filled flexible open-cell foams is qualitatively the same as that of liquid-filled foams, the mechanisms responsible for this behaviour are different. Considering the

specimen arrangement shown in Fig. 4.6 it is assumed[11,18] that at low frequencies the polymer matrix deforms uniformly in the direction of compression and the gas is forced to flow in the transverse, $\pm x$, directions. The fluid flow results in an energy dissipation which increases with increasing frequency and contributes to the loss modulus of the system. At high frequencies the viscous interaction between the fluid and the matrix is large and the fluid is more readily compressed than forced to flow. This is essentially an elastic process so that the storage modulus increases and the energy loss decreases. Although some transverse deformation of the matrix occurs, it was shown by Gent and Rusch that the contribution to the dynamic mechanical behaviour from this mechanism is small. Using a mass-balance equation, these workers were able to establish a relationship for the compressive stress required to produce the fluid flow. Combining this with the stresses related to the homogeneous deformation of the matrix and the compression of the fluid they showed for gas-filled foams that

$$E'(\omega) = E'_m (1 + 1/\alpha - 8\psi_1/\alpha) \qquad (4.17)$$

$$d(\omega) = (\alpha d_m + 8\psi_2)/(\alpha + 1 - 8\psi_1) \qquad (4.18)$$

where

$$\psi_1 = \sum_{n=1,3,5}^{\infty} [n^2\pi^2/\{(n^2\pi^2)^2 + (\alpha\gamma)^2\}]$$

$$\psi_2 = \alpha\gamma \sum_{n=1,3,5}^{\infty} [1/\{(n^2\pi^2)^2 + (\alpha\gamma)^2\}]$$

The parameter α, which is given by

$$\alpha \approx \zeta E'_m/E_g \qquad (4.19a)$$

where E_g is the bulk modulus of the gas, is analagous to the parameter β for incompressible fluids. γ is given by eqn (4.7).

As with liquid-filled foams, the small-strain dynamic mechanical behaviour of air-filled foams is governed by three dimensionless parameters. For compressible fluids, these are γ, α and d_m. The variation of $E(\omega)$ and $d(\omega)$ predicted by eqns (4.17) and (4.18) is the same as that shown in Figs. 4.8 and 4.9, except that for equivalent values of α and β the value of $d(\omega)_{max}$ for air-filled systems is slightly lower than that for liquid-filled systems. This is because under comparable conditions the compression of the gas results in a lower volume flow rate than that which obtains in a liquid-filled system.[11]

4.3.3. Comparison of Predicted and Measured Behaviour

The validity of the equations derived above has been investigated by comparing the values of the parameters α, β and γ measured experimentally for different foam–fluid combinations with those predicted from independent measurement using eqns (4.19b), (4.8) and (4.7). Procedures for measuring the permeability K of the foam have been described in the literature,[1,19] and the porosity ζ has been determined using the water-absorption method. The geometry of the test pieces were such that they approximated to infinitely long rectangular blocks. This condition was assumed to be true when $L > 4W$. For smaller specimens, $L < 4W$, the flow path was controlled by sealing two opposite faces of the block using a thin impermeable rubber skin. This can be done by spreading a thin layer of a compounded rubber latex onto a glass plate, placing the foam block on this layer and curing the compound at the appropriate temperature. When the foam block is carefully stripped from the plate a thin, uniform, impermeable skin is obtained. This skin affects the static compressive stiffness of the test piece. However, in many cases this effect is small, and in principle it can be accounted for when comparing theory with experiment.

At low frequencies $\gamma/12$ tends to zero and eqn (4.10) simplifies to

$$d(\omega) = d_m + \gamma/12 \qquad (4.19b)$$

so that a graph of $d(\omega)$ as a function of linear frequency f should be a straight line with gradient

$$\gamma/12f = (\pi \eta W^2)/(6\zeta E'_m K) \qquad (4.20)$$

and intercept d_m. The coefficient K is the permeability of the foam corresponding to the first term in the fluid-flow equation (4.1). Equation (4.19b) has been found to be valid for a wide range of foam–fluid systems. Some typical data,[11] for a rubber latex flexible foam filled with liquids having different values of η, are shown in Fig. 4.10. Plots such as these allow the determination of an empirical value for the parameter $\gamma/12f$. Values obtained in this way from two series of measurements are plotted as a function of $(\eta W^2)/(\zeta E'_m K)$ in Fig. 4.11.[4,23] The open circles are for PUR flexible-foam test pieces having approximately the same value for W but different ζ, E'_m and K. The closed circles are for a rubber latex foam with different values of W. It is seen that apart from three data points the straight line relationship predicted by eqn (4.20) was followed reasonably well. The gradient of the line, which is approximately 0.4 ± 0.05, is in

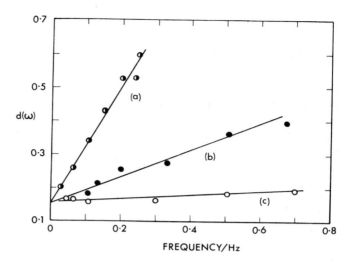

FIG. 4.10. Variation of the loss tangent $d(\omega)$ as a function of frequency $(f \ll f_{max})$ for a rubber latex foam filled with liquids of different viscosity; (a) $\eta = 50$ cP, (b) $\eta = 8.6$ cP, (c) $\eta = 0.9$ cP (from Rusch[11]). The test piece dimension W was the same for each set of data.

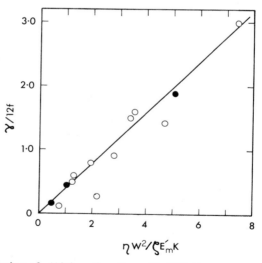

FIG. 4.11. Variation of $\gamma/12f$ as a function of $(\eta W^2)/(\zeta E'_m K)$ for different foams filled with silicone oil, $\eta = 48$ cP. Open circles; PUR flexible foams with different ζ, E'_m and K.[24] Closed circles; rubber latex foams with test pieces having different values of W.[4]

reasonable agreement with the value predicted by the Gent–Rusch theory, $\pi/6 = 0.53$.

An empirical value for β, or α in the case of air-filled foams, can be obtained by fitting eqn (4.12), which relates $d(\omega)_{max}$, $(\gamma/12)_{max}$ and β, to the experimental data at the flow damping maximum using the value of $\gamma/12f$ determined previously. Rusch[11] has reported that for liquid-filled systems with $L \approx 4W$ the dependence of β on h/W is in accord with the behaviour predicted by eqn (4.8). However, for test pieces having rubber skins, marked differences between the measured and predicted behaviours are observed.[3,4,11] An example of this is shown in Fig. 4.12 in which the value of β for a PUR flexible foam filled with silicone oil is plotted as

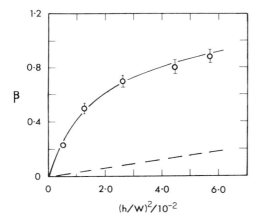

FIG. 4.12. Variation of β for a PUR flexible foam filled with silicone oil as a function of $(h/W)_2$. The broken line is the behaviour predicted by eqn (4.8) assuming $E'_m/E'_t = 1.0$.[4]

a function of $(h/W)^2$. These data were obtained for test specimens with the same W but different h. The broken line is the behaviour predicted by eqn (4.8), assuming that $E'_m = E'_t$. It is seen that the values of β were much larger than expected and that the form of the relationship between β and $(h/W)^2$ was different to that indicated by eqn (4.8).

The discrepancy between the predicted and measured values of β has been attributed to the failure of these test pieces to conform to the Gent–Rusch model. At high frequencies, the model assumes that the transverse deformation of the matrix takes place in the $\pm x$ directions only. Although the flexible rubber skins on opposing faces of the foam block ensure that the fluid flow is approximately unidirectional, they do not restrict the transverse deformation of the matrix. Thus deformation can take place in both x and z directions, so that the effective stiffness associated with the transverse deformation (and hence $E_t(\omega)$) will be smaller than expected. As a result, the value of β will be larger than that predicted by eqn (4.8), as shown by the data of Fig. 4.12. Gent and Rusch have reported[18] experimental values of α for air-filled foams which were larger than those predicted by eqn (4.19a). They attributed this, in part, to the fact that the simple equations for pneumatic damping given in Section 4.3.2 do not take into account the transverse deformation of the matrix. More rigorous equations are given by Rusch.[11]

The dynamic mechanical behaviours of liquid- and air-filled foams measured experimentally over a wide frequency range have been compared with those predicted by the equations given above using the empirically determined values of $\gamma/12f$, α and β. Data for an air-filled PUR flexible foam has been given in Fig. 4.5. The continuous line is the behaviour predicted by eqns (4.17) and (4.18). It is seen that in general there was good agreement between theory and experiment. The discrepancy at high frequencies was attributed to the frequency dependence of the mechanical properties of the polymer matrix and the dissipative nature of the transverse deformation, as explained previously.

With air-filled systems exhibiting fluid-flow effects at low frequencies, it is difficult to determine unequivocally the frequency dependence of the properties of the matrix. This problem does not exist with liquid-filled foams, and it is possible to determine $E'_m(\omega)$ and $d_m(\omega)$ over the frequency range of interest in the absence of flow phenomena. Data for a PUR flexible foam filled with silicone oil are compared with the behaviour predicted by eqns (4.15) and (4.16) in Fig. 4.13. These equations take account of inertia effects and the frequency dependence of the properties of the matrix polymer. It is seen that there was good agreement between theory and experiment. With many foam–fluid combinations motional inertia effects are negligible in the vicinity of the flow damping maximum and they only become significant at high frequencies. Data for a PUR foam filled with silicone oil are shown in Fig. 4.14. The continuous lines are the in-phase stress–strain ratio and the phase angle predicted for this system using eqns (4.15) and (4.16). The fluid-flow damping maximum occurred at about 1 Hz. The behaviour represented by the data of Fig. 4.14 is for a system passing through a resonance condition.

Temperature influences the dynamic behaviour of fluid-filled foams

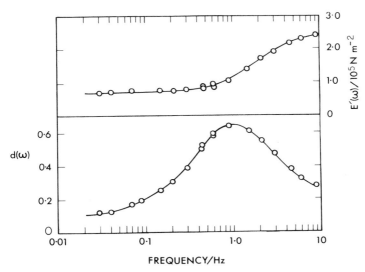

FIG. 4.13. Comparison of the storage modulus $E'(\omega)$ and the loss tangent $d(\omega)$ of a PUR flexible foam filled with silicone oil as a function of frequency with the behaviour predicted by eqns (4.15) and (4.16) (from Hilyard and Kanakkanatt[4]).

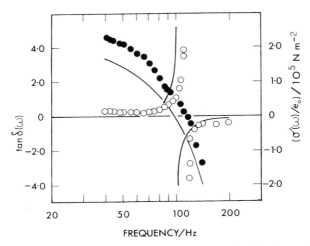

FIG. 4.14. The high-frequency behaviour of a PUR flexible foam filled with silicone oil.[4] Closed circles; the in-phase stress–strain ratio $\sigma'(\omega)/e_0$. Open circles; the phase angle $\delta(\omega)$. Both as a function of the deformation frequency.

in several ways. At temperatures well above the glass transition of the matrix polymer, the frequency for maximum flow damping in liquid-filled systems is shifted to higher frequencies as the temperature is increased. When $E'_m(\omega)$ and $d_m(\omega)$ are independent of frequency the value of $d(\omega)_{max}$ is independent of ω_{max}. This behaviour is shown in Fig. 4.15 and is in accord with the Gent–Rusch theory.

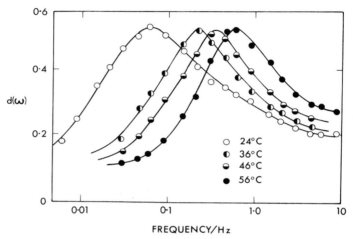

FIG. 4.15. The effect of temperature on the fluid-flow damping peak for a polyether PUR flexible foam filled with glycerin.[24]

For systems in which $d_m(\omega)$ and β are independent of temperature (and frequency) it is predicted from eqn (4.11) that $(\gamma/12)_{max}$ should be constant. Under these conditions, it can be seen from eqn (4.7) that the frequency f_{max} for maximum fluid-flow damping should vary according to

$$f_{max} = \text{const.} \{E'_m(\omega)/\eta\} \qquad (4.21)$$

so that for $T \gg T_g$ the shift of f_{max} is governed primarily by the temperature coefficient of the liquid filling the foam. Figure 4.16 shows f_{max} as a function of temperature for a PUR flexible foam filled with silicone oil. The continuous line is the behaviour predicted by eqn (4.21) using the data at $T = 20°C$ as the reference. By judicious choice of matrix polymer and liquid it should be possible to design a system which provides high mechanical damping at a particular frequency which is relatively insensitive to temperature changes.

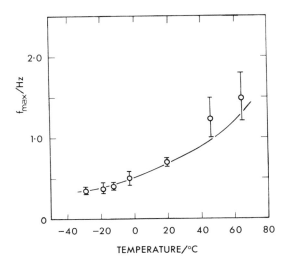

FIG. 4.16. The variation of the frequency for maximum fluid flow damping f_{max} as a function of temperature in a PUR flexible foam filled with silicone oil.[24] The line is the behaviour predicted by eqn (4.21) using 20°C as the reference temperature.

At temperatures approaching the glass transition temperature of the matrix polymer, the behaviour is more complex because two relaxation processes are involved; viscoelastic and fluid flow. Under these conditions the dynamic behaviour can be described properly only by means of a property surface.[3] $E'(\omega), T, f$ and $d(\omega), T, f$ surfaces for an open-cell PUR flexible foam, $T_g \approx -46°C$, filled with silicone oil are shown in Figs. 4.17 and 4.18. The effect of the fluid flow process on $E'(\omega)$ can clearly be seen at 40°C but it became less important as the polymer matrix glass transition was approached. The temperature dependence of $d(\omega)$ is more difficult to explain.

For this particular combination of liquid and polymer matrix the values of f_{max} and $d(\omega)_{max}$ were not strongly dependent upon temperature for temperatures in the range $-10°C$ to $40°C$. However, as the temperature was reduced from 40°C there was initially a small shift of f_{max} to lower frequencies. When the temperature was reduced through $-10°C$, the shift of f_{max} reversed so that f_{max} moved to higher values. This reversal can be attributed to the matrix modulus E'_m becoming more temperature dependent than the viscosity η of the liquid as the glass transition of the matrix polymer was approached (refer to eqn (4.21)). The damping peak

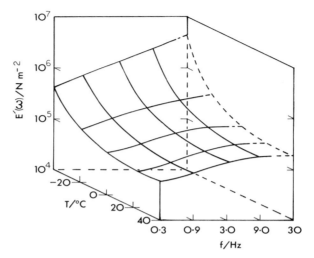

FIG. 4.17. The $E'(\omega)$, T, f surface for a PUR flexible foam, $T_g \approx = -46°C$, filled with silicone oil. Data taken from Kanakkanatt.[3]

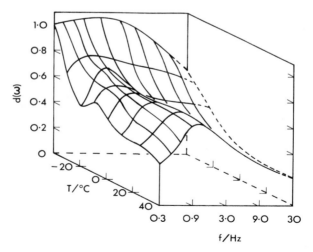

FIG. 4.18. The $d(\omega)$, T, f surface for the system of Fig. 4.17.

at approximately $-10°C$ in the 0·3 Hz isochronal section is a result of this movement of the fluid-flow damping peak, first to lower frequencies and then to higher frequencies. It is seen that at $-40°C$, which was close to the matrix T_g, the flow process had little influence on the mechanical damping of the system.

4.4 VIBRATION ISOLATION USING OPEN CELL FLEXIBLE FOAMS

4.4.1 Experimental Results

It is common practice to use resilient elastomeric mountings to protect structures from mechanical vibrations. The most commonly encountered situations are when a rigid foundation is isolated from the forces generated by a vibrating machine and when a fragile structure is isolated from the movement of its supporting surface. The second of these is shown in Fig. 4.19 where the isolating mount is represented as an elastomer of complex modulus $E^*(\omega)$, the structure as the mass M and the mechanical

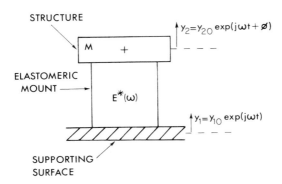

FIG. 4.19. Schematic representation of a simple vibration isolation system employing a resilient mount material with modulus $E^*(\omega)$.

vibration as y. The effectiveness of such an arrangement is usually described by the absolute transmissibility which is defined as the magnitude of the displacement or the acceleration ratio; $|T(\omega)| = |y_2/y_1| = |\ddot{y}_2/\ddot{y}_1|$ where y_2 represents the motion of the isolated mass.[25] Although transmissibility curves for specific cellular polymers have been given in the literature (e.g. Pizzirusso[26] and Hatae[27]) few systematic investigations of their vibration isolation behaviour have been reported. Data for some reticulated PUR flexible foams of different average cell diameter are shown in Figs. 4.20 (i) and (ii).[28] The test pieces were rectangular blocks of height 20 mm and side approximately 60 mm. The amplitude of the forcing acceleration $\ddot{y}_{10}$ was maintained constant during the frequency sweep. The experimental arrangement has been described in the reference given above.[28] The purpose of the investigation was to examine how the cell size (which governs the permeability of the foam and hence the frequency at which fluidic effects are observed), the value of the supported mass and the amplitude of the vibration excitation influence the transmissibility response.

The curves in Fig. 4.20 (i) show the transmissibility of a PUR flexible foam with small cell diameter for different values of the isolated mass M. It is seen that the resonance frequency f_r of the mass-isolator system

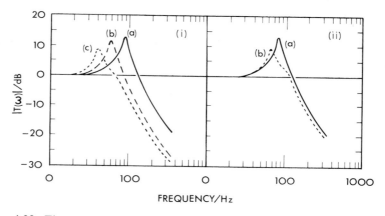

FIG. 4.20. The transmissibility response of air-filled reticulated PUR flexible foams with $\phi \approx 0.03$, (i) $d_{av} \approx 0.25$ mm; (a) $M = 0.047$ kg, (b) $M = 0.107$ kg and (c) $M = 0.12$ kg. (ii) $d_{av} \approx 0.29$ mm, $M = 0.107$ kg for different excitation levels (a) $\ddot{y}_{10} = 0.8$g and (b) $\ddot{y}_{10} = 4.0$ g. Reproduction of original chart records.[28]

decreased with increasing value of M, as would be expected. However, the variation of f_r did not follow the simple equation $f_n = (1/2\pi)\sqrt{K/M}$ where K is the stiffness of the isolator. The deviation from this relationship was most marked with foams of small cell size. This was attributed to the influence of the fluidic process on the stiffness and damping of the test piece as the resonance frequency was reduced to lower values. This conclusion is supported by the observation that for this foam specimen the maximum value of the transmissibility, $|T(r)|$, decreased with decreasing f_r indicating an increase in the damping of the isolator material. For the foam with largest cell diameter, $d_{av} = 0.77$ mm, in which fluid-flow effects were expected to be negligible in this frequency range, the frequency for maximum transmissibility followed quite closely the simple expression given above and the value of $|T(r)|$ was essentially independent of f_r.

The influence of the level of the forcing excitation on the transmissibility of a foam in the same series to test specimens is shown in Fig. 4.20 (ii). It is seen that both $|T(r)|$ and f_r decreased with increasing amplitude, $\ddot{y}_{10}$, of the forcing acceleration. For this test piece f_r decreased continuously from about 91 Hz to 48 Hz and $|T(r)|$ decreased from about 18 dB to 9·5 dB as the excitation level was increased from 0·2 g to 4·0 g. This behaviour can be attributed in part to the non-linear stress–strain behaviour of the material system. Another feature of the transmissibility response is the distortion of the curves at $f > f_r$ observed for large supported masses, Fig. 4.20 (i) (c), and large excitation levels, Fig. 4.20 (ii) (b). The origin of this behaviour is not known. It is interesting to note that although in Fig. 4.20 (ii) (b) the behaviour in the amplification region, $|T(\omega)| > 1.0$, was dependent on the excitation level the transmissibility in the isolating region, $|T(\omega)| < 1.0$, was essentially the same.

Two important features of the transmissibility response of an isolator are the maximum value of $|T(\omega)|$, which occurs at the resonance frequency of the mass-isolator system, and the value of $|T(\omega)|$ in the vibration-isolation region. With conventional systems, such as spring and dashpot and solid elastomeric isolators, a low transmissibility in the isolating region can only be achieved at the expense of a high transmissibility in the amplification region. Various ways of optimising the response of an isolator have been described, see for example Gee-Clough and Waller.[29] In principle it should be possible to achieve superior isolating performance using fluid-filled foams, since these material systems can be designed to produce a definite degree of damping in a specific frequency range, i.e. the frequency range where vibration amplification occurs.

4.4.2. Theoretical Considerations

With a single degree of freedom system in which the mounting is a linear elastic material of modulus $E^*(\omega) = E'(\omega)\{1 + jd(\omega)\}$ the displacement ratio for a sinusoidal excitation is given by[25]

$$\frac{y_2}{y_1} = \frac{1 + jd(\omega)}{1 - \omega^2 M/kE'(\omega) + jd(\omega)}$$

where k is a form factor relating the stiffness $S^*(\omega)$ of the mount to the modulus of the material, $S^*(\omega) = kE^*(\omega)$. It is usual to express equations of this type in terms of the natural frequency ω_n of the system where ω_n is the value of ω at which, in the absence of damping, the displacement y_2 becomes infinitely large. Letting the value of $E'(\omega)$ at $\omega = \omega_n$ be represented by $E'(n)$, then $\omega_n = (kE(n)/M)^{1/2}$ and the transmissibility at frequency ω is given by

$$|T(\omega)| = \frac{\{1 + d^2(\omega)\}^{1/2}}{\{[1 - (\omega/\omega_n)^2\{E'(n)/E'(\omega)\}]^2 + d^2(\omega)\}^{1/2}} \quad (4.22)$$

The main difference between this expression and that obtained for a simple spring and dashpot model[30] is that the damping, represented by $d(\omega)$, and the modulus $E'(\omega)$ are both frequency dependent. Various isolating systems having linear and non-linear stiffness and different types of damping have been considered by Snowdon[25] and Crede and Ruzicka.[31]

The influence of fluidic process on the transmissibility of open-cell flexible foams can be studied theoretically by introducing the relationships for $E'(\omega)$ and $d(\omega)$ given by eqns (4.9) and (4.10), respectively, into eqn (4.22). Although these expressions are for foams filled with incompressible fluids, the discussion given above indicates that qualitatively the material response of a gas-filled foam is approximately the same as that of a liquid-filled foam. The primary difference betwee the two situations is that for a given foam and specimen geometry the frequency at which fluidic effects are observed is much higher in gas-filled materials than in liquid-filled materials.

An investigation of the vibration-isolation behaviour of these systems has been made[32] by choosing a value for the parameter β, eqn (4.8), and then calculating the corresponding value of γ_{max} using eqn (4.11). This gives the value of the parameter γ at the frequency for maximum fluid flow-damping. It is convenient to work in terms of the frequency ratio ω/ω_{max}. This can be done by expressing γ in the form $\gamma = \gamma_{max}(\omega/\omega_{max})$, which is obtained from eqn (4.7). Substituting into eqns (4.9) and (4.10) allows the calculation of the loss tangent and the storage modulus of the isolator material system as a function of ω/ω_{max}. In order to apply eqns

(4.9) and (4.10) to eqn (4.22) it is necessary to relate ω/ω_{max} to ω/ω_n. The simplest arrangement is to assume that the frequency for maximum fluid-flow damping is the same as the natural frequency of the mass-isolator system, i.e. $\omega_{max} = \omega_n$. When this is done the frequency ratio ω/ω_{max} can be replaced by ω/ω_n and there is only one independent variable.

Transmissibility curves predicted for liquid filled foam systems typical of those described in Section 4.3.3 are shown in Fig. 4.21 for different values of β. The parameter β governs the maximum value of the mechani-

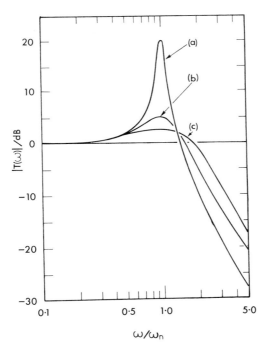

FIG. 4.21. The variation of the transmissibility $|T(\omega)|$ with frequency ratio ω/ω_n for liquid-filled foam mounts predicted by eqns (4.22), (4.9) and (4.10). (a) $\beta = 1 \cdot 0 \times 10^3$, (b) $\beta = 0 \cdot 50$ and (c) $\beta = 0 \cdot 20$. The loss tangent of the polymer matrix $d_m = 0 \cdot 10$.[32]

cal damping resulting from the fluid flow process, see eqn (4.12). The behaviour for $\beta = 1 \times 10^3$, curve (a), corresponds to the situation where fluid-flow effects are negligible and it represents the transmissibility response of the polymer matrix. The loss tangent of the matrix polymer $d_m = 0.10$. The equations used in this analysis assume linear elastic behaviour and consequently are applicable only for small material deformations. The behaviour predicted by curve (a) of Fig. 4.21 is similar to that observed for the PUR flexible foam of intermediate cell size at small excitation levels (curve (a); Fig. 4.20 (ii)).

Decreasing the value of β corresponds to increasing the damping of the isolator system. For curves (b) and (c) in Fig. 4.21, the fluidic effects have a significant influence on the mechanical properties of the isolator material, and it is seen that the increased damping considerably reduces the transmissibility in the amplification region. An important difference between the response of these systems and that of the classic Hookean spring–Newtonian viscous dashpot isolator is the frequency at which the value of the transmissibility is unity for $\omega/\omega_n > 1.0$. With the spring and dashpot isolator, this occurs when $\omega/\omega_n = \sqrt{2}$ and is independent of the degree of damping. For fluid-filled foams the value of ω/ω_n for $|T(\omega)| = 1.0$ increases with increasing damping and, in this respect, the behaviour is similar to that of an elastically coupled viscous damped vibration isolating system described previously in the literature.[31]

It can also be seen for situations (b) and (c) that the transmissibility in the vibration isolating region, $|T(\omega)| < 1.0$, was larger than that for the polymer matrix. This is due to the enhanced value of the elastic modulus $E'(\omega)$ at frequencies above that for maximum fluid flow damping, see for example Figs. 4.5 and 4.13. This behaviour is undesirable, since it degrades the performance of the isolator under operating conditions. A lower value of $|T(\omega)|$ in the isolating region can be achieved while maintaining a relatively low value in the amplification region by optimising the design of the isolator system. One way of doing this is to arrange for the fluid-flow damping maximum to occur at a frequency below the natural frequency, i.e. $\omega_{max} < \omega_n$. In general terms this can be expressed as $\omega_{max} = a\omega_n$ where the coefficient $a \leq 1.0$. The behaviour predicted for a liquid-filled foam isolator with $a = 0.5$ is compared with that determined experimentally[25] for two solid rubber mounts in Fig. 4.22. The transmissibility of the foam system in the isolating region is approximately the same as that of the rubber mounts, but in the amplification region it is considerably less. It should be possible to reduce the modulus enhancement at high frequencies and hence improve further the performance of a flexible foam isolator by replacing the Newtonian liquid with a shear-thinning liquid. So far no studies of foams filled with non-Newtonian fluids have been reported.

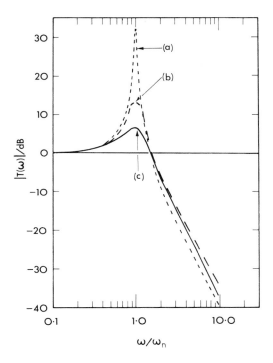

FIG. 4.22. Comparison of the transmissibility of (a) a solid natural rubber and (b) a solid butyl rubber isolating mount measured experimentally[25] with that predicted (c) for a liquid-filled open-cell foam with $\beta = 0.5$, $d_m = 0.10$ and $a = 0.5$.[32]

ACKNOWLEDGEMENT

The author wishes to thank K. C. Rusch, S. V. Kanakkanatt, J. Beadman and the editors of *The Journal of Cellular Plastics, Rubber Chemistry and Technology, Journal of Physics* and *The Journal of Sound and Vibration* for permission to use material contained in their publications.

APPENDIX

From eqns (4.3) and (4.14), the compressive stress σ_{ft} associated with the fluid-flow process and the transverse deformation of the system when the transverse motional inertia forces and the frequency dependence of the properties of the matrix polymer are included is

$$\sigma_{ft} = \frac{E'_m(\omega)(\gamma/12)^2\beta' e_0 \sin \omega t}{(1 - \beta''\gamma/12)^2 + (\beta'\gamma/12)^2}$$
$$+ \frac{E'_m(\omega)(\gamma/12)(1 - \beta''\gamma/12)e_0 \cos \omega t}{(1 - \beta''\gamma/12)^2 + (\beta'\gamma/12)^2}$$

(A1)

where

$$\begin{aligned}
\beta^* &= \beta' + j\beta'' \\
\beta' &= \beta(1 - \chi)/\{(1 - \chi)^2 + d_m^2(\omega)\} \\
\beta'' &= -\beta d_m(\omega)/\{(1 - \chi)^2 + d_m^2(\omega)\} \\
\beta &= 3\{E'_m(\omega)/E'_t(\omega)\}(h/W)^2 \\
\chi &= (\omega^2\rho h^2)/\{5E'_t(\omega)\} \\
\gamma &= (\omega\eta W^2)/\{\zeta E'_m(\omega)K(\omega)\} \\
d_m(\omega) &= E''_t(\omega)/E'_t(\omega) = E''_m(\omega)/E'_m(\omega)
\end{aligned}$$

(A2)

REFERENCES

1. JONES, R. E. and FESMAN, G. (1965) *J. Cell. Plast.*, **1**, 200.
2. HILYARD, N. C. and DJIAUW, L. K. (1971) *J. Cell. Plast.*, **7**, 33.
3. KANAKKANATT, S. V. (1969) *PhD. Thesis, The University of Akron*, Ohio, U.S.A.
4. HILYARD, N. C. and KANAKKANATT, S. V. (1970) *J. Phys. D: Appl. Phys.*, **3**, 906.
5. FERRY, J. D. (1962) *Viscoelastic Properties of Polymers*, John Wiley, New York.
6. COOPER, S. L. and TOBOLSKY, A. V. (1966) *J. Appl. Polym. Sci.*, **10**, 1837.
7. WHITTAKER, R. E. (1971) *Shoe Materials Progress* (S.A.T.R.A.), **3**, 165.
8. HULL, G. K. (1976) *SPI Cell. Plast. Div. Cell. Plast. Conf., 4th and Annu. Conf., 19th Montreal Que., Nov. 15–19, 1976*. Publ. by Technomic Publ., Westport, Conn., 1977, p.76.
9. KONTSKY, J. A., HIEN, N. V. and COOPER, S. L. (1970) *Polym. Lett.*, **8**, 353.
10. FIELDING-RUSSEL, G. S. and FITZHUGH, R. L. (1972) *J. Polym. Sci., A-2*, **10**, 1625.
11. RUSCH, K. C. (1966) *PhD. Thesis, University of Akron*, Ohio, U.S.A.
12. CRITCHFIELD, F. E., KOLESKE, J. V. and PRIEST, D. C. (1972) *Rubber. Chem. Tech.*, **45**, 1467.
13. PATTEN, W., SEEFRIED, C. G. and WHITMAN, R. D. (1974) *J. Cell. Plast.*, **10**, 276.
14. SEEFRIED, C. G., WHITMAN, R. D. and POLLART, D. F. (1974) *J. Cell. Plast.*, **10**, 171.

15. HEPBURN, C. and REYNOLDS, R. J. W. (1975) *The Chemistry and Technology of Polyurethanes. Molecular Behaviour and the Development of Polymeric Materials*, A. Ledwith and A. M. North (Eds), Chapman and Hall, Chap. 7.
16. AGGERWAL, S. L. (Ed.). (1970) *Block Polymers*, Plenum Press, New York.
17. DICKIE, R. A. (1973) *J. Appl. Polym. Sci.*, **17**, 45.
18. GENT, A. N. and RUSCH, K. C. (1966) *Rubb. Chem. Tech.*, **39**, 389.
19. GENT, A. N. and RUSCH, K. C. (1966) *J. Cell. Plast.*, **2**, 46.
20. LIBER, T. and EPSTEIN, H. (1969) *ASME Vibr-46*, April, 1969.
21. GENT, A. N. and LINDLEY, P. B. (1959) *Proc. Inst. Mech. Eng.*, **173**, 111.
22. HILYARD, N. C. (1970) *J. Acoust. Soc. Am.*, **47**, 1463.
23. STOKEY, W. F. (1976) *Vibration of Systems having Distributed Mass and Elasticity. Shock and Vibration Handbook*, C. M. Harris and C. E. Crede (Eds.), 2nd Edn, McGraw-Hill, Chap. 7.
24. HILYARD, N. C. and KANAKKANATT, S. V. (1970) *J. Cell. Plast.*, **6**, 87.
25. SNOWDON, J. C. (1968) *Vibration and Shock in Damped Mechanical Systems*, John Wiley, New York, Chap. 2.
26. PIZZIRUSSO, J. (1973) *Package Eng.*, **18**, 54.
27. HATAE, M. T. (1976) *Shock and Vibration Handbook*, C. M. Harris and C. E. Crede (Eds.), 2nd Edn, McGraw-Hill, Chap. 41.
28. BEADMAN, J. (1980) Project Report, Sheffield City Polytechnic.
29. GEE-CLOUGH, D. and WALLER, R. A. (1968) *J. Sound Vib.*, **8**, 364.
30. BLAKE, R. E. (1976) *Shock and Vibration Handbook*, C. M. Harris and C. E. Crede (Eds.), 2nd Edn, McGraw-Hill, Chap. 2.
31. CREDE, C. E. and RUZICKA, J. E. (1976) *Shock and Vibration Handbook*, C. M. Harris and C. E. Crede (Eds.), 2nd Edn, McGraw-Hill, Chap. 30.
32. HILYARD, N. C. (1974) *J. Sound. Vib.*, **32**, 71.

Chapter 5

SHOCK MITIGATION—MATERIAL BEHAVIOUR

N. C. HILYARD
*Department of Applied Physics, Sheffield City Polytechnic,
Sheffield, UK*

5.1 INTRODUCTION

As with the vibration-isolation problems discussed in the preceding chapter, shock-isolation situations can usually be divided into two categories. These are when the aim is to limit the shock-induced stresses in critical components of the protected item or to limit the forces transmitted to the structure which supports the device within which the shock originates. Although the objectives of these situations are different their analysis using the classical equations of mechanics is much the same. It is clear that the effectiveness of a shock-isolating system is directly related to the force-deflection characteristics of the device, or the material, under impact-loading conditions.

In practice, mechanical shock is a complex transient motion involving a combination of decaying vibrations at different frequencies. However, in many situations the deformation of the isolator is in the compression mode and in special cases simplification of the motion is permissible. The idealisation of a shock motion as a simple velocity step defined by $\dot{y} = 0$ at $t < 0$ and $\dot{y} = v_i$ at $t = 0$, where v_i is the speed at the instant of impact, can form the basis for designing a shock isolator and evaluating its effectiveness.[1]

An understanding of how the mechanical behaviour of an isolator influences the overall shock response, or how an isolating material behaves under impact-loading conditions can be obtained by considering the lumped parameter model shown in Fig. 5.1(a). This consists of an isolator, a material or a device, for which the stiffness S is a function of

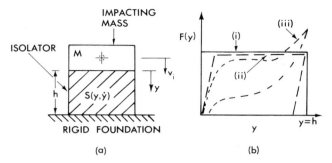

FIG. 5.1. (a) The idealised lumped parameter model of a shock-mitigating system with deformation and rate-dependent stiffness S. (b) Compressive force–deformation diagrams for (i) the 'ideal' shock mitigator (ii) the 'semi-ideal' shock mitigator and (iii) a typical flexible cellular polymer.

the compressive deformation y mounted on a rigid foundation. In more general terms the stiffness may also depend upon the strain rate, $\dot{y}$. The height of the isolator is h. At time $t = 0$ a mass M moving with speed v_i impacts upon the upper surface of the isolator creating a uniform stress over the cross-section the initial rate of deformation c_i being equal to v_i/h. The equation of motion is

$$M\ddot{y} + F(y,\dot{y}) = 0 \qquad (5.1)$$

At times $t > 0$, the motion continues in the positive y direction until the work W done on the isolator system is equal to the kinetic energy W_i of the body at the instant of impact. The influence of changes in the gravitational potential energy are usually ignored in these analyses. At this instant $\dot{y} = 0$ and $y = y_m$, where y_m is the maximum deformation of the isolator, so that

$$W_i = \int_0^{y_m} F(y,\dot{y}) \, dy \qquad (5.2)$$

After this the potential energy stored in the isolator is returned to the impacting body, which rebounds in the negative y direction with kinetic energy W_0. This parameter is given by eqn (5.2) using the function $F(y,\dot{y})$ and the limits of integration appropriate to the unloading part of the compression cycle. If the mass is not bonded to the isolator, it can be assumed that contact between the mass and the isolator is lost when the net force on M, in the $-y$ direction, is zero.

Shock-isolating systems take many different forms ranging from complex reactive or non-reactive mechanical devices,[2,3] moulded materials of complex shape to simple pads or sheets of isolating material. In this chapter we are concerned with the behaviour of specimens of cellular polymer with simple geometry, i.e. pads or sheets, subjected to impact loading in the compression direction. In the following chapter, specific shock-mitigating systems are considered.

5.2 DESCRIPTION OF PROPERTIES

Shock-mitigating materials can be divided into two broad categories; elastic (i.e. flexible or resilient) and non-elastic (i.e. brittle or crushable).[4,5] Materials that sustain less than an arbitrarily selected amount of permanent deformation when deformed during impact are placed in the first category, and materials that sustain larger amounts are placed in the second. The choice of a particular category depends on whether the isolator material is to be subjected to a single impact or to multiple impacts.

The effectiveness of an isolator material has been described in a variety of ways. For packaged goods probably the most important parameter is the maximum acceleration, $\ddot{y}_m = a_m$, experienced by the packaged object.[4] In order to meet design specifications this must be less than the 'fragility index' of the object, defined as the maximum permissible acceleration that the object can withstand. Under certain circumstances, however, other parameters may be of importance. These include the maximum deformation y_m of the isolator material, which is relevant when there are space limitations, and the rebound energy W_0 which is important in some personal-protection situations. The second parameter is related to the material properties by

$$E_d/E_s = (W_i - W_0)/W_0 \qquad (5.3)$$

where E_d, the energy dissipated in the material, is given by the area enclosed by the force-deformation diagram during the deformation cycle and E_s is the stored elastic energy. The ratio E_d/E_s may be considered as the energy-absorbing capacity of the isolator material and it is analogous to the loss tangent measured under small oscillatory strains (see Chapter 4). The rebound resilience, R, of the material is related to the E_d/E_s ratio by

$$R = 1/(1 + E_d/E_s)$$

As far as the 'fragility' requirement is concerned, an 'ideal' isolating material is one that produces the smallest maximum acceleration for a given impact energy. This is achieved by an isolator for which the force is constant, independent of the deformation, over the entire thickness h of the material.[6] To satisfy the minimum rebound-energy criterion the area under the force-deformation diagram during the unloading part of the deformation cycle should be zero. However, all isolator materials would be expected to have an initial Hookean region in the stress–strain response and a finite area under the force-deformation diagram during unloading. The force-deformation diagrams for 'ideal' and 'semi-ideal' systems and a typical flexible cellular polymer are shown diagrammatically in Fig. 5.1(b).

Many methods have been used for the presentation of information needed for the design of shock mitigating systems. In early work[4] acceleration data for specific materials of different thickness are reported for different impact energies (expressed in terms of drop height and mass). More generalised approaches for the description of isolator materials have been suggested by several workers.[5-7] According to Woolam[6,7] the acceleration efficiency J of a material can be defined as the ratio of the maximum acceleration, a_m, experienced by the object with the specified material to that which would be experienced by the object when using an 'ideal' isolator, as described above, of the same height. For an ideal isolator $F(y,\dot{y}) = \text{constant} = F_m$, where F_m is the maximum force experienced by M. Since W_i for the ideal material is $W_i = Mv_i^2/2 = F_m y_m$ and $y_m = h$, we have for the acceleration efficiency of a specific material

$$J = a_m/(v_i^2/2h) \tag{5.4}$$

For the 'ideal' isolator the value of J is unity and for real materials $J > 1$.

With cellular polymer isolating materials, which have force-deformation diagrams similar to that shown in Fig. 5.1(b), the maximum compressive deformation, and hence the maximum force experienced by the impacting object, is directly related to the kinetic energy of the object at the instant of impact. This depends upon two quantities, M and v_i. By using the impact-energy density, $U = Mv_i^2/2Ah$, where A is the area over which the force is distributed, it is possible to construct a family of curves such as those shown diagrammatically in Fig. 5.2. These describe the acceleration efficiency of the material as a function of the independent variable U for different values of the initial strain rate c_i. Although, in general, such curves exhibit rate effects with many isolator materials these are small,[6] so that for design purposes the performance of a material

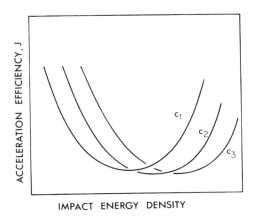

FIG. 5.2. Variation of the acceleration efficiency of a shock-mitigating material as a function of the impact energy density for different initial strain rates c_i.

may be represented by a single J–U curve. Experimental data for two reticulated PUR flexible foams are shown in Fig. 5.3.[8] The data were obtained using the experimental arrangement described below using

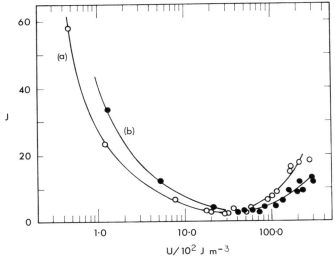

FIG. 5.3. Variation of the acceleration efficiency J as a function of the impact-energy density U for two open-cell PUR flexible foams of density ratio $\phi \approx 0.03$ for a range of impact speeds and masses. (a) Average cell diameter $d_{av} \approx 0.25$ mm. (b) Average cell diameter $d_{av} \approx 0.77$ mm.

four different masses (in the range 0·5–2·5 kg) at different impact speeds. The foams had approximately the same density ratio, $\phi \approx 0.03$, the same cell structure geometry, but different cell size. It can be seen that the data superposed reasonably well giving continuous J–U curves, indicating the validity of using U as an independent variable, and that the curves exhibited broad minima. For flexible cellular polymers the minimum value of J typically lies in the range 2·0 to 4·0.[6–8]

The important features of a J–U curve are the width of the J–U minimum in the optimum energy-absorbing region and the minimum value of J. These are governed by the size of the plateau region of the compression stress–strain curve, the slope of the curve in this region and the level of the stress. For design purposes, the value of the impact-energy density for minimum J is also required.

5.3 PREDICTION OF SHOCK-MITIGATION BEHAVIOUR FROM STATIC STRESS–STRAIN CURVES

Because of the wide availability of compression stress–strain diagrams measured under quasi-static conditions, there has been much interest in the prediction of the behaviour of shock-mitigating systems employing cellular polymers using this type of information. This can be done in two ways, by graphical analysis[9] or by using mathematical models.[5,8,10] For the second procedure it is necessary to have an analytic expression for $F(y)$, or $\sigma(\varepsilon)$, so that the integral $\int F(y)dy$ can be evaluated. Several different forms of non-linear expression relevant to shock-isolating materials have been described in the literature.[1] These include $F(y) \alpha \tan(\pi y/2h)$, $F(y) \alpha \tanh(y/y')$ where y' is a constant depending upon the properties of the material and the geometry of the specimen, and $F(y) = F_0 + \{1 - \exp(-ay)\} F_{\lim}$ where F_o, $F_{\lim}$ and a are constants.[3]

Rusch[5] has put forward a procedure which makes use of the expression $\sigma(\varepsilon) = E_f \varepsilon F(\varepsilon)$. This expression and the relationship between the parameters E_f and $F(\varepsilon)$ and the composition and cell geometry of the foam have been discussed in Chapter 2B. Rusch proposed that the behaviour under impact-loading conditions can be described by three dimensionless variables. The first of these is the energy-absorbing efficiency $K = a_{\text{ideal}}/a_m$, which is the inverse of J defined above, and takes values between zero and unity. This parameter is related to the maximum strain in the material, ε_m, by the equation

$$K = \int_0^{\varepsilon_m} F(\varepsilon)\,d\varepsilon / \{\varepsilon_m F(\varepsilon_m)\} \tag{5.5}$$

The second variable is the maximum stress divided by the modulus of the foam, $\sigma_m/E_f = I/K$, where

$$I/K = \varepsilon_m F(\varepsilon_m) \tag{5.6}$$

The third, and independent, variable is the impact-energy density divided by the foam-compression modulus, $U/E_f = I$, which is given by

$$I = \int_0^{\varepsilon_m} \varepsilon F(\varepsilon)\,d\varepsilon \tag{5.7}$$

He also proposed that the strain function $F(\varepsilon)$ during compressive deformation can be related to the strain by

$$F(\varepsilon) = a\varepsilon^{-p} + b\varepsilon^q \tag{5.8}$$

where the coefficients a, b, p and q are empirically determined material constants assumed to be independent of the strain rate. Using eqns. (5.6), (5.7) and (5.8), he showed that for $\varepsilon \gg \varepsilon_b$, where ε_b is the buckling strain,

$$I = \{a/(2-p)\}\varepsilon_m^{(2-p)} + \{b/(2+q)\}\varepsilon_m^{(2+q)}$$

and

$$K = I/\{a\varepsilon_m^{(1-p)} + b\varepsilon_m^{(1+q)}\}$$

Curves showing the variation of K and I/K as a function of I can be generated from these equations by selecting appropriate values for ε_m. The shape of the K–I curve is the inverse of the J–U curve (Figs. (5.2) and (5.3)) and exhibits a maximum value with increasing value of I. The detailed shape of the curve is governed by the parameter $F(\varepsilon)$, which is dependent on the matrix geometry, but is independent of the foam modulus E_f. Schwaber and Meinecke[10] have found that the strain function defined in eqn (5.8) does not satisfactorily describe all cellular polymer systems.

By analysing static stress–strain data for different types of cellular polymers Rusch was able to identify the material parameters that control the energy-absorbing efficiency K and as a result specify an optimum material. An optimum system was defined as one for which the value of K is large and the maximum in the K–I curve is broad. According to this analysis, an optimum material will have a compression stress–strain behaviour (refer to Chapter 2B) which exhibits a small value of $F(\varepsilon)_{min}$, a large difference between the strain at $F(\varepsilon)_{min}$ and the buckling strain ε_b, and zero slope of the stress–strain diagram in the plateau region, i.e. $\log F(\varepsilon)/\log \varepsilon = -1\cdot 0$. The criteria are best fitted by a foam which has a brittle matrix, because these have a plateau region which is wider and flatter than that of resilient or ductile foams, a large cell size, a narrow cell-size distribution and a minimum number of reinforcing membranes between the cells.

Unfortunately there are few experimental data available in the literature to verify these theoretical predictions. However, Benning[11] has observed for polyethylene foams that large cells, or anisotropic cells with maximum dimension orientated in the direction of compression, are required for the best shock mitigation. Also it can be seen from the data of Fig. 5.3, which is for flexible foams, that although the two foams had approximately the same minimum value for J, when considered on a linear U scale rather than logarithmic the width of the J–U minimum for the foam with the larger cell size was greater than that for the foam with the smaller cell size.

5.4 MECHANICAL BEHAVIOUR OF FLEXIBLE FOAMS UNDER IMPACT LOADING

5.4.1 Stress–Strain Relationships: General Considerations

It has been reported that the shock-mitigating behaviour of an isolator system predicted from material stress–strain data measured under static conditions does not always agree with that found in practice[10,12–14] so that for package-design purposes this type of data has limited usefulness. This discrepancy can be attributed to a number of factors.

The mechanical properties of the base material, i.e. modulus and damping, may be rate sensitive so that with the high strain rates experienced during impact loading the material properties may be different to those measured under essentially static conditions. However, these effects are expected to be negligible with rigid foams and small with flexible foams except at temperatures close to their glass transition.[5,14] In some situations the dynamic action of the fluid enclosed by the polymer matrix, usually air, influences the mechanical response of the cellular polymer,[12,13,15–17] as described in Chapter 4. The magnitude of this effect depends on several material and specimen parameters including the average cell size, the deformation rate and the dimensions of the test piece. Another factor which should be taken into consideration is that the deformation processes in the polymer matrix under impact loading may be different to those for quasi-static loading.

Impact and static stress–strain diagrams for an open-cell PUR flexible foam are compared in Fig. 5.4. These results were obtained for a reticulated PUR foam with $\phi \approx 0.03$, mean cell diameter $d_{av} \approx 0.77$ mm and

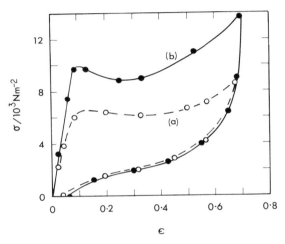

FIG. 5.4. Compressive stress–strain diagrams for a reticulated PUR flexible foam $\phi \approx 0.03$, $d_{av} \approx 0.77$ mm. (a) Quasi-static loading conditions. (b) Impact-loading conditions, $c_i = 53$ s^{-1}.

$E_f \approx 1 \times 10^5$ Nm^{-2} using the low-energy impact apparatus shown in Fig. 5.5. The specimen thickness $h = 3.0$ cm, loaded area $A = 32$ cm^2 and the initial strain rate $c_i = 53$ s^{-1}. The impact data were obtained under conditions such that flow phenomena were absent, as explained below. It can be seen that during the loading part of the compression cycle the proportional limit and the level of the stress in the plateau region for impact conditions were significantly larger than those for static conditions. In addition, the impact value of the initial compression modulus was about 20% larger than the static value. During the unloading part of the compression cycle, the stress–strain behaviour was essentially the same for the two loading conditions. Although this behaviour was typical of all

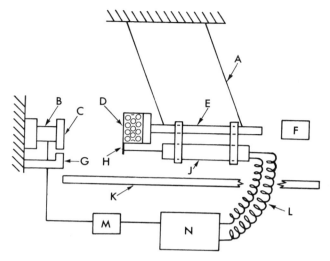

FIG. 5.5. The low-energy impact apparatus. A, suspension; B, piezoelectric force transducer; C, anvil; D, foam sample; E, pendulum tube; F, holding magnet; G, displacement stop; H, plate; J, displacement transducer; K, graduated scale; L, flexible leads; M, charge amplifier; N, storage oscilloscope.

specimens tested,[8] with some materials small differences in the static and impact curves during unloading were observed.

Similar stress–strain behaviour has been reported by Nagy et al.[18] and Schwaber and Meinecke[10] for other cellular polymers, including PUR foams. However their testing procedures were such that fluid-flow effects were not excluded. These workers also found for their systems that the level of the stress and the slope of the stress–strain curve in the plateau region increased continuously with increasing strain rate.

Schwaber and Meinecke proposed that the dependence of the stress on the strain and the strain rate is separable so that the stress can be presented by an equation of the form

$$\sigma = F(\varepsilon)\, E_0(v)$$

where the strain function $F(\varepsilon)$ is rate independent and the function $E_0(v)$ depends upon rate. For their particular cellular systems, which were SBR latex coated PUR foams, the rate-dependent parameter could be represented over limited ranges of strain rate by the expression

$$E_0(v) = E(k) v^r$$

where $E(k)$ is the value of the modulus $E(v)$ for some arbitrary value of the deformation speed, v, and r is a rate constant. The rate dependence of $E_0(v)$ reported by Schwaber and Meinecke is similar to the frequency dependence of the small-strain dynamic modulus of an air-filled open-cell PUR flexible foam shown in Fig. 4.5 in that at low and high rates $E_0(v)$ was only weakly dependent on rate, but at intermediate values of rate the system passed through a transition region in which $E_0(v)$ increases rapidly with increasing rate. The extent of the transition region was approximately one decade in strain rate for the two situations, so that the transition observed by Schwaber and Meinecke in the modulus–deformation speed curve might be attributed to fluid-flow phenomena. The weak dependence of the modulus on rate outside of the transition region can be attributed to the viscoelastic response of the matrix polymer. Cousins[14] has used the empirically determined Schwaber–Meinecke rate function and a strain function $F(\varepsilon) = (1 - \varepsilon)^{-m}$ to predict the impact-mitigating behaviour of rate-dependent materials.

By analysing large-strain stress–strain diagrams obtained using controlled deformation rates Nagy et al.[18] were also able to factorise the compressive stress into two parts; a rate-independent strain function $F(\varepsilon)$ and a rate-dependent function. However, in their case, the rate-dependent function was not independent of strain. The relevant expression is

$$\sigma = F(\varepsilon)(\dot{\varepsilon}/\dot{\varepsilon}_0)^{n(\varepsilon)}$$

The parameter $\dot{\varepsilon}_0$ is a reference strain rate and the exponent $n(\varepsilon) = a + b\varepsilon$, where a and b are constants. It was found that both $F(\varepsilon)$ and $n(\varepsilon)$ were dependent upon the cellular system. The dependence of this rate function on strain is also indicative of fluid-flow phenomena as explained in Section 5.3.

5.4.2 Test Procedure

The work just described shows that in order to obtain an understanding of the mechanisms responsible for the material behaviour of cellular polymers, it is necessary to separate out the behaviour due to the matrix and that due to fluid flow. The theory of Gent and Rusch for the stiffness and damping of flexible open-cell foams under small oscillatory strains (see Chapter 4) can be used to establish equations showing how fluidic effects influence the mechanical behaviour under impact loading (see Section 5.3.4).

Using these equations, it is possible to design experiments in which fluidic effects are negligibly small so that the impact behaviour of the polymer matrix can be investigated in the absence of flow phenomena.[8] This involves selecting a range of initial strain rates, v_i/h, appropriate to the size of the specimen and the physical parameters of the foam, such as modulus and permeability.

In the following discussion we will consider first the behaviour of the polymer matrix and then describe how fluid flow modifies this behaviour. The data presented were obtained using a low-energy impacting system, shown diagrammatically in Fig. 5.5. The force experienced by the specimen was measured using a piezoelectric transducer (B) and the deformation using an LVDT (J). Force-deformation diagrams were displayed on a storage oscilloscope (N). The impact and rebound energies were measured using the graduated scale (K).

5.4.3 The Cellular Matrix

Mechanical Properties

The compressive stress–strain behaviours of a reticulated PUR flexible foam measured under impact and quasi-static loading conditions have been compared in Fig. 5.4. In this investigation[8] it was found that the impact stress–strain diagrams for specimens of different loaded area were approximately the same so that the differences in the static and impact behaviour were not related to the fluid phase.

Examination of the cell structure of a flexible foam during compression at slow rates shows that buckling of the cell elements is confined initially to regions adjacent to the surface over which the external load is applied.[8,11] Under these conditions, the cell elements in the interior of the specimen will be deformed elastically, whereas those at the loaded boundaries will be buckled. The effect of the buckled boundary region on the stress–strain behaviour of rigid foams has been described by Phillips and Waterman[19] and its influences in the measured value of the initial-compression modulus has been discussed by Menges and Knipschild in Section 2A.3.2. With flexible foams, the buckled region propagates through the test specimen as the compressive strain is increased until a state of uniform cell deformation is reached. Thus under impact-loading conditions with high initial-strain rates, we are concerned with the transmission of both buckling and elastic deformations.

It is possible here to compare the deformation mechanisms in a cellular polymer with those in a ductile material loaded beyond its yield point.[20] Under impact conditions the stress is transmitted as two waves, an elastic wave with speed $c_0 = \sqrt{(E/\rho_0)}$, where E is the elastic modulus and ρ_0 the density of the material and a plastic wave with speed $c_p = \sqrt{(P/\rho_0)}$, where P is the slope of the stress–strain diagram post yield. The speed of plastic-wave propagation is often an order of magnitude less than that of the elastic wave. Transferring this concept of two deformation waves to cellular polymers it may be assumed that the buckling wave travels at a much slower speed than the elastic wave. Consequently with high initial compressive-strain rates, the buckled boundary region will propagate only a short distance before the stress in the interior has reached the critical buckling value. Thus under impact-loading conditions the deformation in the cellular material will be more uniformly distributed throughout the specimen and a higher modulus, an enhanced rectilinear region in the stress–strain diagram and a higher level of stress in the plateau region would be expected. The material properties observed under these conditions will be the inherent properties of the cellular polymer unaffected by the boundary region.

Because of these differences between the static and impact stress–strain behaviours differences between other material properties would be expected. This is in fact the case. The variation of the energy absorbing capacity, E_d/E_s, as a function of the maximum compressive strain, ε_m, is shown for a reticulated PUR flexible foam in Fig. 5.6. The impact data, represented by the closed circles, were obtained for a single impacting mass. The static data were calculated from the static stress–strain diagram. It can be seen that E_d/E_s initially increased rapidly with increasing ε_m. This was followed by a region with smaller gradient and subsequently a region of higher gradient. This behaviour can be related to the three regions of the compressive stress–strain curve. It can also be seen that the energy-absorbing capacity under impact loading was significantly larger than the static value. In addition it was found in this series of experiments that E_d/E_s was not governed uniquely by ε_m or U. Small rate effects were observed when using different combinations of values for the impacting mass M and speed v_i.[8]

Model Representation and the Prediction of Behaviour
An alternative approach to that of Rusch has been put forward for predicting the behaviour of a shock-mitigating system using material stress–strain data.[8] This analysis is based on the expression $\sigma(\varepsilon) = \sigma_{25} F(\varepsilon)$ where σ_{25} is the stress at 25% compression and $F(\varepsilon)$ is an empirically determined, but analytic, strain function. This form of expression was chosen because of the wide use of σ_{25} to describe the mechanical behaviour of

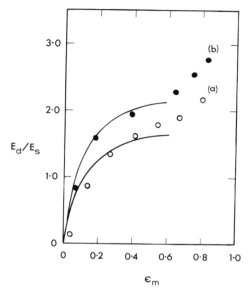

FIG. 5.6. Damping capacity E_d/E_s of a reticulated PUR flexible foam as a function of maximum compressive strain ε_m for (a) quasi-static loading and (b) impact-loading conditions. Continuous lines generated using eqns (5.3), (5.11) and (5.12) and material parameters appropriate to the two loading conditions.

flexible foams. Unlike the Rusch approach, the analysis allows the prediction of the energy absorbing capacity. However because of the particular strain function chosen it can only be applied in situations where the maximum strain does not exceed the limit of the plateau region.

For PUR flexible foams with uniform cell structure and size the force $F(y)$ during the compression part of the cycle can be related approximately to the deformation y by

$$F(y) \approx F_{25}\{1 - \exp(-y/\lambda)\} \tag{5.9}$$

where $F_{25} = A\sigma_{25}$ and λ, which is a curve-fitting constant, is the value of y when $F(y) = 0.63 F_{25}$. This expression presumes that the stress–strain curve in the plateau region is parallel to the strain axis, which is approximately true for uniform PUR foams. During the unloading part of the cycle

$$F(y) \approx (F_m/\pi)\cos^{-1}(1 - 2y/y_m) - \alpha F_m \tag{5.10}$$

where y_m is the maximum deformation and F_m the corresponding force. The parameter α is given by

$$\alpha = (1/\pi)\cos^{-1}(1 - 2y_0/y_m)$$

where y_0 is the deformation that is not recovered on the completion of the deformation cycle, i.e. when the force has returned to zero. With PUR flexible foams the ratio y_0/y_m is approximately constant so that to a reasonable approximation the parameter α can be assigned a constant value independent of y_m.

Combining eqns (5.2) and (5.9) gives

$$W_i \approx F_{25}\left[y_m + \lambda\{\exp(-y_m/\lambda) - 1\}\right] \tag{5.11}$$

Using the appropriate limits of integration it can also be shown for $y_m \gg y_0$ that the rebound energy is given by

$$W_0 \approx (1 - \alpha)F_m y_m/2 \tag{5.12}$$

By using empirically determined values for F_{25}, α and λ and assigning values to y_m, the corresponding values of F_m, W_i and W_0 can be calculated. These quantities can then be used to determine the values of the parameters E_d/E_s and $J = F_m/W_i h$ for different maximum strains ε_m and impact energy densities U.

The behaviour predicted by these equations using material parameters obtained from static and impact stress-strain diagrams is shown in Figs. 5.6 and 5.7. It can be seen that there was reasonable agreement between the predicted behaviour and that determined experimentally over the range for which the strain function was approximately valid, $0 < \varepsilon_m \leq 60\%$. It was also found in this investigation for a series of foams of approximately the same density and cell structure geometry that the E_d/E_s ratio at 50% compression was essentially independent of cell size. In all cases the impact value of E_d/E_s was larger than the static value.

From these results, and those of other workers, it is clear that in the absence of flow phenomena the material properties of flexible cellular polymers measured under impact loading conditions are different to those determined under quasi-static conditions. This can be attributed in part to the viscoelastic, and hence rate-dependent, nature of the matrix polymer. However, it appears that one of the major factors contributing to this disparity is the difference in the cell deformation mechanisms.

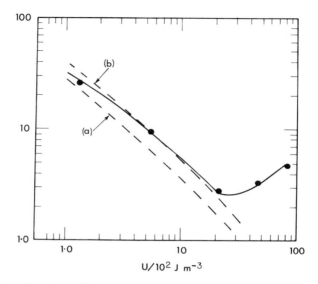

FIG. 5.7. Acceleration efficiency J of a reticulated PUR flexible foam as a function of the impact energy density. Broken curves predicted from material data measured under (a) quasi-static conditions and (b) impact conditions.

5.4.4 Fluidic Effects

Influence on the Compression–Deflection Characteristics
An important characteristic of a shock mitigating system is the width of the minimum in the J–U curve, i.e. the optimum shock-mitigation region. For rate-independent cellular polymers, this is governed by the extent of the plateau region in the stress–strain curve. This is controlled by the cell-structure geometry. Rusch[5] has indicated that the optimum region can be expanded if the effective compressive stiffness of the system increases with increasing deformation rate. This can be achieved by operating the cellular polymer close to its glass transition temperature or by designing the system such that the flow of the fluid enclosed by the polymer matrix influences the overall mechanical behaviour. The first situation is of little practical importance because of the temperature sensitivity of the arrangement.

There have been two detailed studies of the influence of fluidic effects on the mechanical behaviour of cellular polymers under impact loading conditions. The first of these[16,17] considered air-filled systems and the second[21] liquid-filled systems. With gas-filled foams it is difficult to identify which phase, gas or polymer, is responsible for the observed behaviour. This problem does not exist with liquid-filled systems, and by suitable choice of polymer matrix and liquid, fluidic effects can be observed at relatively low impact speeds. This facilitates the experimental investigation of these phenomena and enables the determination of the parameters that govern the mechanical response of the system. Qualitatively the behaviour of gas-filled and liquid-filled systems should be similar, as demonstrated by the studies of Gent and Rusch[22] using small oscillatory strains.

In order to discuss the shock-mitigation behaviour of fluid-filled systems it is necessary to show first how the presence of fluid-flow phenomena modify the compressive stress–strain behaviour under large strain dynamic loading conditions. Force–time and deformation–time traces for an open-cell PUR flexible foam (air-filled, but with no fluid-flow effects) are compared with those for the same matrix filled with a liquid in Fig. 5.8. These data were obtained using the low energy-impact appa-

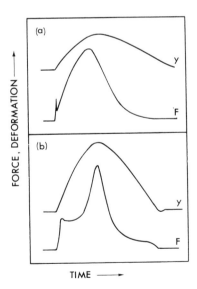

FIG. 5.8. Force (F)–time and deformation (y)–time traces for a reticulated PUR flexible foam (a) liquid filled and (b) air filled (with no fluid-flow effects).

ratus shown in Fig. 5.5. It can be seen for the liquid filled foam that the shape of the F–t curve is different to that of the polymer matrix, the maximum in the F–t curve does not coincide with the maximum in the y–t curve, as it does for the matrix, and at the instant of impact there is a short-duration impulse. The size of this impulse is small at low initial-strain rates but increases with increasing impact speed. The mechanism responsible for the impulse has not been identified but it is probably due to the propagation of a high-speed elastic wave through the liquid medium. The speed of propagation of this wave would be significantly larger than that of the elastic wave transmitted through the polymer matrix.[20]

Model Representation and the Prediction of Behaviour

As mentioned above, there have been two theoretical and experimental investigations of the influence of the fluid phase on impact behaviour of flexible open-cell foams. These theoretical analyses employed a common approach in that they both used lumped parameter phenomenological models to represent the different deformation mechanisms. One such model, based upon the Gent–Rusch theory[22] for the small strain dynamic mechanical behaviour (see Section 4.3) is shown in Fig. 5.9. This model assumes that the force acting on the impacting mass M is made up of two parts; $F_1(y)$ due to the polymer matrix which, although assumed to be indendent of rate, is the force which obtains under impact loading condi-

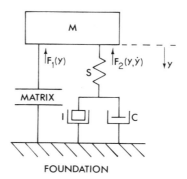

FIG. 5.9. A lumped parameter model representing a fluid filled cellular polymer: S, the elastic stiffness associated with the transverse deformation of the matrix; I, flow inertia; C, viscous flow, $F_1(y)$, the force associated with the deformation of the matrix in the direction of compression.

tions, and $F_2(y,\dot{y})$ due to the fluidic processes which is both deformation and rate dependent. The equation of motion is

$$M\ddot{y} + F_1(y) + F_2(y,\dot{y}) = 0 \tag{5.13}$$

A similar model was used by Liber and Epstein[16,17] in their study of the shock-mitigating behaviour of air-filled cellular polymers. The components of their model represent the non-linear response of the polymer matrix, the retarded elastic response of the matrix and the air-flow process. All of these components were considered to be deformation and rate dependent. At high deformation rates, they assumed, like Gent and Rusch,[22] that the gas is more readily compressed than forced to flow.

In order to predict the impact behaviour of a liquid-filled foam using the model of Fig. 5.9, it is necessary to express $F_1(y)$ and $F_2(y,\dot{y})$ as analytic functions of y and $\dot{y}$.[21] The non-linear matrix force, $F_1(y)$ was assumed to be given by eqns (5.9) and (5.10) with the parameters F_{25}, λ and α determined under impact-loading conditions. The fluidic processes, $F_2(y,\dot{y})$ were analysed as follows.

Applying the superposition principle, described by eqn (4.6), the deformation y of the specimen in the direction of compression is made up of two parts

$$y = y_f + y_t \tag{5.14}$$

where y_f is the deformation resulting from the flow of fluid and y_t is the deformation (in the direction of compression) due to the transverse deformation of the matrix. The second deformation component can be expressed as

$$y_t = F_2/S$$

where S is the compressive stiffness of the specimen associated with the transverse deformation mechanism. The relationship between the specimen coefficients and the materials properties and the specimen geometry are given in the Appendix. The variable y_f is related to F_2 by

$$F_2 = C\dot{y}_f + I\dot{y}_f^2 \,(\text{sgn } \dot{y}_f) \tag{5.15}$$

where C is a viscosity coefficient, which is related to the permeability of the foam $K(y)$, and I is a flow inertia coefficient, which is a function of the parameter $B(y)$, see Section 4.3. The function sgn $\dot{y}_f$ takes a value of 1 for the loading part of the compression cycle and -1 for the unloading part.

Both $K(y)$ and $B(y)$ are functions of the compressive deformation of the foam.[16,21,23] It has been found experimentally[21,23] that $K(y)$ decreases approximately in proportion to the compressive strain, y/h, as predicted by Gent and Rusch,[23] over a large range of strains so that $K(y)$ can be expressed as

$$K(y) = K_0 \{1 - k(y/h)\} \tag{5.16}$$

where K_0 is the permeability in the undeformed state. The constant of proportionality, k, determined empirically is, in general, larger than that predicted by the Gent–Rusch equation.[21,23,24] For the PUR flexible foams which were the subject of the study described here $k \approx 1.3$. Gent and Rusch have also shown that the parameter $B(y)$ is much less strain dependent than $K(y)$. Since for the PUR systems used in this study the forces resulting from flow inertia were small compared with those due to viscous flow the parameter $B(y)$ was assumed to be constant, independent of y, and was assigned a value equal to that for the undeformed specimen, i.e. $B(y) = B_0$.

By substituting $y_f = y - y_t$ and $y_t = F_2/S$ into eqn (5.15) and ignoring a term $F_2^2 (I/S^2)$, which is small, gives

$$F_2 + \dot{F}_2 \{(C/S) + 2(I/C)\dot{y}\} \approx G\dot{y} + I\dot{y}^2 (\operatorname{sgn} \dot{y}) \tag{5.17}$$

(Editor's note: There are typographical errors in the equation given in the original article.) This equation allows the calculation of the force F_2 acting on the impacting mass due to the fluid-flow processes over a large range of strains. By combining eqns (5.9), (5.10) and (5.17) relationships can be established which can be used to predict the motion of M during the loading and unloading part of the cycle, and hence determine the acceleration experienced by M and the energy absorbed during the compression cycle. These equations are; for the loading phase

$$\dddot{y} + \frac{\ddot{y}}{\tau I'} + \frac{I}{I'\tau M}\left[1 + \frac{2F_{25}\exp(-y/\lambda)}{\lambda S}\right]\dot{y}^2 +$$

$$\left[\omega^2 + \frac{F_{25}\exp(-y/\lambda)}{\lambda M}\right]\frac{\dot{y}}{I'} + \frac{F_{25}}{\tau M I'}\left[1 - \exp(-y/\lambda)\right] = 0 \tag{5.18}$$

and for the unloading phase

$$\dddot{y} + \frac{\ddot{y}}{\tau I'} - \frac{I}{I'\tau M}\left[1 + \frac{4F_m}{\pi S y_m \chi(y)}\right]\dot{y}^2$$

$$+ \left[\omega^2 + \frac{2F_m}{\pi M y_m \chi(y)} \right] \frac{\dot{y}}{I'} + \frac{F_m \cos^{-1}(1 - 2y/y_m)}{\pi I' M \tau}$$

$$- \frac{\alpha F_m}{I' M \tau} = 0$$
(5.19)

The parameters used in eqns (5.18) and (5.19) are defined in the Appendix. Introducing appropriate values for these the equations can be solved numerically using the initial conditions $\ddot{y} = y = 0$ and $\dot{y} = v_i$ at $t = 0$. The point of transition from the loading to the unloading equation is obtained by searching for the condition $\dot{y} \leq 0$ during the loading phase.

The impact behaviour of liquid-filled foams has been investigated experimentally using the low energy apparatus described above with the specimen arrangement given in Fig. 5.10. Thin latex rubber skins, about 0·3 mm thick, were applied to the surfaces of a rectangular block of foam. The block was then bonded to an aluminium plate and the top part of the

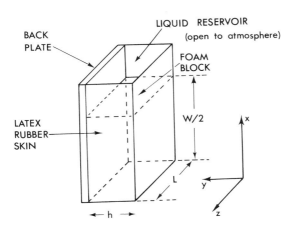

FIG. 5.10. The sample arrangement used for the investigation of liquid-filled foams using the low-energy impact apparatus of Fig. 5.5.

foam block removed leaving the latex skin in place. The open cavity so formed acted as a flexible liquid reservoir. The cellular system was filled with the liquid and entrapped air removed by placing the specimen under vacuum. The liquid level in the reservoir was maintained approximately 2 mm above the surface of the foam. If there is no deformation of the bottom and the side latex walls in the $-x$ and the $\pm z$ directions, respectively, during compression in the y direction, the fluid is constrained to flow in the $+x$ direction. This behaviour conforms to the idealised fluid-flow pattern assumed in the Gent–Rusch theory. For these investigations the liquid-filled sample was attached to the plate C, Fig. 5.5, and not the pendulum as was the case for the air-filled foams.

Experimental data obtained using this arrangement are given in Figs. 5.11 and 5.12. The continuous lines are the behaviour predicted theore-

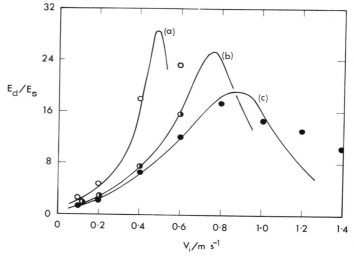

FIG. 5.11. Variation of the damping capacity E_d/E_s of a water-filled open-cell PUR flexible foam as a function of the impact speed v_i for different impacting masses (a) $M = 3\cdot45$ kg, (b) $M = 1\cdot05$ kg, (c) $M = 0\cdot65$ kg. Continuous lines generated using eqns (5.18) and (5.19) and those given in the Appendix.

tically. Figure 5.11 shows the variation of the energy-absorbing capacity, E_d/E_s, as a function of the impact speed v_i for an open-cell PUR flexible foam, with porosity $\zeta \approx 0\cdot98$, filled with water at 22°C. A complete set of parameters characterising this foam, which is the same as that for the data of Figs. 5.6, 5.7 and 5.12 are given in the original publications[8,21] (foam B). It can be seen that the E_d/E_s ratio increased rapidly with increasing impact speed and, from the closed circles, that it passed through

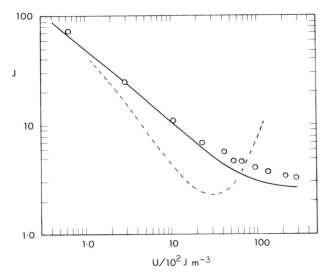

FIG. 5.12. Variation of the acceleration efficiency J as a function of the impact-energy density for the foam–liquid system of Fig. 5.11, $M = 0.65$ kg. Continuous line generated using eqns (5.18) and (5.19). Broken line, the behaviour of the polymer matrix, air filled, with no fluidic effects.

a maximum value. This maximum in the $E_d/E_s - v_i$ curve has been observed with other liquid-filled PUR foams with similar physical and structural characteristics. The energy-absorbing behaviour of the liquid-filled foam differs from that of the matrix (with no flow effects, see Fig. 5.6) in two major respects; (i) except for low values of v_i (or ε_m), E_d/E_s for the liquid-filled system is much larger than that for the polymer matrix and (ii) E_d/E_s for the matrix increases monotonically with increasing ε_m, it does not pass through a maximum. In addition, although the energy-absorbing behaviour of the matrix is dependent on the value of the impacting mass, this dependence is small. For liquid-filled systems, the mass dependence of the energy-absorption behaviour is large, as can be seen from Fig. 5.11. These observations indicate that for the conditions obtaining in these experiments, fluid-flow effects were dominating the mechanical behaviour. The variation of the acceleration efficiency J as a function of the impact-energy density U for the same foam–liquid system is shown in Fig. 5.12. This behaviour can be compared with that of the polymer matrix, which is indicated by the broken curve. It is seen for the liquid-filled system that J decreased less rapidly with increasing U than for the matrix. Although it was not possible to define the minimum in the J–U curve for the liquid-filled system using the experimental conditions available it was apparent that the minimum value of J would occur at a much higher value of U and that the minimum in the J–U curve would be broader than that for the matrix (in the absence of fluid flow effects), as predicted by Rusch.[5]

The equations given above and in the Appendix have been used to predict the shock-mitigating properties of these liquid-filled foam systems. In order to facilitate the comparison between the large-strain impact behaviour and the small-strain oscillatory response the coefficient C which governs the viscous-flow process and is important at low rates has been expressed in terms of the parameter γ. The coefficient S, which is related to the transverse deformation mechanism, and is important at high rates, was expressed in terms of the parameter β. The parameters γ and β are defined by eqns (4.7) and (4.8), respectively. This was done primarily because of the understanding of the significance and the applicability of these parameters obtained from the small oscillatory strain studies of liquid-filled systems. For the prediction of the impact behaviour, all coefficients and constants were determined from the material properties and the geometry of the specimen except for the value of β, which was taken as a curve-fitting constant. The parameters defining the stress–strain behaviour of the matrix (i.e. F_{25}, λ and α) were obtained for specimens with latex walls under impact-loading conditions, and account was taken of the effect of the liquid in the mechanical properties of the matrix polymer.

The behaviour predicted for the liquid-filled foam under impact-loading conditions is shown as the continuous lines in Figs. 5.11 and 5.12. It is seen, from Fig. 5.11, that for impact speeds less than that for the damping maximum, there was reasonable agreement between the predicted and the observed behaviour, thus justifying the assumptions made in establishing the model. However, above the damping maximum, the predicted E_d/E_s ratio was less than that determined experimentally. The value of β determined by curve fitting (the same value was used for Figs. 5.11 and 5.12) was about one third of that predicted by eqn (4.8). The reason for this discrepancy is not known, but it is probably related to the fact that with the sample arrangement used in these investigations, the fluid flow was multidirectional and not unidirectional as assumed in the Gent–Rusch model. The empirically determined value of β obtained from small oscillatory strain measurements was also different to that predicted. This was also attributed to the sample arrangement not conforming to the constraints of the theoretical model (see Section 4.3.3).

Some insight into the processes that occur during the large-strain impact deformation of a liquid-filled foam can be obtained by considering the experimental data and the theoretically predicted behaviour shown in Figs. 5.11 and 5.12. The theoretical curves terminate at a maximum compressive strain of 60%, which is approximately the limit of the plateau region for this PUR flexible foam, and is the maximum strain for which the strain functions defined by eqns (5.9) and (5.10) are valid. From Fig. 5.11 it is seen that maximum damping occurs before the end of the matrix plateau is reached. The data given in the original paper shows that the deformation at maximum E_d/E_s lies in the range 30–45% compression, the strain at maximum E_d/E_s increasing with increasing impacting mass. It can also be demonstrated that the impact-energy density for maximum E_d/E_s increases with increasing value of the impacting mass.

By using the mathematical equations it can be shown that for a given set of material and specimen parameters (such as matrix modulus, foam permeability, liquid viscosity, specimen dimensions) one of the most important processes governing the impact behaviour of a fluid-filled foam is the dependence of the effective permeability on the compressive strain, as expressed by eqn (5.16). The value of the permeability not only determines the energy absorption at low rates, it also determines the rate at which the transition from viscous flow to transverse elastic deformation takes place. It is this transition that gives rise to the energy-absorption maximum, both for small-strain oscillatory deformations and large-strain impact deformations. The shift in the E_d/E_s peak to lower impact speeds with increasing impact mass, shown in Fig. 5.11, and the sharpening of the E_d/E_s-v_i peaks predicted theoretically are indicative of this strong deformation dependence of the flow process in fluid-filled foams. Above this transition the behaviour becomes dominated by transverse elastic deformation mechanisms where the fluid does not flow through the matrix. Thus below the damping maximum the behaviour is controlled by a dissipative flow process and above by an elastic process. During this transition there is a continuous increase in the effective modulus of the system until an equilibrium value is reached (see for example Fig. 4.13). Under extreme conditions there is no flow of fluid through the matrix and the energy dissipation is essentially that due to the polymer matrix. Although the situation is analogous to that which obtains under small-strain oscillatory deformations with increasing frequency it is more complex in that the mechanical behaviour of the material is governed by the initial strain rate and the impact-energy density rather than the frequency alone. By comparing the data of Fig. 5.12 with that of Fig. 5.11, curve (c), it can be seen that the increase in the elastic response has only a small effect on the minimum value of the acceleration efficiency and gives rise to a flattening of the J–U curve to produce a broader minimum.

Although these theoretical and experimental studies have been made on liquid-filled foams the results are in general agreement with the Gent–Rusch model so that they should be applicable to any fluid-filled open-cell system. The behaviour of cellular polymers filled with compressible fluids will differ in detail with those filled with incompressible fluids, but the essential features of the mechanical behaviour under impact-loading conditions should be the same. For example, replacing the liquid by a gas will reduce considerably the value of the parameter γ (in proportion to the ratio of the viscosities of the two fluids), so that a gas-filled cellular polymer will exhibit qualitatively similar impact behaviour to that of the liquid-filled polymer but at much higher initial strain rates and impact-energy densities.

ACKNOWLEDGEMENT

Figures 5.3 to 5.12 are based on original data or diagrams contained in *Journal of Cellular Plastics* (1971), 7, 33–42 and 84–90 and are published courtesy of Technomic Publishing Company Inc., Westport, CT 06880, USA.

APPENDIX

The parameters of the model given in Fig. 5.9 are related to the material properties and the geometry of the foam specimen by

$$S = E_t W^3 L / 6 h^2$$
$$C = \eta W^3 L / \{ 24 \zeta K_0 (1 - ky/h)(1 - y/h) \}$$
$$I = \rho W^4 L / \{ 64 \zeta^2 h^2 B(y)(1 - y/h)^2 \}$$

The material and geometric parameters used in the above equations are defined in Chapter 4.

The parameters used in eqns (5.18) and (5.19) are

$$F_m = F_{25}\{1 - \exp(-y_m/\lambda)\}$$
$$\tau = C/S$$
$$\omega = (S/M)^{\frac{1}{2}}$$
$$\chi(y) = [1 - (1 - 2y/y_m)^2]^{1/2}$$
$$I' = 1 + (2I/C)\dot{y}$$

REFERENCES

1. NEWTON, R. E. (1976) *Shock and Vibration Handbook*, C. M. Harris and C. E. Crede (Eds), 2nd Edn, McGraw-Hill, Chap. 31.
2. ESHELMAN, R. L. (1973) *Isolation of Mechanical Vibration, Impact and Noise*, ASME, AMD, Vol. 1, 221.
3. MERCER, C. A. and REES, P. L. (1971) *J. Sound Vib.*, **18**, 511.
4. HATAE, M. T. (1976) *Shock and Vibration Handbook*, C. M. Harris and C. E. Crede (Eds), 2nd Edn, McGraw-Hill, Chap. 41.
5. RUSCH, K. C. (1970) *J. Appl. Polym. Sci.*, **14**, 1433.
6. WOOLAM, W. E. (1968) *J. Cell. Plast.*, **4**, 79.
7. WOOLAM, W. E. (1968) *J. Cell. Plast.*, **4**, 334.
8. HILYARD, N. C. and DJIAUW, L. K. (1971) *J. Cell. Plast.*, **7**, 33.
9. MEINECKE, E. A. and CLARK, R. C. (1973) *Mechanical Properties of Polymeric Foams*, Technomic Publishing Company Inc., Chap. 4.
10. SCHWABER, D. M. and MEINECKE, E. A. (1971) *J. Appl. Polym. Sci.*, **15**, 2381.
11. BENNING, C. J. (1969) *J. Cell. Plast.*, **5**, 40.
12. JONES, R. E., HERSCH, P., STIER, G. G. and DOMBROW, B. A. (1959) *Plastics Technol.*, **5**, 55.
13. HUMBERT, W. E. and HANLON, R. G. (1962) *Package Engineering*, 7, 79.
14. COUSINS, R. R. (1976) *J. Appl. Polym. Sci.*, **20**, 2893.
15. JONES, R. E. and FESMAN, G. (1965) *J. Cell. Plast.*, **1**, 200.
16. LIBER, T. and EPSTEIN, H. (1969) *Shock Vib. Bull.*, **40**(5), 291.
17. LIBER, T. and EPSTEIN, H. (1969) *ASME Vibrations Conference*, Philadelphia, Paper No. 69-VIBR-46.
18. NAGY, A., KO, W. L. and LINDHOLM, U. S. (1974) *J. Cell. Plast.*, **10**, 127.
19. PHILLIPS, P. J. and WATERMAN, N. R. (1974) *Polym. Eng. Sci.*, **14**, 67.
20. JOHNSON, W. (1972) *Impact Strength of Materials*, Edward Arnold, Chap. 5 and Chap. 2.
21. HILYARD, N. C. (1971) *J. Cell. Plast.*, **7**, 84.
22. GENT, A. N. and RUSCH, K. C. (1966) *Rubber Chem. Tech.*, **39**, 389.
23. GENT, A. N. and RUSCH, K. C. (1966) *J. Cell. Plast.*, **2**, 46.
24. RUSCH, K. C. (1966) *PhD Thesis, University of Akron*, USA.

Chapter 6

SHOCK MITIGATING SYSTEMS

M. A. MENDELSOHN
*Westinghouse Electric Corporation, Pittsburgh,
Pennsylvania, USA*

6.1 INTRODUCTION

As described in the preceding chapter, the shock-mitigating properties of cellular materials are dependent on several factors, including both the chemical composition and the cellular geometry. In studying foams, it is difficult to isolate composition and cell-structure variables. Changes in the chemical composition of foams can affect, in addition to properties of the polymer, factors such as cell size and shape, the relationship between the thickness of cell diaphragms and ligaments and the openness of cells. By employing structured coatings or mouldings, these interpretative difficulties can be eliminated and the effects of both geometry and chemical composition can be controlled and evaluated independently.

In this chapter we will first review an investigation[1,2] of the chemical and engineering properties of polyurethane isolator pads. For greater detail the reader is referred to these references. The effects of chemical variables on the shock–mitigating properties of polyurethane flexible foams and phenolic rigid foams are then considered.

6.2 STRUCTURED MATERIALS

6.2.1. Mechanical Test Procedures

The mechanical tests performed in this study consisted of measuring the static and dynamic compression–deflection (C–D) characteristics and vibrational damping on specimens about 6 in. wide and 4 in. long (Fig. 6.1). Unless otherwise specified, strut heights employed in this study were approximately 1·5 in.

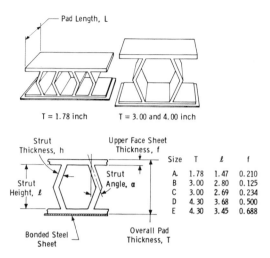

FIG. 6.1. Laboratory specimens of cast urethane shock pads.

In many cases, absolute values of data in the various graphs are not directly comparable because numerous changes in pad design were employed throughout this study.

The static C–D tests are performed in an Instron Universal Test Machine. A pad is loaded to the bottomed deflection, normally at about 65–70% compression, and unloaded without pause for three or four successive cycles. Almost no discernible difference in the C–D data results from additional loading cycles (Fig. 6.2). Cross-head rates of 2·0, 2·0 and 5·0 in. min^{-1} were employed for pads having a thickness of 1·78, 3·00 and 4·30 in., respectively.

The term pseudoset is used as a measure of the rate of recovery or resilience characteristics of the pad. Slowly recoverable materials have high values. Pseudoset is defined as the displacement between the start of the first and fourth cycle loading strokes.

Dynamic C–D tests were run in a sand-drop, shock-test machine shown in Fig. 6.3. A weight guided by a ball bushing bears against the test specimen. A deflectometer connected to the weight permits the linear deflection to be determined, while an accelerometer mounted on the weight is used to measure the dynamic compressive stress. The deflection and acceleration traces are then recorded on an oscillograph. An adjustable drop-table height, which permits test velocities up to 120 in. sec^{-1} pro-

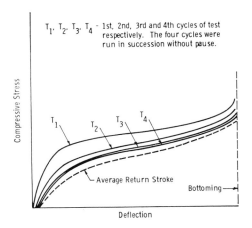

FIG. 6.2. Effect of loading cycle on compressive stress–strain characteristics.

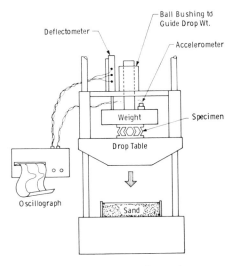

FIG. 6.3. Drop table for obtaining dynamic compression data.

vides the proper initial-strain rate. Initial-strain rates employed for structured castings and foams were 2500 in. in.$^{-1}$ min^{-1} and 1500 in. in.$^{-1}$ min^{-1}, respectively.

The damping quality factor, Q

$$Q = \frac{1}{2C/C_c} \text{ (where } C/C_c = \text{damping ratio)}$$

is measured in a servo-hydraulic test apparatus (Fig. 6.4). The 4·30 and 3·00 in. pads (sizes B–E, Fig. 6.1) were preloaded to 0·30 in. compression and subjected to a 10 Hz ± 0·02 in. sinusoidal deflection. Figure 6.5 shows the resulting load–unload hysteresis loop from which Q is calculated. The same test was used for the 1·78 in. (size A) pads except that the precompression was 0·15 in.

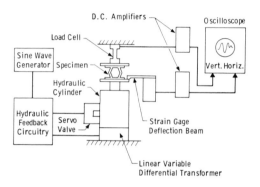

FIG. 6.4. Damping measurement apparatus.

6.2.2. Polymer Processing

The polyurethane polymers (Adiprene L100, L167 and L315 are isocyanate-terminated prepolymers, manufactured by Du Pont, containing polyoxytetramethylene chain segments and having equivalent weights of approximately 1020, 660 and 450, respectively. Conap DP4736, manufactured by Conap, Inc., is a polyoxyisopropylene-based prepolymer having an isocyanate equivalent weight of approximately 800) were prepared by treating isocyanate-terminated prepolymers and combinations of prepolymers with the chain extender, MOCA [4,4'-methylenebis (2-chloroaniline)], at elevated temperatures.

The studies involving only geometrical factors utilised pads prepared from Adiprene L100 unless otherwise stipulated.

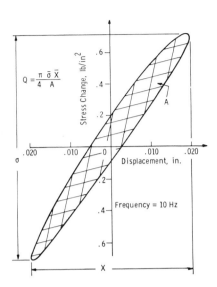

FIG 6.5. Typical hysteresis loop for determining damping.

The quantity of MOCA employed for single prepolymers and for their combinations was adjusted to give a NCO:NH$_2$ ratio of 1·0:0·95. This ratio was selected since lower concentrations of MOCA gave materials which exhibit excessive drift in their properties during aging due to the high NCO:NH$_2$ ratios, whereas greater MOCA concentrations resulted in properties which were too rate sensitive.

The procedure below describes a typical preparation of a polyurethane casting at a nominal processing temperature of 80°C. A solution consisting of 70 parts Adiprene L167 and 30 parts Adiprene L315 is heated to 78–80°C while 21·0 parts MOCA are brought to 124–126°C. Degassing of the Adiprene solution and MOCA at a pressure of 1–2 Torr for about 5 and 1 min, respectively, follows. When the temperature of the prepolymer blend has dropped to 70 ± 2°C, it is stirred together with the MOCA, which is at 122–124°C. After agitating for about 2 min, the resultant solution, at a temperature of 79–81°C, is poured into the mould which was preheated to the same temperature and then cured at that temperature for a prescribed length of time. After completion of the oven cure, demoulded samples were permitted to equilibrate to ambient, ~ 23°C/50% RH for at least seven days and then tested.

6.2.3. Effect of Column and Cell Geometry on Mechanical Behaviour

Strut Shape

The simplest structure, consisting of straight columnar struts, gives the C–D relationship shown in Fig. 6.6. While displaying a high resistance to buckling, it exhibits highly rate-sensitive dynamic stress values as evidenced by the dynamic spike. Qualitatively, one may consider that the large compressive load under dynamic conditions develops before appreciable buckling occurs since the strut does not have time to 'fly out'. Delicate objects surrounded by isolator pads having a straight column configuration may not receive adequate shock protection under dynamic loading conditions. As shown by the dashed line in Fig. 6.6, by prebuckling the columns, the dynamic spike is greatly reduced.

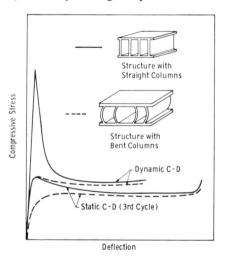

FIG. 6.6. C–D curves for columnar structure.

In many applications, it is desirable that the C–D curve should not display a negative slope, i.e. the compressive stiffness should increase continually with deflection. This can be provided by increasing the strut angle (Figs. 6.1 and 6.7). Furthermore, as the strut angle is increased, the rate sensitivity to loading is diminished. However, even with an angle as great as 20°, a small dynamic spike is evident, giving a slightly negatively sloped dynamic C–D curve at low deflections. Use of greater strut angles will

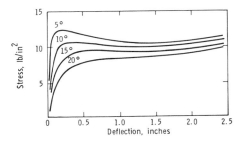

FIG. 6.7. Effect of strut angle on 1st cycle C–D characteristics.

eliminate the negative dynamic C–D slope, but has the disadvantage of requiring thicker struts and therefore more material in order to obtain the same load-bearing capability. As a result, the deflection to bottoming is reduced. Compensating for the decrease in compressive stiffness by using a more rigid polymer will not always solve this problem, since stiffer materials tend to exhibit greater rate sensitivity.

Under both dynamic and static loading, use of a notched bent strut will give a large reduction in compressive stiffness at low deflections and minimise the C–D plateau effect by imparting a steeper slope to the stress–strain curve (Fig. 6.8). The low initial modulus is an obvious consequence of the notch. As the deflection increases, the walls of the strut just adjacent to the notch begin to bottom against each other, providing an increase in the load. The dynamic C–D curve more closely parallels that of the static curve. Overall, this type of design imparts a low rate sensitivity to the structure. (Throughout this discussion, the term 'rate sensitivity' will refer to the ratio of dynamic to static stress at identical deflection.)

Load-bearing properties have been related to strut thickness and height for structures of the type shown in Fig. 6.1 by the Kim–Rudd relationship:

$$P_i = C_i \left[\frac{h_i}{l_0}\right]^{n_i}$$

where h_i is the strut thickness, l_0 is the original strut height, P_i is the line load at an arbitrary axial displacement, C_i is the variable coefficient and n_i is the variable exponent. If strut thickness and resultant loads are the variables, the above equation can be rewritten to give:

$$n_i = \frac{\log P_i - \log P_0}{\log h_i - \log h_0}$$

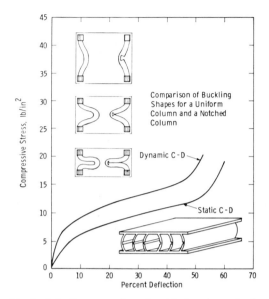

FIG. 6.8. Typical C–D curves for a buckled and notched structure.

where P_0 is the reference line load and h_0 is the reference strut thickness.

The value of n_i can be determined by measuring the slope of a dimensionless plot of a reduced line load function, $\log(P_i/P_0)$, against a reduced strut thickness function, $\log(h_i/h_0)$.

$$\log(P_i/P_0) = n_i \log(h_i/h_0)$$

Converting line load to stress, S, a relatively simple relationship has been found to be quite convenient.

$$\log(S) = n \log(h_i) + C$$

A study was made in which this equation was applied (Fig. 6.9). One of the face sheets of the pads was bonded to a rigid surface, whereas the other was free to deform (Fig. 6.10). By plotting S against $\log(10\,h)$, values of n in the range of $2·40 - 2·52$ were obtained for C size (strut height = 2·69 in.) pads prepared from several different polymer formulations. Interestingly, application of the same treatment to the B size (strut height = 2·80 in.) pads gives values of n ranging from 2·07 to 2·12. At first, it may appear puzzling why the very slightly shorter strut should

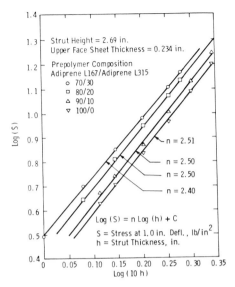

FIG. 6.9. Relation between strut thickness and compressive stress.

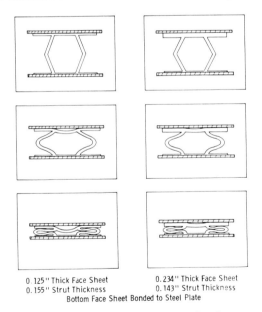

0.125" Thick Face Sheet 0.234" Thick Face Sheet
0.155" Strut Thickness 0.143" Strut Thickness
Bottom Face Sheet Bonded to Steel Plate

FIG. 6.10. Effect of pad geometry on mode of compression.

display such a significantly greater load-bearing dependency on thickness; however, this difference is explanatory when considering that the bending of the face sheet affords an important contribution to the C–D characteristics of the pads. The pads using a 2·69 in. strut have a face sheet thickness, 0·234 in., almost twice that of the 2·80 in. strut pads whose face sheet thickness is 0·125 in. As the thickness of the strut increases, the thickness of the face sheet becomes more important since at the very low strut thicknesses, the ratio of face-to-strut thickness approaches a value where the struts buckle easily while the face does not significantly distort. At low strut thickness such as 0·125 in., the effect of strut height, 2·69 in. vs. 2·80 in., is smaller on an absolute and percentage basis than for the heavier struts. Thus, pads having a thinner face sheet exhibit less sensitivity of C–D characteristics to strut thickness due to the relatively greater contribution that the distortion of the face sheet makes to the deformation of the pad.

The pad configurations of this study utilise opposed struts in order to avoid the problem of lateral shifting of the face sheets relative to one another during compressive loading. Unidirectional alignment of struts in a large pad would be expected to alter the mode of the rippling of the face sheets and thus change their quantitative contribution to the C–D characteristics of the pads. This difference is, however, minimised with increasing thickness of the face sheet.

From the first equation, one may write:

$$S_1/S_2 = (l_2/l_1)^{n_i} = (2·80/2·69)^{2·50} = 1·106$$

The value of 2·5 for n was preferred to 2·1, since this represents the condition of less distortion of the face sheet. Thus, if the face sheet were not a factor $[f(h) \ll g(f)$, or if both face sheets were bonded], one could expect a strut having a height of 2·69 in. to have approximately 10·6% greater load bearing than a 2·80 in. strut.

One could be faced with the problem of wishing to maintain certain C–D characteristics while altering the height of the pad. A brief study of this problem indicated that when the ratio of strut height to thickness was maintained, essentially the same C–D characteristics were obtained when deflections are compared on a percentage basis. This, of course, assumes that the relative participation of the face sheet is not altered. In Fig. 6·11, a comparison of the stress value at the same percentage deflection for two pairs of pads shows an essentially equal stress for structures having the same strut height-to-thickness ratio.

SHOCK MITIGATING SYSTEMS

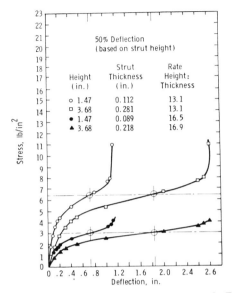

FIG. 6.11. Effect of strut height to thickness ratio on C–D characteristics.

Circular and Hexagonal Structures

Other structures such as multirow, circular and hexagonal-shaped cells were examined, since these structures are particularly economical to prepare because standard stock may be employed for mould inserts.

Circular configurations display static characteristics similar to those of the columnar bent strut, except that they exhibit bottoming at lower deflections. However, under dynamic loading, the webs between the circles behave much like straight columns giving the dynamic 'spike' effect. By slightly precompressing the circular structure to provide an initial bow, the spike is reduced.

Hexagonal structures give dynamic and static C–D responses similar to those of the bent-notched columns; however, inherently inefficient packing also causes these pads to bottom at considerably lower deflections than those of the columnar-cell configuration.

6.2.4. Effect of Polymer Composition on C–D Characteristics

In spite of the previously described effects of cell geometry on C–D characteristics, the chemical composition of the pad also strongly influences the shape of the compressive stress–strain curve.

Figure 6.12 shows the C–D characteristic of a polyoxyisopropylene-based pad. In spite of the notched bent strut, there is a very strong

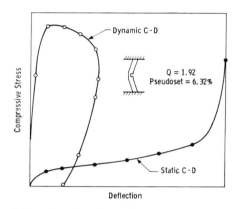

FIG. 6.12 C–D and damping data for polyoxyisopropylene–MOCA structure.

dynamic effect. The low Q value of 1·9 indicates that this material provides excellent vibrational damping and the high pseudoset of 6·3% shows that the pad is not very resilient. By comparison, a pad prepared from a polyoxytetramethylene-based prepolymer (Adiprene L100) using the same mould exhibited the C–D characteristics similar to those shown in Fig. 6.8. A Q value of 8·6 indicated poor damping and a pseudoset of 2·5% showed that the polyoxytetramethylene-based pad exhibits moderately rapid recovery from compression.

A study of a series of compositions is summarised in Fig. 6.13. As the concentration of polyoxyisopropylene is increased, the material displays a reduction in the static compressive stiffness despite the slight increase in the substituted urea concentration of the polymer. (The urea concentration was higher because of the lower equivalent weight of the polyoxyisopropylene, Conap DP4736). This effect could be attributed to disruption of the alignment of the polymer network by the increasing quantity of polyoxyisopropylene chain segments. The resultant increase in the random character of the network diminishes the intermolecular attractive forces. The dynamic stiffness also decreased with increasing polyoxyisopropylene, but less rapidly than the static stiffness, until a mini-

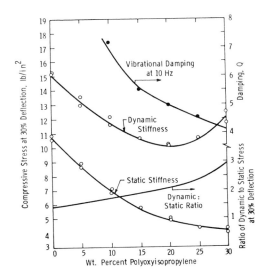

FIG. 6.13. Properties of polyoxytetramethylene–polyoxyisopropylene MOCA extended polyurethanes.

mum was reached at about 20 weight percent, followed by a rapid increase. Throughout the entire region studied (Fig. 6.13), the rate sensitivity, determined by the ratio of dynamic to static properties, increased with polyoxyisopropylene concentration. The increased rate sensitivity appears to be a consequence of the increased difficulty of rapid realignment of the displaced hydrogen bonds and polymer chains that can result from the steric contribution of the protruding methyl groups of the polyoxyisopropylene segments. In addition, the slight increase of urea linkages with increasing concentration of the lower equivalent weight polyoxyisopropylene prepolymer contributes to higher rate sensitivity. As expected, the vibrational damping increased (decrease of Q values) with increasing rate sensitivity.

These results suggest that the two types of prepolymers can be blended to give materials having intermediate rate sensitivity, damping and resilience properties.

The same effects as described above have been observed in comparing one-shot polyurethanes prepared from polyoxytetramethylene and polyoxyisopropylene-based polyols.

The effect of the $NCO:NH_2$ ratio was studied by varying the MOCA concentration (Fig. 6.14). The polymers were prepared by treating

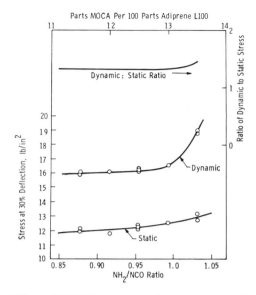

FIG. 6.14. Effect of MOCA concentration on compressive stiffness.

L100 with MOCA at a processing temperature of 80°C. Results showed that the static compressive stiffness rose slowly with increasing MOCA content throughout the range studied. The dynamic stiffness followed the same pattern until a 0·98 NH_2:NCO ratio was reached, and then rose rapidly. The rate sensitivity appeared to be insensitive to the MOCA below this ratio, and then increased. The higher rate sensitivity at the upper MOCA concentrations is consistent with the expected increase of the glass transition temperature of the material. At lower MOCA concentrations, increased numbers of biuret and perhaps allophanate crosslinks are formed and the concentration of urea linkages and aromatic content is reduced. This is consistent with the reduction in compressive stiffness with decreasing MOCA content. The effects of the increase in biuret and allophanate crosslinks which tend to increase rate sensitivity and the resultant reduction in urea linkages which gives lower rate sensitivity appear to be of about equal importance in the NH_2: NCO ratio interval from 0·88 to 0·98.

In most applications where low rate sensitivity is desirable, 12·5 parts MOCA (0·95 NH_2:NCO) per 100 parts L100, appear optimal. At this concentration, problems associated with a drift of properties during ageing are minimal. Use of 0·98 NH_2:NCO is too close to the region where the dynamic stiffness is highly sensitive to MOCA content and from a practical standpoint, gives an inadequate provision for error during manufacturing.

As the equivalent weight of the prepolymer decreases, the increased concentration of urea and urethane linkages results in an increased compressive stiffness, rate sensitivity, vibrational damping and decreased resilience. The magnitude of the effect on C–D properties is shown in studies of the Adiprene L100–Adiprene L315 system (Fig. 6.15) and the Adiprene L167–Adiprene L315 system (Figs. 6.9 and 6.16).

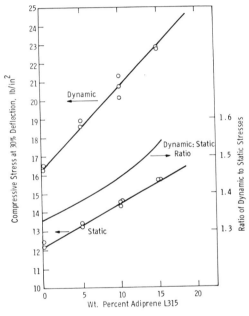

FIG. 6.15. Effect of composition of Adiprene L315–Adiprene L100 solution on C–D characteristics.

6.2.5 Effect of Processing Temperature

A study was performed in which a series of polyurethane castings were processed at temperatures ranging from 105°C to 135°C (Figs. 6.17 and 6.18). In another investigation, the effect of processing tempera-

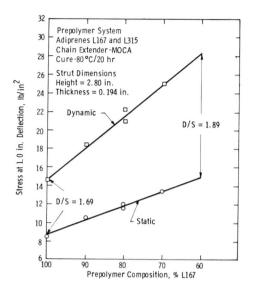

FIG. 6.16. Effect of polymer composition on C–D characteristics at 1·0 in. deflection.

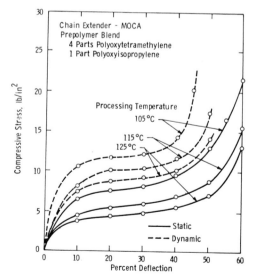

FIG. 6.17. Effect of processing temperature on C–D characteristics.

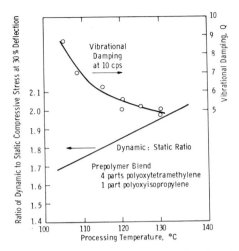

FIG. 6.18. Effect of processing temperature on vibrational damping and rate sensitivity of C–D characteristics.

ture over the range of 80–120°C for a series of polyoxytetramethylene–polyoxyisopropylene prepolymer blends was determined (Fig. 6.19). As the processing temperature increased, the static and dynamic compressive stiffness decreased. The rate sensitivity showed a slight

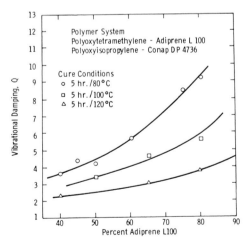

FIG. 6.19. Effect of prepolymer composition on vibrational damping.

increase and the Q decreased. Higher reaction temperatures give increased concentrations of biuret and allophanate linkages which interfere with the hydrogen bonding. Although crosslinks generally tend to increase the rigidity of many polymers, the few additional crosslinks formed at the higher temperatures interfere with the spatial alignment of a vastly larger number of hydrogen bonds and thus can cause a net reduction in the intermolecular forces of the urethane polymer. The decrease of Q with increasing processing temperature is believed to be a consequence of the additional random crosslinks. Recovery of the orientation of distorted polymer chains, so as to regain the intermolecular attractive forces, is hindered when the concentration of the allophanate and biuret crosslinks has increased. From the type of data of Figs. 6.17–6.19 processing temperatures can be employed to 'fine tune' the selection of polymers having the desired damping and rate sensitivity characteristics.

6.2.6 Summary of Geometrical and Chemical Composition Effects

Use of structured elastomers permits selection of geometrical designs, polymer systems and processing conditions which will provide predictably a complex set of desired chemical and physical characteristics. It is now feasible to engineer shock-mitigating structures with polymeric materials that are comparable in sophistication to mechanical systems

6.3 FLEXIBLE POLYURETHANE FOAMS

6.3.1 General Considerations

There has been a tremendously wide use of flexible polyurethane foam in packaging applications. However, in many cases, the foam was not developed to meet specific shock-mitigating requirements—its selection is frequently haphazard. However, the military and furniture industries have displayed considerable interest in engineering properties, such as C–D characteristics, glass transition temperature, breathability, etc. Because of the 'expanding' applications a huge body of technical and trade literature exists, provided to a large extent by corporations involved in manufacturing foam products and especially in supplying its raw materials, which relates properties of foam to its chemical composition.

(The compositional and structural requirements of polyurethane foams for comfort cushioning are discussed in detail in Chapter 3 above. For further background reading on polyurethane technology, the reader is referred to Saunders and Frisch two-part volume[3,4].)

Since summarising this vast literature would require much more space than available, we will review a single investigation[5,6] of chemical structure–property relationships in moderate depth in order to share the philosophy employed in developing a flexible foam to meet certain load-bearing and shock mitigating requirements.

The foam that will be described was required to have specific C–D properties, good resilience or recovery from distortion, a moderate tensile strength, low glass transition temperature, high hydrolytic stability and good fatigue resistance.

6.3.2 Properties and Their Measurement

The load-bearing testing of the flexible foam differs slightly from that of the structured pads. An Instron Universal Tester or a Southwark Tate Emery Testing Machine with automatic cycling controls and a recorder is used to measure the load-bearing capacity of the foam (C–D test). The specimen, usually 6 in. × 6 in. × 2 in., which had been permitted to age for at least seven days at 50% RH and $25 \pm 2°C$, is centred between two parallel flat plates that cover the entire sample area and a total load of one pound is applied to the specimen surface. The distance between plates is measured to the nearest 0·01 in., and the sample is loaded to 25% of its initial height and then unloaded without pause. A total of four deflection cycles, applied at the rate of $1 \pm 0·1$ in. min^{-1} per in. of initial thickness, are imposed on the sample with essentially zero time lag between cycles. The zero point on the chart is determined from the first loading curve. Pseudoset is defined as the distance on the abscissa (Fig. 6.20) where the stress value for the fourth-cycle loading curve begins departing from zero. Compression–deflection data which are reported are obtained from the fourth loading cycle. Typical raw data obtained directly from the tester that illustrate the loading characteristics and the hysteresis loops are shown in Fig. 6.20.

The specified properties of the foam require that the C–D curve (fourth cycle) fall within the limits shown in Fig. 6.20. In addition, the unloading value for deflections of 20, 30, 40 and 50% should not be less than 25%, nor more than 85% of the loading values.

Tensile strength of the foam is required to exceed 30 lb in.$^{-2}$ when

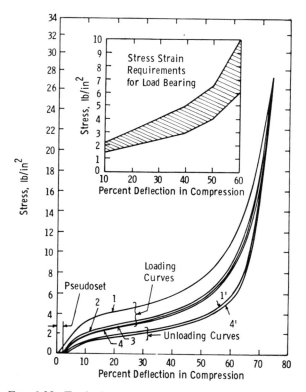

FIG. 6.20. Typical stress–strain data for four-cycle test.

tested in accordance with ASTM-D1564-62T. It is desirable that the elongation exceeds 70%. Another requirement is that compression set values (ASTM-D1564-62T, Method B) expressed as percentage of original height, do not exceed 8%.

While the subject of hydrolytic stability is beyond the scope of this chapter, the requirements pertaining to it are stringent.

Breathability is defined here as the pressure drop observed when passing an air flow of 5 ft^3 min^{-1} through a hollow cylindrical specimen, 6 in. outside diameter, 1 in. inside diameter and 2 in. high, while it is compressed to 1 in. in height. Air is introduced at the centre and is permitted to pass through only in a radial direction due to the external fixture that contains two parallel plates that press on the faces of the sample. Specifications require that the foams impart a maximum pressure drop of 4 lb in.$^{-2}$

Typical results for most of the foams described here are of the order of 1–2 lb in.$^{-2}$

Since the shock protection provided by the foam at high impact rates, such as 12 ft sec^{-1}, is correlatable with the glass transition temperature by a time–temperature superposition relationship, it was specified that the glass transition temperature should not exceed $-40°C$. Dilatometric determinations were performed by evacuating the sample and replacing the gas with iso-octane, which does not appear to cause noticeable swelling of the foams described here. The filled dilatometer is cooled to $-80°C$ and the temperature is permitted to rise at $0.5°C$ min^{-1}. Typical T_g values for most of the foams described here are about $-55°C$ to $-60°C$.

While we are primarily concerned with load-bearing characteristics, certain limitations dictate other characteristics of the foam. In order to conform to the required C–D envelope (Fig. 6.20), it is necessary that the pseudoset on the fourth cycle be small, generally less than 3%. Since the testing apparatus operates essentially continuously for the four cycles, foams having slow rates of elastic recovery cannot keep up with the movement of the platens. A typical slow-recovery foam may have recovered only 90% of its original thickness between the third unloading curve and start of the fourth load cycle and, thus, show a zero value for the stress at 10% compression. Materials exhibiting moderate rates of recovery may have pseudosets of about 5%. If such a material was to have a sufficiently high modulus as to meet the 10% deflection requirements, it would probably exceed the upper stress limits at the higher deflections. Foams having greater resiliences display lower pseudoset values, thus providing an inherent advantage in meeting the requirements shown in Fig. 6.20. While compression set stimultaneously measures several different properties, lack of resilience generally gives high values which exceed specifications. Since the foam is used to surround certain instruments and protect them from vibration, it is apparent that use of materials that do not retain rapid-recovery charcteristics results in development of rattle spaces.

Good breathability is an important requirement, since it permits air to escape sufficiently rapidly at high-frequency vibrations and under shock conditions. If the gas cannot escape, its *PVT* characteristics will tend to raise the stress values beyond the permissible limits. (The effect of gas flow on the dynamic mechanical behaviour is considered in some detail in Chapters 4 and 5 above.)

An arbitrary requirement was that the foam retain at least 50% of its initial-stress and vibrational-modulus characteristics after 40 000 fatigue cycles. Fatigue cycling was carried out using an apparatus similar to that shown in Fig. 6.4 on 4 in. × 4 in. × 2 in. specimens at a precompression of 5% and a stroke of ± 10% about the precompressed height. The fatiguing frequency of 0·5 Hz resulted in a temperature rise of about 4 to 5° F. Periodically, the large-deflection cycling was interrupted and the sample subjected to a one minute run at lower amplitude and higher frequency simulating vibrational conditions. The vibration tests were run at 3·5 Hz with the foam precompressed 5% and at a double amplitude of 0·040 in. A typical running fatigue loop obtained at the start of testing and after 40 000 cycles, the outstrokes of the fatigue loop at various test intervals, and the typical behaviour of a fourth-cycle C–D test before and after 40 000 fatigue cycles are shown in Fig. 6.21.

There were no requirements regarding cell structure or size, except that the structure should be free of striations and be uniform in order to avoid discontinuities in properities. Foam prepared according to the fomulations in Table 6.1, which met the required physical properties, has a 'bread' structure when prepared with a low-pressure, high-shear Martin Sweets Machine, and a 'caviar' structure when a high-pressure, low-shear Hennecke UBT machine is used. The results reported in this study are on foams prepared by a Martin Sweets Machine on variations of the basic formula shown in Table 6.1 and having the properties summarised in Table 6.2.

6.3.3 Effect of Composition on Mechanical Behaviour

Selection of Formulation

A polyurethane foam is a highly complex heterogeneous form of matter; therefore, it is impossible completely to isolate effects of chemical variables. Studies of sensitivities of physical properties to chemical compositions, complex as they are, are much more amenable to understanding in the case of casting resins. Additional physical variables that control the properties of the foam are imposed as a consequence of change of chemical composition. For example, the rheology of the system, interfacial tensions, gas solubility, rates of diffusion of the forming gas, number of nucleating centres, etc., are often altered with chemical changes in the formulation. Furthermore, chemical changes alter the kinetics and thermodynamics of all of the phases of the foam formation. The above factors affect the resultant density and cell structure of the foam and therefore influence many of the physical properties. Therefore,

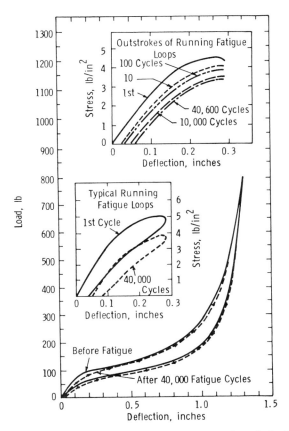

FIG. 6.21. Typical fourth-cycle C–D tests obtained before fatiguing and after 40 000 fatigue cycles.

chemical structure sensitivity studies that are described here should be considered qualitative in nature.

The major portion of this chapter is concerned with the effect of various chemical variations on the C–D or load-bearing characteristics of the foam. In this study, the load bearing is considered to be a function of density and intrinsic compressive stiffness. The latter term has been created in an attempt to separate the effect of density from that of the actual stiffness of the polymeric structure, since changes of density occur with variations in chemical composition. Chemical structure of the

TABLE 6.1
Typical Formulation of Protective Foam

Stream		Parts of components[a]	Total parts of stream[a]
1	Polyoxypropylene triol (POPT)[b] (mol.wt. approximately 4000)	35	45.2
	Tribasic fatty acids (TA)[c]	10	
	N-Ethylmorpholine (NEM)	0.2	
2	Ethylene glycol (EG)	5.00	5.8 + X
	Hydroquinone	0.80	
	Water	X	
3	Tall oil fatty acids[d]	0.2025	0.27
	Stannous octoate (SnOct)	0.0675	
4	Polymethylene polyphenylisocyanate (PMPPI)[e]	18.8	30.3
	Tolylene diisocyanate (TDI)[f]	11.5	

[a] X = quantity of H_2O added adjusted to give desired C–D properties. Most common values are 0.38–0.41 parts water.
[b] Polyoxypropylene triol, Voranol CP-4000; manufactured by Dow Chemical Company. Similar results are expected with the use of other 4000 mol. wt. poly (propylene ether triols).
[c] Tribasic fatty acids: trimer acids, Empol 1040, manufactured by Emery Industries, consists of approximately 90% of C_{54} tribasic acids and 10% of C_{36} dibasic acids.
[d] Tall oil fatty acids: Acintol FA-1, manufactured by Arizona Chemical Company. Other fatty acids can be substituted.
[e] Polymethylene polyphenylisocyanate; Mondur MR, manufactured by Mobay Chemical Company. Similar results are expected from PAPI, manufactured by the Carwin Company. Materials believed to have an average functionality of about 2.75.
[f] Tolylene diisocyanate consists of approximately 80% of 2,4-tolylene diisocyanate and 20% of 2,6-tolylene diisocyanate.

polymer is thus largely reflected in the intrinsic compressive stiffness; however, this concept is nevertheless, largely qualitative since the nature of the cell structure and its effect has not been isolated.

In order to employ this concept, a relationship between load bearing and density was developed empirically for a series of foams having, except for different catalyst concentrations, the same chemical composition, as shown in Table 6.1. This provides foams having similar structures over a range of densities. A plot of density versus load bearing for the 10% and 60% deflection points is shown in Fig. 6.22. Even though the points

TABLE 6.2
Typical Properties of Protective Foam[a]

Tensile strength	30–60 lb in.$^{-2}$
Elongation	70–100%
Compression set	4–7%
Glass transition temp.	$-55°$ – $-60°$C
Breathability	1–2 lb in.$^{-2}$
Density	7·0–8·5 lb/ft^{-3}
Load bearing	Within limits of Fig. 6.20
Pseudoset	1·5–3·0%
Cream time	30–40 sec
Rise time	2·5–3·5 min
Hydrolytic stability[b]	Very high

[a] Values pertain to measurements described in text.
[b] A discussion of hydrolytic stability is beyond the scope of this chapter. However, requirements are almost as stringent as those described for pads.[7]

display some scatter, they nevertheless, provide density–load bearing envelopes suitable for making rough comparisons with foams of slightly different chemical compositions.

Figure 6.23, on which the envelope obtained in Fig. 6.22 is superimposed, provides an example of the intrinsic compressive stiffness con-

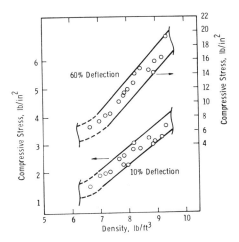

FIG. 6.22. Correlation between density and load bearing for foams of same formulation except for different catalyst concentrations.

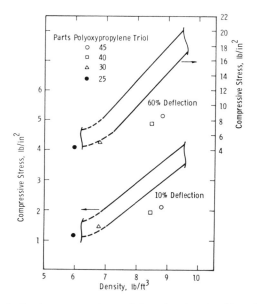

FIG. 6.23. Effect of polyoxypropylene triol on density–load bearing relationship.

cept. This graph shows that as the concentration of the polyoxypropylene triol increases, the density and load bearing increase; however, the inherent stiffness of the polymer decreased as evidenced by the points falling below the envelope. This treatment is also illustrated for the effect of isocyanate index. For the illustrations of this treatment to other formulation variables, the reader is referred to Mendelsohn and Rosenblatt.[7]

Unfortunately, this treatment is subject to some uncertainty because variations in catalyst concentration can affect both the cell structure and rigidity of the polymer chains. In addition to possible surface activity, the catalysts influence factors such as formation of allophanate and biuret crosslinks, sequence of reactions, degree of completion of reactions, etc. Nevertheless, catalyst variations are relatively one of the most innocuous means of isolating the effects of density while maintaining essentially the same structure of water-blown foams. (Even foams that are expanded with volatile inert gases, such as the fluorocarbons, are slightly affected because of surface activity of the blowing agent and its cooling effect, which can give some alteration of the reaction sequences due to differences in activation energies.)

In order to meet the previously described required properties, one cannot employ conventional formulations. Firstly, castor oil based formulations which have found success in some shock-protection and high load-bearing applications could not be used here since they lacked the required resilience. Polyester-based foams can be prepared so as to have good load-bearing and recovery properties; however, they must be eliminated owing to their poor hydrolytic stability. Conventional flexible polyethers generally give load-bearing properties that are too low for this application, whereas the rigid polyether formulations are obviously lacking in their cushioning qualities and resilience. As a result, a combination of a very low and a high molecular weight polyol (Table 6.1) was selected.

High molecular weight polyethers, as the sole polyol source, give foams in the moderate density range that are too low in load bearing. Employment of lower equivalent weight polyethers gives increased load bearing but imparts poor resilience. Use of ethylene glycol and other very low molecular weight polyols improves the rigidity of the foam and yet gives good resilience when used in combination with polyethers having molecular weights above about 3500. The added stiffness can be attributed to the large increase in hydrogen bonding and concentration of rigid domains. A considerable portion of the polyether groups, which comprise the soft domains, are replaced with the highly polar aromatic urethanes formed by the reaction of the glycol and isocyanates. Large concentrations of hydrogen bonds generally give polymers having slow rates of recovery. The high resilience obtained despite the large number of hydrogen bonds might be attributed to the long uninterrupted polyether chain segments between the hydrogen bonds. This is consistent with findings that intermediate molecular weight polyethers, used in place of the combination of ethylene glycol and high molecular weight polyethers and having the same total NCO requirements, give about the same total number of hydrogen bonds but yet have poorer resilience.

Trimer acids were incorporated into the formulation in order to provide a high percentage of open cells. It is likely that this effect is of a kinetic nature based on the different rates of reaction of carboxylic and various hydroxyl groups with isocyanates. Carboxylic groups enter into the reaction at a slower rate than that of the water and most of the hydroxyls of the polyols. Thereby, CO_2 can be generated from the acids after the foam has been partially formed and cause many of the newly formed cells to burst.

Density of the foam greatly affects the load-bearing characteristics.

While all of the ingredients have some effect on the density, it is particularly convenient to control it by adjusting the water and catalyst content. Increased quantities of water and amine catalyst give decreased density, whereas the density increases with the SnOct (stannous octoate) concentration. These variables can be adjusted so as to provide a foam having the desired load-bearing properties; however, such adjustments are highly specific for given temperatures and machine conditions.

In order to improve control of metering, the SnOct can be diluted with a carrier, such as tall oil fatty acids. Other liquid fatty acids or unreactive carriers, such as dibutyl phthalate or perchloroethylene can also be used, provided that they do not exhibit undesirable surface effects.

The ratio of polymethylene polyphenylisocyanate to tolylene di-isocyanate can be adjusted to control the stiffness of the polymer chains. An increased ratio of tolylene di-isocyanate gives a foam having a higher elongation and greater flexibility owing to fewer crosslinks.

Addition of hydroquinone greatly reduced the scorch and yet had only a minimal effect on the physical properties of the foam. Buns greater in volume than 4 ft^3 generally attain their maximum exotherm temperature of about 150–160°C from 1 to 4 hrs after pouring. If buns of flexible open-cell foam without an antioxidant or free-radical acceptor are cut open shortly after the foam is prepared, the entering air oxidises the foam, causing a rather severe darkening of the material. Autoxidative degradation is expected when one considers the high temperature resulting from the exotherm and that the substrate consists largely of a polyether containing tertiary hydrogen atoms.

Almost all of the commercially available silicone surfactants were studied and none of them contributed towards meeting the required properties. In general, most surfactants gave a slightly lower density and finer cells. Since the foam is normally quite breathable and highly open celled, reduction in cell size by surfactants resulted in an increased pressure drop for flow, and thus reduced the breathability. Actually, the formulation can be considered to contain its own surfactant since the ethylene glycol has miscibility with water and the organic ingredients. Surfactant activity could also result from the combination of the amine catalyst and fatty acids. In this particular set of foams, the surfactant also was not required to stabilise the rising foam just after the pour. The lack of requiring an added surfactant is, however, quite unusual.

Catalyst

A detailed study was made of the effect of catalyst concentrations on load bearing and density. Since the amine catalyst relatively favours the NCO–H_2O reaction that generates the CO_2, increased concentration gives lower density and load bearing. The tin catalyst increases the rate of the NCO–OH reaction, relative to the water reaction, thus causing the polymer to gel faster. Therefore, use of more SnOct gives a foam that can set before it attains the maximum expansion and thereby gives denser and higher-load-bearing foams. These effects are complex, since each of the catalysts catalyses all of the pertinent reactions and their concentrations also affect the exotherm while the foam is forming.

It is shown in Figs. 6.24 and 6.25 that the load bearing and density increase in a nonlinear manner with the tin concentration and decrease

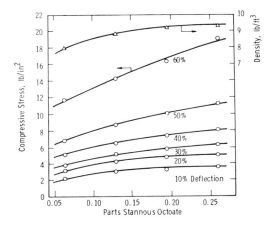

FIG. 6.24. Effect of stannous octoate concentration on load bearing and density. NEM concentration, 0·2 parts.

with increasing amine content. Since the stress values at increasing deflections are more sensitive to density, they showed the greatest dependency on the catalyst variations. It can readily be seen that very similar load-bearing properties can be obtained with a variety of catalyst compositions, since both the ratio of the partially competing catalysts as well as their absolute concentration and possible synergistic interactions govern the final properties. For example, foams having densities of 7·4 lb ft^{-3} and stress values of 2·0 and 8·8 lb in.$^{-2}$ at 10 and 60% deflections, respectively, were obtained using catalyst combinations of 0·40 NEM with 0·083 SnOct and also from 0·80 NEM with 0·13 SnOct. Similarly, foams having densities of about 7·9 and corresponding stress values of 2·5 and 11·0 were found for 0·33 NEM with 0·130

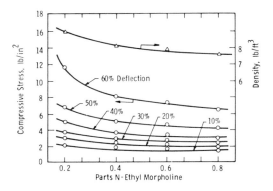

FIG. 6.25. Effect of N-ethylmorpholine concentration on load bearing and density. Stannous octoate concentration, 0·065 parts.

SnOct and for 0·80 NEM with 0·195 SnOct. The correspondence of both load bearing and density at different concentration levels of catalyst suggests that the previously described complications in developing the intrinsic compressive stiffness envelopes may be minimal.

Isocyanate Concentration

Results of a study of the effect of isocyanate index, defined as 100 times the ratio of isocyanate employed to that theoretically required to react with all of the active hydrogen compounds, are shown in Fig. 6.26. The

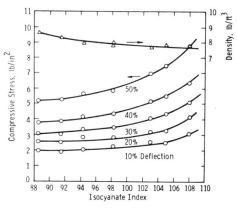

FIG. 6.26. Effect of isocyanate index on load bearing and density. Water, 0·38 parts; SnOct, 0·0675 parts.

load bearing increases and density decreases slightly with increasing isocyanate (NCO) level. The density effect can be attributed to increased availability of NCO for the CO_2 generating reactions. Since the load bearing of this series of foams increases at decreasing densities, it is apparent that the increased quantities of NCO give a stiffer polymer. This effect is further illustrated in Fig. 6.27 in which the compressive stress–density relationship is shown against the 10 and 60% deflection parameters. Foams prepared from formulations having higher NCO

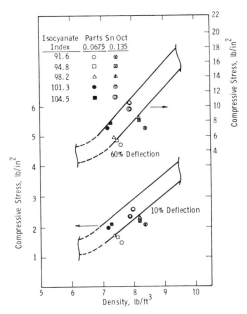

FIG. 6.27. Effect of isocyanate index at 0·27 and 0·54 parts SnOct on density–load bearing relationship.

indices gave stress values near the top of the envelope, whereas low-index foams clearly showed a reduced intrinsic compressive stiffness. At the lower NCO indices, the polymer chain length is shorter due to the deficiency of the available NCO to react with all of the hydroxyl containing compounds. As the index increases, greater intrinsic compressive stiffness results due to increased molecular weight, formation of additional crosslinks due to allophanate and biuret linkages, crosslinks resulting from the increased quantity of the polyfunctional polymethylene polyphenylisocyanate, and a higher concentration of aromatic groups in the polymer chains.

Ethylene Glycol

In view of the above discussion, the results obtained with varying quantities of ethylene glycol (EG) are somewhat enigmatic. From Fig. 6.28, it can be seen that the load-bearing capacity and density increase with EG content to a peak at about 5 parts EG, then decrease with increased

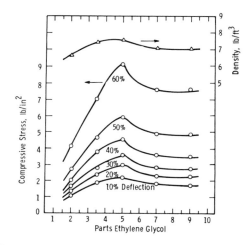

FIG. 6.28. Effect of ethylene glycol concentration on load bearing and density.

glycol content. The relative intrinsic compressive stiffness of the foams containing 9, 7 and 5 parts EG are about the same and then show a decrease for 2·0 and 3·5 parts EG.

As the quantity of EG is raised, the rates of foam gelation are increased due to more crosslinks and greater exotherm. These additional crosslinks can consist of allophanate and biuret linkages which result from the required increased quantity of isocyanate. The polymethylene polyphenylisocyanate component provides crosslinking through urethane linkages. However, the higher exotherm and greater isocyanate concentration also favour the blowing reactions. Hence, there are two opposing trends, and over the range of concentrations studied, the overall gelation process appears to be optimised with respect to the blowing reactions when the quantity of EG is between 4 and 5 parts where the density is at a maximum.

On increasing the quantity of EG from 2 to 5 parts, the increased hydrogen bonding is largely responsible for the stiffening of the polymer chains. As the hydrogen bonding is increased further, it becomes more difficult for the polymer chains to align themselves so as to take advantage of the additional hydrogen bonding. Increased crosslinking resulting at higher EG concentrations also interferes with the ability of the chains to align.

Since the EG is more reactive and is the source of about six times the hydroxyl equivalents (based on a formulation with 5 parts EG) compared to the polyoxypropylene triol, the product consists of a block-type polymer containing rigid and soft domains. Thus, long polyether chains are interspersed between multiconsecutive segments of compact and highly polar urethane-containing groups. Hence, the hydrogen bonding occurs in clusters or bunches rather than being uniformly distributed along the polymer chains. If the number of hydrogen bonds in the bunches is increased, for example, from that corresponding to use of 5 parts EG to 9 parts EG, there might not be an appreciable difference in chain stiffening, since the clusters could have been close to their maximum rigidity at 5 parts EG. One might also consider that the polyether containing segments which comprise a large portion of the volume of the polymer are relatively unaffected and behave in a manner analogous to a buffer.

Another possible contribution to the lack of stiffening of the polymer chains at the upper range of the EG concentration studied is that there may be more unreacted groups since the quantity of material of low equivalent weight is increased. This is a familiar occurrence in many types of polymeric systems where the ends of the many but relatively low molecular weight, entangled polymer chains are somewhat restricted in their motions due to the required coordination with other segments of the network. In addition, the greater exotherm resulting when more EG is used may contribute to the dissociation of some of the already reacted secondary hydroxyl groups from the polyether.

That the load-bearing capacity of this series of foams increased with EG content from 2 to 5 parts is attributed to both the increase in bulk density of the foam and stiffness of the polymer chains. The subsequent decrease in load-bearing capacity with increasing the EG from 5 to 9 parts results from the decrease in density. The tensile strenght also peaks at about 5 parts EG, reflecting primarily the increased density. Elongation was essentially the same for 9 to 5 parts EG and then rose rapidly with decreasing EG concentration due to the fewer crosslinks.

Polyoxypropylene Triol

Load-bearing capacity and density increase with increasing polyoxypropylene triol (POPT) content (Fig. 6.29). This can be attributed to a bulk or dilution effect, since the ratio of the quantity of gas evolved to the total mass of the foam decreased with increasing POPT

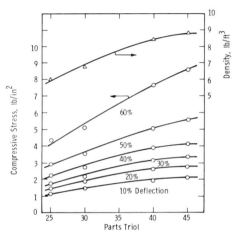

FIG. 6.29. Effect of polyoxypropylene triol (MW ≈ 4000) concentration on load bearing and density.

concentration. Since the POPT is primarily responsible for the flexibility of the polymer chains, it can be seen that the intrinsic compressive stiffness decreases with increasing POPT content (Fig. 6.23). Of the two above opposing trends, dilution and softness, that of dilution appears to predominate with respect to overall load-bearing capability. The general trend for both elongation and tensile strength is to increase with the POPT. Even though the calculated crosslink density increases slightly with the POPT ($\overline{M}_c$ 1450, 1400, 1300 and 1250 for 45, 40, 30 and 25 parts POPT, respectively), this apparently is overshadowed by the decrease in hydrogen-bond concentration, as indicated by the generally reduced intrinsic compressive stiffness. Higher elongation values at greater POPT concentrations are also consistent with this view. The concomitant increase in tensile strength is attributable to the increase in density.

Water

As the concentration of water is increased with a concomitant adjustment in isocyanate to maintain a constant NCO index, the density and load bearing decrease (Fig. 6.30) as a result of the additional blowing effect. Since the variation in water concentration represents minor changes in the total active hydrogen content, the effect on the rigidity of the polymer is relatively small, showing only a slight increase at the upper concentration of water.

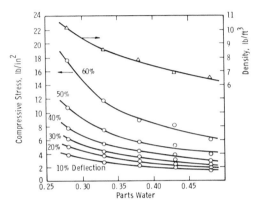

FIG. 6.30. Effect of water concentration at constant isocyanate index on load bearing and density.

Trimer Acids

Load-bearing capacity and density decrease with increasing trimer acids (TA) concentration (Fig. 6.31). Since the reaction product of an organic acid with NCO groups generates CO_2, an increase in the TA content gives more blowing and hence lower-density foams. Increased TA gives a more open-celled foam, as evidenced by the increase of breathability with the TA concentration (Fig. 6.32). This phenomenon is attributed to the previously described kinetic effect. It appears that the intrinsic compressive stiffness of the foam is almost independent of the TA concentration over the range of 5–15 parts. Omission of TA causes a major change in the rheological characteristics and cell structure of the foam. The fundamental difference in structure was apparent, since normal linear shrinkage for most of the foams discussed in this paper was about 1.5% whereas about 5% shrinkage occurred when TA was omitted. As a result, the apparent decrease of intrinsic compressive stiffness of the zero TA foam is not amenable to an explanation based solely on chemical structure sensitivity concepts.

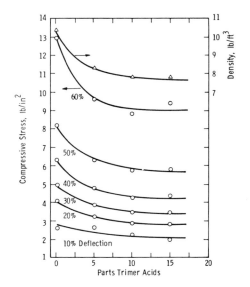

FIG. 6.31. Effect of trimer acids concentration on load bearing and density.

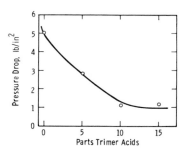

FIG. 6.32. Effect of trimer acids concentration on breathability.

Isocyanate Composition
The effect of the composition of the isocyanate stream was studied by varying the ratios of the tolylene diisocyanate (TDI) and the polymethylene polyphenylisocyanate (PMPPI) components. Over the range studied, the increased crosslink density, obtained by a higher ratio of PMPPI to TDI, resulted in increased load-bearing properties (Fig. 6.33).

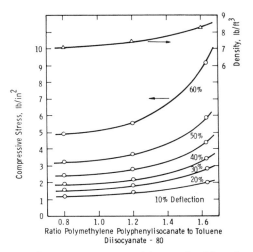

FIG. 6.33. Effect of isocyanate composition on load bearing and density.

The intrinsic compressive stiffness of the foam was about the same, within experimental error, for the two lower PMPPI ratios and showed the expected increase for the weight ratio of 1·63 parts PMPPI per part TDI. Elongation decreased, as expected, with an increase in crosslink density. As the PMPPI ratio is increased further, beyond about 2·5, the foams tend to become friable and begin losing load-bearing capability, as measured on the fourth loading cycle, due to rupture of cell ligaments.

6.3.4 Summary

Since the foams represent heterogeneous non-ideal systems, the interpretations were qualitative, and some were speculative in nature. A more fundamental study would involve ideal systems employing thin films of polymeric materials. However, since the physics and chemistry of foam formation and structure cannot be divorced, it was considered that the results of a structure–sensitivity study of actual foams would be a useful contribution to the field of applied foam technology. A further extension of this work that would help elucidate many of the discussed phenomena is a detailed study of the polymeric structure and end-group analyses of the material of actual foams.

Since density is of major importance in controlling load-bearing capacity, an attempt has been made to determine qualitatively the contributions of the stiffness of the polymeric structure independent of the bulk density of the foam. This led to the development of the concept of intrinsic compressive stiffness accompanied with discussions of the effects of chemical structure on it.

Within the concentration limits studied, increased isocyanate content, higher functionality of the isocyanate mixture, greater polyoxypropylene triol concentration, and increased tin catalyst concentration give increased load-bearing capability to the foam. The latter three cause an increase in the bulk density of the foam, whereas the density decreases slightly with increasing isocyanate index. Increased quantity of trimerised fatty acids gives a reduction in compressive stress and density and higher breathability. The density and load-bearing capacity decrease with increasing water content and amine catalyst concentration, and appear to reach a peak midway in the studied concentration range of the ethylene glycol.

6.4 PHENOLIC RIGID FOAMS

6.4.1 General Considerations

A rigid phenolic foam was developed to protect delicate and massive equipment from shock resulting from explosive detonations. In contrast to the elastomeric pads and foam which are designed to protect equipment over a great many deflection cycles, this foam absorbs energy as it is destructively crushed.

In such an application, it is necessary that the compressive strength of the foam exceed a certain minimum value in order that a sufficient quantity of energy be absorbed and that the equipment be adequately supported. However, the foam must exhibit crushing at a stress level that will not impart excessive loading to the apparatus that it is protecting. Furthermore, in order to avoid the build-up of excessive pressure on shock impact, it is also essential the foam meet certain minimum gas permeability requirements.

Since a matrix of a brittle foam is broken rather than flexed during compression, its C–D curve displays a considerably flatter and wider stress plateau than obtainable with a flexible foam. The essentially constant stress up to the ultimate strain offers the desired property of high-energy absorption (area under the load–compression curve) while preventing the attainment of undesirably high stress. High permeability towards gases is required to prevent an excessive increase in stiffness of the foam that could result from air trapped inside the matrix during dynamic loading.

The influence of the physical characteristics of the foam matrix on the C–D and energy absorption properties has been investigated by Rusch.[8–10] His work is reviewed in Section 2B.3 and 5.2, and is consistent with the observations of Mendelsohn *et al.*[1,2] in that the C–D behaviour of cast cellular polyurethanes are highly sensitive towards the shape of the cells.

Since in addition to being rigid, the foam was required to be highly open celled, a phenolic material was selected. Other rigid foams, such as polyurethanes and isocyanurates, were not employed, since their structures are essentially closed cell.

Thus a detailed study[11–13] of the effects of variations in the chemical formulation and process conditions was performed on a series of phenolic foams whose basic formulation is described in Table 6.3.

The processing, described in detail in Mendelsohn *et al.*,[11] is designed to prepare 40–45 kg buns of foam using a batch process in which there is a sequential addition of the ingredients under high-speed, high-shear mixing conditions. First, the resins, stored at close to 0° C, are combined with the surfactants and brought to the nominal processing temperature, usually 18° C. Next the blowing agent and then the acids are added under carefully controlled mixing conditions. After about 20 sec have elapsed from the time of the initial introduction of acid, the mixture has been poured into a large cardboard carton and permitted to rise. The material creams within several seconds and completes its rise usually over the next minute.

Within 2–4 hr of foam preparation, the skin plus about 2 in. of material is trimmed from the bun. The trimmed buns are baked in a vented forced-air circulation oven for 8 hr at 100°C.

6.4.2 Mechanical Tests and Requirements

C–D Properties

The C–D properties that were specified during the initial phase of this development are shown in Fig. 6.34. At low or essentially static rates of compression, the foam exhibits stress–strain properties lying between the upper and lower static limit curves. Under a high strain rate of 1500 in. in.$^{-1}$ min.$^{-1}$ the C–D characteristics of the foam are delineated by the space between the upper-dynamic-limit and lower-static-limit curves.

After the foam has been post-baked, the block is trimmed to $24 \times 18 \times 14$ in. Samples $6 \times 6 \times 2$ in. are then cut from various sections

TABLE 6.3
Basic Formulation of Phenolic Foam

Ingredients[a]	Parts by weight
Phenolic resin BRL 2760	80
Phenolic resin BRL 2759	20
Tween 60	0·25
Span 80	0·50
Niax 113 or Freon 113	12
47·5 wt % sulphuric acid	25
85 wt % phosphoric acid	15

[a] Bakelite BRL-2760 is a one-step heat-reactive liquid phenolic resin manufactured by Union Carbide Corporation. Its suggested use is for preparation of medium-density rigid phenolic foam. Properties listed by Union Carbide include: viscosity at 25°C, 2350–3125 cP, pH 6·0–6·8, and nonvolatile matter (78–81)%.

Bakelite BRL-2759 also is a one-step heat-reactive liquid phenolic resin manufactured by Union Carbide Corporation. Its suggested use is for low-density rigid phenolic foam. Listed properties include the following: viscosity at 25°C, 400–800 cP, pH 6·0–6·3, and specific gravity at 25°C, 1·210–1·225.

Tween 60, polyoxethylene sorbitan monostearate, is manufactured by ICI America, Inc., Atlas Chemical Division. Its listed properties include: viscosity 600 cP at 25°C, acid number 2·0 max, saponification number 45–55, hydroxyl number 81–96, and the HLB number 14·9.

Span 80, sorbitan mono-oleate, is manufactured by ICI America, Inc., Atlas Chemical Division. It has the following properties: viscosity ~ 1000 cP at 25°C, acid number 8·0 max, saponification number 145–160, hydroxyl number 193–210, and hydrophile-lipophile-balance (HLB) number 4·3. (Higher HLB numbers indicate greater hydrophilic character. Lipophilic surfactants are assigned HLB numbers below 9·0, whereas ones that are hydrophilic are given numbers above 11·0. Those in the range of 9–11 are intermediate.)

Niax 113 or Freon 113, 1, 1, 2-trichloro-1, 2, 2-trifluoroethane, fluorocarbon blowing agent, boiling point 47·6°C, is manufactured by Union Carbide under trade name Niax or DuPont as Freon.

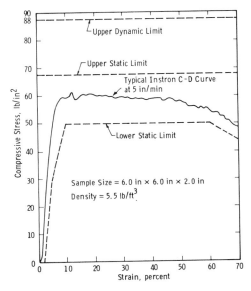

FIG. 6.34. Requirements and typical stress–strain properties.

of the bun and crushed in an Instron Universal Test Machine at a rate of 5 in. min^{-1} producing an autographic record as shown in Fig. 6.34.

Permeability

A hollow cylindrical specimen having uniform wall thickness 3·5 in. inside diameter × 5·5 in. outside diameter × 2·0 in. height, is cut from a 2 in. thick sample taken from near the centre of the bun. The pressure differential of air flowing at a rate of 17 ft^3 min^{-1} from the centre of the specimen in an outward radial direction is measured. It is required that the foam have a sufficiently open-cell structure so that the pressure drop will not exceed 31 Torr.

6.4.3 Effect of Composition and Process Variables

Variations in chemical composition were made by varying the concentration of a particular ingredient while otherwise maintaining the formulation described in Table 6.3 and employing a nominal processing temperature of 18°C.

Compressive stress S at 20% deflection was arbitrarily selected to represent the load-bearing capability of the foam. Pressure drop ΔP of air flowing through the foam was employed as a measure of the breathability or permeability. Since an increase in the density D or volume fraction polymer in the foam matrix tends to give a reduced permeability by providing less space for air to pass through, a figure of merit $\Delta P/D$ relating to the openness of the cell membranes is plotted against the investigated variable. An analogous relationship, termed a permeability-load-bearing function $\Delta P/S$ is similarly plotted. The latter is particularly useful for design purposes, since it suggests a potential means for optimising the relationship between breathability and C–D characteristics.

Reaction Rate and Density
Although the rate of reaction is not a design property of the completed foam, it nevertheless is important since it determines whether manufacture of a certain foam is feasible, affects the cost of processing equipment, and influences the properties of the finished product.

Excessively fast rates of reaction are to be avoided due to the difficulty in filling a mould without pouring foam onto already creaming and rising material. Even for a continuous conveyor process a very fast reacting material will have the disadvantage of requiring a very large capacity foam machine. A reaction rate that is too slow permits separation of the mixed ingredients before their reaction with the result that the foam does not possess uniform properties throughout the bun.

One cannot predict with assurance what effect the overall reaction rate will have on the foam's density, load bearing and permeability. This is illustrated by the effect of increasing the concentration of sulphuric acid which increases the rate of the polymerisation reactions and the exothermic temperature. Since, in this case, the higher rate of gelation is relatively more important than the more vigorous expansion of the blowing agent, the resultant foam has a greater density accompanied by higher load-bearing and reduced breathability properties. However, in contrast, increasing the rate of reaction by employing a higher process temperature favours the expansion of the blowing agent relative to the rate of polymerisation. Thus, foams prepared at higher temperatures display reduced densities, lower load-bearing capabilities and higher permeabilities.

Density has a very strong influence on the load-bearing properties of the foam (Figs. 6.35 and 6.36, and Table 6.4). Its dependence on compressive strength σ has been expressed by previous investigators[9,14,15] by the

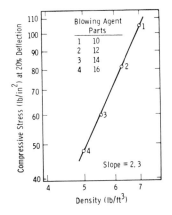
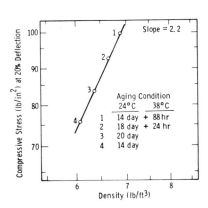

FIG. 6.35. Effect of blowing agent on density and compressive stiffness of foam.

FIG. 6.36. Effect of ageing of phenolic resin intermediates on density and load bearing of foam.

TABLE 6.4
Factors Affecting Dependency of Compressive Stress on Density

Variable	Range of variation	Slope of plot of log stress/log density
Blowing agent	10–16 parts	2.3
H_2SO_4 (47.5 wt. %)	15–35 parts	2.7
H_3PO_4 (85 wt. %)	0–30 parts	3.3
Tween 60	0.125–1.00 parts	2.3
Span 80	0–2.0 parts	1.6
Total surfactant (Tween 60/Span 80 = 0.5)	0–3.0 parts	2.1
BRL 2759	0–20 parts	3.2
Ageing of resins	(see Fig. 6.36)	2.2
H_2O (change in H_2O concentration from standard formula)	(−2.5)–(+12.5) parts	2.1
Temperature	15–24°C	3.5

empirical relation

$$\sigma \propto D^a$$

in which D is the bulk density of the foam and a is a constant. Values of the constant a in the range of 1·4–1·6 have been reported for rigid polyurethane foams (see Section 1.3.2). Generally, the load-bearing capability of phenolic foams herein discussed is much more dependent on density.

A single log–log plot of the compressive stress versus density for all of the foams prepared in this study displayed such a wide scatter of the data points that a 'best straight line' drawn through the points had little meaning. However, smooth curves resulted from stress versus density plots in which the density was affected by a single composition or process variable (see Figs. 6.35 and 6.36 for typical stress–density plots). The exponential dependency of compressive stress on density obtained from the slopes of such curves is shown in Table 6.4.

The large differences obtained in the exponential relationships suggest that variations in at least most of the process and composition variables affect, in addition to density, other intrinsic properties of the foam. This is not surprising in view of the highly involved phenomena occurring during the formation of the foam. For example, in addition to previously described factors, such as the rate phenomena for sulphuric acid are its highly complex surface actions in which the most conspicuous effects consist of a diminution of cell size with increasing acid concentration. The acid also affects the structure and composition of the polymer chains giving a more rigid material as its concentration increases.

The relationship between load bearing and density has been related qualitatively to cell structure. Comparing the relative cell sizes of the respective members of a series, one finds that when the cell size diminishes with increasing density, the exponent, a, tends to be large. The smaller exponents result when the cell size tends to increase with density. This is evidenced by the marked increase in coarseness of the cell structure with density as the concentrations of water and Span 80 are increased. In contrast, the relatively high exponential dependencies of compressive stiffness on density are obtained from the previously described series in which the increasing density, resulting from higher concentrations of phosphoric acid, sulphuric acid, phenolic resin BRL 2760, and lower processing temperature, is accompanied by a diminution in cell size. An example of this effect is shown in Table 6.5 which lists the cell sizes for the highly exponentially dependent temperature and phosphoric acid series and the far less dependent Span 80 series.

Water Concentration

An example of the effect of reaction rate on properties of the finished product is provided by adjusting the concentration of water in the formulation. In a series of experiments, the concentration of water was

TABLE 6.5
Effect of Processing Temperature and Composition on Cell Size

Temperature (°C)	Parts H_3PO_4	Parts Span 80	Cells in.$^{-1}$
15	15	0·5	24
18	15	0·5	21
21	15	0·5	18
24	15	0·5	13
18	0	0·5	10
18	7·5	0·5	—
18	15	0·5	29
18	22·5	0·5	38
18	30	0·5	40
18	15	0	80
18	15	0·25	42
18	15	0·5	26
18	15	1·0	19
18	15	1·5	14

increased by adding it to either the resin or acid components and decreased by employing acid solutions more concentrated than the standard. Since water dilutes the catalyst and reactants and absorbs heat from the exothermic reaction, the rate of reaction decreases with increasing water content (Fig. 6.37).

An increase in the density and compressive stress values with added water (Fig 6.38) indicates that the expansion processes are more severely retarded than those involving the polymerisation of the resin. Analogously to the effect of increasing the concentration of sulphuric acid, permeability decreases with increased water content (Figs. 6.38 and 6.39). However, the relatively slight effect on the permeability-load-bearing function up to about 10 parts added water indicates a means to obtain higher load bearing while experiencing only a minor diminution in breathability.

Blowing Agent
Variation of concentration of blowing agent 1,1,2-trichloro-1,2,2-trifluoroethane, was explored over the range of 10–16 parts (Figs. 6.40 and 6.41). Adjustment of the quantity of blowing agent provides a convenient method of controlling density and load-bearing properties. A decrease in compressive stiffness of close to 60% results from an approximately 60% increase in the quantiy of blowing agent. The expected

252   MECHANICS OF CELLULAR PLASTICS

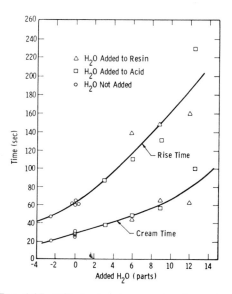

FIG. 6.37. Effect of water on rate of reaction.

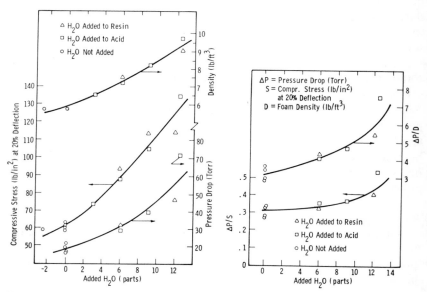

FIG. 6.38. Effect of water on load bearing, permeability and density.

FIG. 6.39. Effect of water on permeability functions.

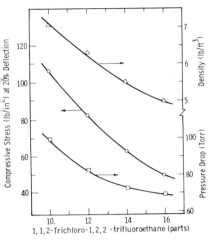

FIG. 6.40. Effect of blowing agent on load bearing, permeability and density.

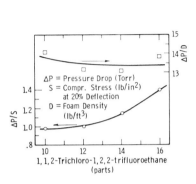

FIG. 6.41. Effect of blowing agent on permeability functions.

decrease in permeability, as indicated by the higher pressure-drop values, follows the increased density at lower blowing-agent concentrations. Over the concentration range studied, the degree of openness of the cells, as indicated by $\Delta P/D$, appears essentially constant; however, a moderate increase of the permeability–load bearing function $\Delta P/S$ occurs with increasing quantities of blowing agent.

Processing Temperature

As the temperature is increased, the load bearing decreases sharply and the cells become increasingly open (Figs. 6.42 and 6.43). The increase in permeability with temperature is attributed to the more vigorous formation and expansion of volatiles which cause rupturing of the cell membranes as well as the reduction of density. The overall rate of reaction increases such that the rise time is decreased by 50%, as the nominal processing temperature is increased from 15°C to 24°C.

Acid Concentration

The rate of reaction increases rapidly with increasing sulphuric acid concentration. At the high concentrations of acid, the resultant high rate of rigidification of the foam matrix relative to the expansile capability of the

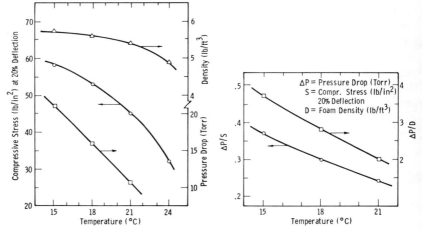

FIG. 6.42. Effect of process temperature on load bearing, permeability and density.

FIG. 6.43. Effect of temperature on permeability functions.

blowing agent results in high density and load bearing and low permeability properties (Figs. 6.44 and 6.45). The combination of rate and surface effects result in a sharp diminution of cell size with increasing acid concentration.

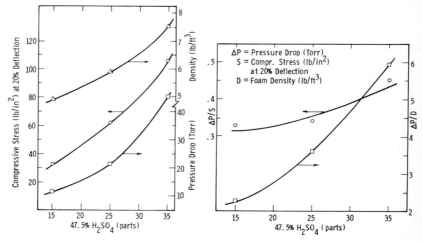

FIG. 6.44. Effect of sulphuric acid on load bearing, permeability and density.

FIG. 6.45. Effect of sulphuric acid on permeability functions.

The effect of variation of the concentration of phosphoric acid on the rate phenomena, as expected from a consideration of the relative strengths of the acids, is less pronounced than that for sulphuric acid. In addition to moderately reducing the time required for gelation of the resin, increasing concentration of phosphoric acid greatly decreases the permeability of the foam while imparting higher load-bearing capability (Figs. 6.46 and 6.47). Phosphoric acid appears to have the effect of reducing the interfacial surface energy of the components of the foaming system. This phenomenon is depicted in Table 6.5 in which the cell size is shown to decrease dramatically with increasing acid concentration.

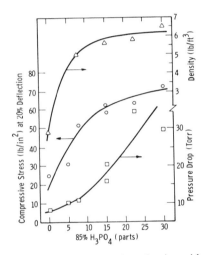

FIG. 6.46. Effect of phosphoric acid on load bearing, permeability and density.

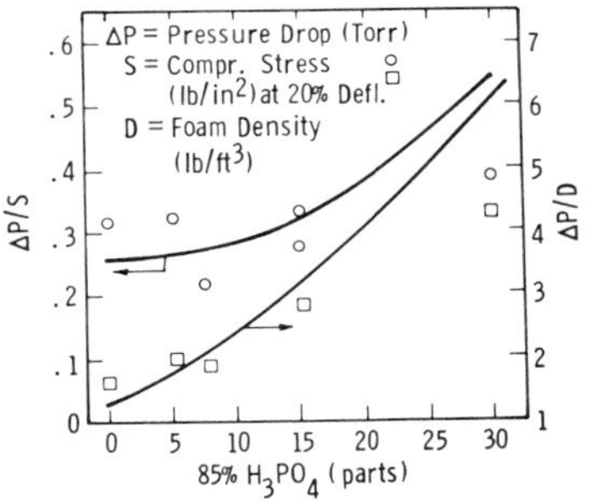

FIG. 6.47. Effect of phosphoric acid on permeability functions.

Composition of Phenolic Resin
Wide variations in foam properties can be obtained by adjusting the composition of the phenolic resin. Union Carbide suggests use of BRL 2759, BRL 2760 and BRL 2761 for preparation of low, medium and high density foams, respectively.[16,17]

Foams prepared from blends containing from zero to 30% BRL 2759 in BRL 2760 showed that as the BRL 2759 is increased, the overall rate of the foaming reaction increases, load bearing and density decrease and breathability is improved (Figs. 6.48 and 6.49). The large increase in

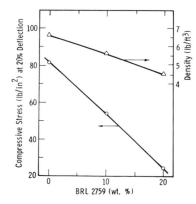

Fig. 6.48. Effect of phenolic resin composition on load bearing and density.

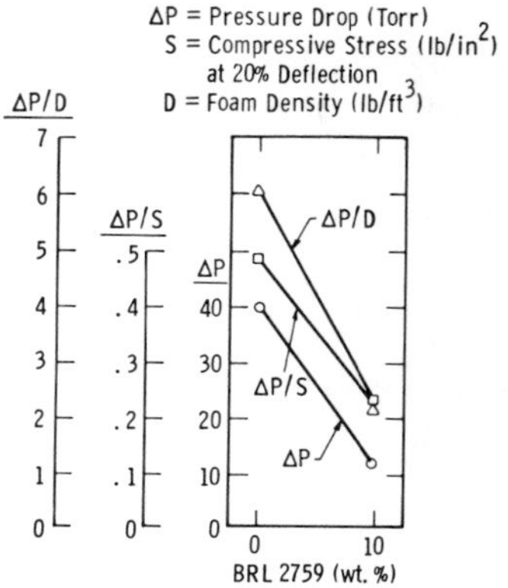

Fig. 6.49. Effect of phenolic resin composition on permeability characteristics of foam.

cell size, 21 and 8 cells per in. for 10% and 20% BRL 2759, respectively, accompanying the reduced density is consistent with the very large dependency of load bearing on density.

The study of the effect of ageing of the phenolic resins was performed by permitting the resin mixture 90% BRL 2760 with 10% BRL 2759, to age at room temperature and at 37°C. As the resin polymerises during storage, it gives more dense and concomitantly higher load-bearing and less breathable foam (Figs. 6.36, 6.50 and 6.51). The importance of storing phenolic resin intermediates at close to 0°C is evident.

Surfactant System

Although the surfactants do not have a great effect on the rate of reaction, they strongly influence the permeability and C–D characteristics of the foam matrix.

Tween 60 stabilises the foam during its formation as evidenced by the collapse of rising foam in which it was omitted while Span 80 was the only surfactant. As its concentration increases from 0·125 to 0·25 parts, while Span 80 is maintained at 0·5 parts, the cell size of the foam diminishes sharply from 17 to 42 cells per in. At higher Tween 60 levels, there is no apparent trend in the size of the cells. While maintaining a Span 80 concentration of 0·5 parts, the compressive stress increases with Tween 60 until 0·37 parts is reached and then diminishes with further increase in the concentration of Tween 60 (Fig. 6.52).

SHOCK MITIGATING SYSTEMS 257

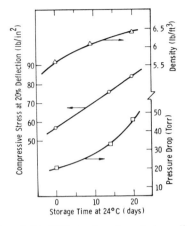

FIG. 6.50. Effect of ageing of phenolic resins on load bearing, permeability and density.

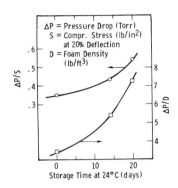

FIG. 6.51. Effect of ageing of phenolic resins on permeability functions.

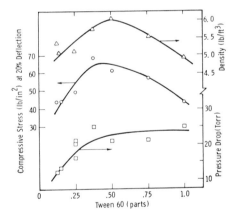

FIG. 6.52. Effect of Tween 60 on load bearing, permeability and density.

The pressure drop of air flowing through the foam matrix increases with Tween 60 reaching an apparent maximum at approximately 0·4 parts and then remains essentially constant (Fig. 6.52). Since the density of the foam decreases at the higher levels of Tween 60, the lack of a corresponding improvement in breathability indicates that the cells are becoming progressively more closed while the concentration of this surfactant is increasing. The plot of the effect of Tween 60 on the openness of the cells and load bearing function (Fig. 6.53) also indicates the increasingly closed-cell structure at higher levels of this surfactant. Apparently a cell structure that provides maximum load-bearing properties is obtained at an intermediate Tween 60 concentration (Fig. 6.52). While the cell

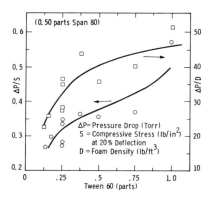

FIG. 6.53. Effect of Tween 60 on permeability functions.

size diminishes and the structure becomes increasingly closed, the density of the matrix increases thereby giving higher compressive stress values. After about 0·4 parts Tween 60, the apparent increasing of the closed-cell content while not continuing the trend of reducing the cell size results in a reversal of the effect of the Tween 60 concentration on load bearing and density. This phenomenon is attributed to the more effective trapping of the blowing agent and other volatiles within the closed cells, thereby giving a net reduction in density and subsequent load bearing of the foam.

Span 80, the lipophilic component of the surfactant system, exhibits behaviour that is at least partially antithetical towards that of Tween 60. Not only is its presence not required to stabilise the foam during rise, but it tends to act as a destabiliser. Formulations employing 0·5 parts and perhaps even less Span 80 require the inclusion of Tween 60 to avoid collapse, whereas a fine-cell foam exhibiting good stability during the crucial rise period was prepared in which 0·25 parts Tween 60 was employed and Span 80 was omitted. Observing that stable foam has been prepared in which both surfactants are omitted, it appears that Tween 60 is necessary to counterbalance the destabilising effects of Span 80.

Foams were studied in which the Span 80 was increased from 0 to 2·0 parts while the Tween 60 was held constant at 0·25. As the concentration of the Span 80 is increased, the density and load bearing increase, the matrix becomes more permeable and the cells become progressively larger (Figs. 6.54 and 6.55, and Table 6.5). Thus, Span 80 can be employed efficaciously to provide the unusual combination of increased breathability and load bearing. The unusual behaviour of the Span 80 is attributed to its proclivity towards providing a highly open-cell

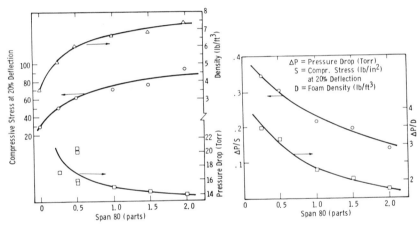

FIG. 6.54. Effect of Span 80 on load bearing, permeability and density.

FIG. 6.55. Effect of Span 80 on permeability functions.

structure. As the foam matrix displays increasingly open-cell character, it loses efficiency in trapping the expanding blowing agent, thereby not rising to its full extent. The relatively low rate of increase of compressive strength with concentration of Span 80, especially above 0·5 parts, is attributed to the corresponding increase in the coarseness of the cell structure. This effect on cell geometry has been shown to give relatively low exponential dependencies of load bearing on density.

A series of runs was made in which the total surfactant concentration was varied while maintaining a constant 2:1 weight ratio of Span 80 to Tween 60. The compressive strength (Fig. 6.56) appears to follow almost an

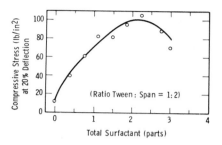

FIG. 6.56. Effect of total quantity of surfactant at constant composition ratio on load bearing.

additive type relationship between the individual variations of each surfactant. Thus, adjustment of both the individual concentrations and the hydrophilic–lipophilic balance of the surfactants provides a powerful means for controlling the physical properties of the foam.

6.5 SUMMARY OF PHENOLIC FOAM EFFECTS

A phenolic foam system has been developed for use in an application requiring energy absorption and shock mitigation. The rigid foam is characterised by having the required combination of high breathability and load-bearing capability.

Compressive stiffness and density of the foam increase with increasing concentrations of either acid, BRL 2760, water, Span 80, and with decreasing concentrations of BRL 2759 and the blowing agent. The load-bearing capability increases passing through a maximum and then decreases with increasing concentration of Tween 60. Increasing process temperature gives a reduction in the density and compressive stiffness of the foam. Breathability or permeability decreases with increasing density of the foam in every case except when density increases with increasing concentrations of Span 80. Employing relatively high concentrations of Span 80 provides the unusual combination of high load bearing and breathability; however, this approach is limited by the resultant instability of the foam during its period of rise.

Data presented on the effects of variations of the processing and composition parameters provide the type of knowledge necessary for adjusting or 'fine tuning' the permeability and C–D characteristics of the foam.

ACKNOWLEDGEMENT

The author wishes to express his gratitude to those who have coauthored with him the referenced articles on structured elastomers, flexible polyurethane foam and rigid phenolic foam. These include R. G. Black, H. J. Connors, J. F. Meier, H. F. Minter, G. B. Rosenblatt, G. E. Rudd and R. H. Runk.

Appreciation is also expressed to the laboratory technicians of the Insulation and Applied Chemical Research Departments and the Physical Testing Section of the Westinghouse Research Laboratories who prepared the materials and performed the measurements.

REFERENCES

1. MENDELSOHN, M. A., RUNK, R. H., CONNORS, H. J. and ROSENBLATT, G. B. (1971) *I&EC Prod. Res. and Dev.*, **10**, 14.
2. MENDELSOHN, M. A., RUDD, G. E. and ROSENBLATT, G. B. (1975) *I&EC Prod. Res. and Dev.*, **14**, 181.
3. SAUNDERS, J. H. and FRISCH, K. C. (1962) *Polyurethane Chemistry and Technology, Part I, Chemistry*, Interscience, New York.
4. SAUNDERS, J. H. and FRISCH, K. C. (1962) *Polyurethane Chemistry and Technology, Part II, High Polymers*, Interscience, New York.
5. MENDELSOHN, M. A., BLACK, R. G., RUNK, R. H. and MINTER, H. F. (1965) *J. Appl. Polym. Sci.*, **9**, 2715.
6. MENDELSOHN, M. A., BLACK, R. G., RUNK, R. H. and MINTER, H. F. (1966) *J. Appl. Polym. Sci.*, **10**, 443.
7. MENDELSOHN, M. A. and ROSENBLATT, G. B. (1979) *Effect of Hydrolytic Degradation on Compression–Deflection Characteristics and Hardness of Structured Polyurethane Castings*, in R. K. Eby (Ed), *Durability of Macromolecular Materials* (ACS Symposium, Series 95), Chap. 11.
8. RUSCH, K. C. (1969) *J. Appl. Polym. Sci.*, **13**, 2297.
9. RUSCH, K. C. (1970) *J. Appl. Polym. Sci.*, **14**, 1263.
10. RUSCH, K. C. (1970) *J. Appl. Polym. Sci.*, **14**, 1433.
11. MENDELSOHN, M. A., MEIER, J. F., RUDD, G. E. and ROSENBLATT, G. B. (1979) *J. Appl. Polym. Sci.*, **23**, 325.
12. MENDELSOHN, M. A., MEIER, J. F., RUDD, G. E. and ROSENBLATT, G. B. (1979) *J. Appl. Polym. Sci.*, **23**, 333.
13. MENDELSOHN, M. A., MEIER, J. F., RUDD, G. E. and ROSENBLATT, G. B. (1979) *J. Appl. Polym. Sci.*, **23**, 341.
14. FERRIGNO, T. H. (1967) *Rigid Plastics and Foams*, 2nd Ed., Reinhold, New York, p. 148.
15. TRAEGER, R. K. (1967) *J. Cell. Plast.*, **3**, 405.

16. UNION CARBIDE CORP. (1976) *Bakelite Phenolic Resins for Rigid Foam,* New York.
17. PAPA, A. J. and PROOPS, W. R. (1973) *Plastic Foams,* Marcel Dekker, New York, Chap. 11.

Chapter 7

STRUCTURAL FOAMS

J. L. THRONE
*Amoco Chemicals Corporation,
Naperville, Illinois, USA*

7.1 INTRODUCTION

Although thermoplastic structural foam moulding was first envisioned by Beyer and Dahl at Dow in the late 1950s,[1] the efforts of Angell at Union Carbide[2,3] in the mid-1960s marked the beginning of a commercialisation of the process. Basically, thermoplastic structural foam (TSF) is characterised by the deliberate addition of an inert gas into molten or softened plastic in a machine resembling an injection moulding machine, then injection of this mixture into the mould cavity in a manner that allows the gas to expand the plastic to fill the cavity. There are many variations of this process,[4] including accumulation and high-speed injection, moving moulds and two-component injection. The final parts have the general characteristics of a cellular core and a solid or near-solid skin (Fig. 7.1). For most single-resin foams, there is a rather gradual density gradient from the nearly uniform density skin to a nearly uniform density foam core.[5]

The advantages of foaming an injection-moulded part include reduced weight at constant part thickness or an increase in stiffness at the same part weight. Some disadvantages to TSF include poor surface quality when compared with an injection-moulded part, uncontrolled density distribution throughout the part, and lowered mechanical properties owing to the deliberate addition of a gas to the resin.

The primary applications for TSF moulded parts have been in furniture (seat backs, chair frames), automotive parts (glove boxes, under-dash electrical boxes, kick panels), industrial parts (underground cable ducts, water meter housings, valve bodies), materials handling

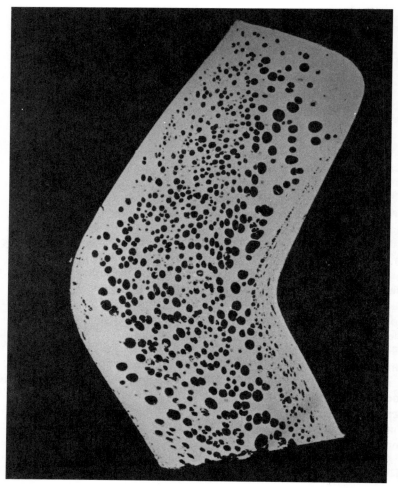

FIG. 7.1. Thermoplastics structural foam cross-section of 5 mm thick polystyrene. Note bubbles in skin area, row of microvoids behind skin and near-spherical bubbles in foam core.

(food containers, pallets, milk cases), equipment cabinets (computer housings, typewriter housings, radio and TV cabinet components) and recreational equipment (water and snow skis, tennis rackets, baseball bats, hockey masks). The primary competitions to TSF are painted sheet metal, wood, glass-reinforced polyester sheet moulding compound and injection-moulded solid thermoplastic resins. By their very nature, TSF are not as strong as the solid resin before foaming. The ability to form thick, lightweight parts has enabled them to perform as structural components. They are, in fact, 'structured' materials, in that the integral skin, which has the greatest strength, is placed at the maximum fibre stress (in a bending mode). More important, since most TSF applications are made of single-component foams, there is no definable shear plane as is typical of the laminated sandwich panel discussed by Hartsock.[6]

For the purpose of this discussion, we consider thermoplastic structural foams (TSF) to have bulk densities not less than 50% of the solid resin densities. Normal operations yield densities that are between 25% and 35% less than that of the bulk-resin densities. Thus mechanical properties of TSF materials are preferentially compared with solid-resin properties rather than the low-density foams described in detail by Benning[7] and Meinecke and Clark.[8] Nevertheless, the concepts used by these authors form a lower limit for explaining the behaviour of TSF materials.

Some terms used below require some explanation here. It has been the tradition in TSF literature to relate all mechanical properties to the 'o' value, which is that of the unfoamed or solid resin. All foams are categorised in terms of their density relative to that of the solid resin. Thus, we shall use the term 'reduced density' in this regard. There are two types of foam densities that are used below. The bulk density is obtained by simply weighing and measuring the foam sample. The bulk reduced density is called ϕ. The foam-core density is obtained by carefully microtoming the sample until the measured density remains constant. The reduced core density is called R'. The skin thickness can be obtained either by observation or by measuring a soft X-ray of the foam cross-section using an isodensitracer.[9] The reduced skin thickness is called e. In bending, the foam stiffness is most important and will be carefully defined below. In tension, the most important property is Poisson's ratio. It is well known that Poisson's ratios for steel, concrete and elastomers are 0·3, 0·1 and 0·4–0·5, respectively.[10] For low-density expanded-polystyrene foams, Benning reports values as low as 0·03,[7] although the models described by Meinecke and Clark yield values in the range 0·35–0·5, regardless of the foam density. This point will be touched on below.

For the record, the need for adequate models for predicting the stress–strain behaviour of structural foams in both bending and tensile modes is most important to the engineer charged with the responsibility of ensuring the performance of a structural foam part. The importance of uniform cellular structure and skin thickness on the overall performance of the structural foam part in bending and tension is also of importance to the processing engineer, since he is faced with selecting a process that will yield a part having the design engineer's desired mechanical properties at the lowest cost.

It is assumed that the reader is familiar with the basic concepts in the bending and stress–strain behaviour of conventional materials. The emphasis throughout this chapter will be on single-component TSF that are produced by injecting a gas–polymer mixture into a mould cavity having a volume substantially larger than the volume of plastic. Where necessary, we will refer to differences in mechanical behaviour of foams produced in other ways, such as through moving mould faces or simultaneous two-component injection moulding. The reader should also be aware of processes for producing structural foam profiles via continuous extrusion. The most successful of these is the Celuka process.[11] The strength of the foam produced this way does not differ appreciably from that produced by injection moulding. Most recently, two-component continuous extrusion of ABS has been commercialised. Again, the differences in the mechanical behaviour of this foam and single-component foams will be noted at the appropriate time. However, attention will be focused on the injection-moulded single-component foam that develops an integral skin owing to the interrelationship between the solubility of the gas in the polymer, the internal pressure in the foam while in the mould, and the rate of heat removal through the mould walls. The interested reader can find the details of this interrelationship elsewhere.[12]

7.2 CELLULAR MORPHOLOGY

It is perhaps instructive to note the dramatic difference in cellular morphology between TSF foams and typical insulation foams. A typical TSF foam cross-section is shown in Fig. 7.1 for a 5 mm thick impact polystyrene foamed to about 25% ($\phi = 0.75$). Compare this with the cellular morphology of a 2 lb ft^{-3} general-purpose polystyrene insulating foam ($\phi = 0.03$) as shown in Fig. 7.2. In general, TSF cells are surrounded by resin, whereas low-density foams appear as dodecahedronal structures characterised by struts and membranes (see Chapters 1 and 2). We would not expect models describing deformation and stress–strain behaviour in low-density foams to be applicable in TSF foams. Nevertheless,

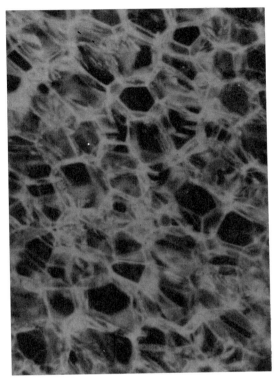

FIG. 7.2. Micrograph of 2 lb ft^{-3} expanded polystyrene foam, showing dodecahedronal cell structure, strut–membrane construction. Compare with TSF in Fig. 7.1. 50×.

some of them can be extrapolated to TSF density range with remarkable success. We discuss these shortly. Figure 7.3 shows several additional aspects of cellular morphology in TSF foams. This scanning electron micrograph of glass-reinforced polybutylene terephthalate shows near-spherical cells toward the centre of the foam sample, elongated cells near the surface, and a near absence of cells in the skin region. Note also that the glass fibres are oriented in the direction of flow (parallel to the surface) and that the glass does not act as a primary nucleating source for a bubble, nor do the fibres form randomly around a bubble, thereby reinforcing it.[13] Thus, we can assume that glass simply reinforces the resin. This assumption allows us to compare reinforced, filled and unfilled resin foams on a reduced basis, where the 'o' value is the

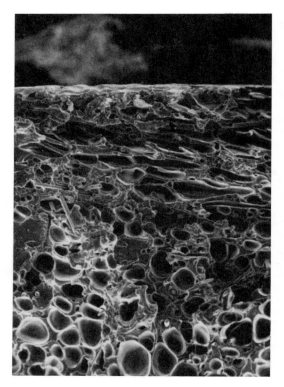

FIG. 7.3. Scanning micrograph of glass-reinforced polystyrene TSF near skin. Note flattened cellular structure in skin area, near-spherical bubbles in core. 70×.

appropriate value for the specific type of resin and reinforcement. This observation is supported by extensive property measurements on many types of TSF foams.[14]

Careful examination of the cellular structure in the skin region frequently shows a layer of extremely fine cells just behind the identifiable skin. Various theories have been proposed about their occurrence, with the most plausible one being that these bubbles are the result of arrested gas dissolution into the surrounding resin.[15]

Frequently, these bubbles appear to be 'markers' indicating the flow direction and flow of the wavefront during filling of the mould cavity. Figure 7.4 shows this phenomenon. In addition, note carefully the irregular nature of the surface of the foam. The folding over and entrap-

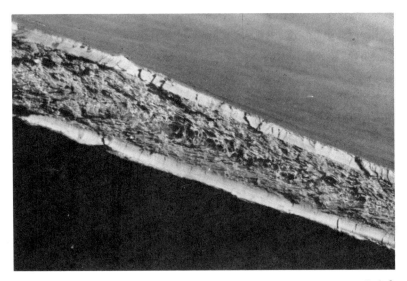

FIG. 7.4. Optical micrograph of GP polystyrene TSF with heavy well defined skin. Bubbles show left-to-right flow direction.

ment of gas at the foam surface will be discussed in detail below when impact strength is discussed.

It has been common practice to relate all foam properties to the values of the unfoamed resin, e.g. the 'o' values. This tacitly assumes that unfoamed resin containing gas has the same property as resin without gas. More important, it assumes that the skin, seen so prominently in Fig. 7.1, has the density of the unfoamed resin and therefore the same physical properties as those of the unfoamed resin. Careful examination of the surface regions in Figs. 7.1, 7.3 and 7.4 show an occasional cell. Thus, we might expect that the skin may not have the properties of the solid resin.

Thermoplastic structural foams are frequently considered to be stress free owing to their method of processing. Normally processing pressures (in the mould cavity) are quite low and the foam is allowed to expand at leisure until the cavity is filled. Cooling times are considerably longer than those for injection-moulded parts, thereby allowing some annealing and relief of internal residual stresses. This point will be discussed in detail, with illustrations to show that the formation of the cellular structure may induce changes in material morphology around the cell and may decrease the impact and fatigue resistance of the part owing to residual stresses around the individual cell.

7.3 UNIFORM-DENSITY CELL BEHAVIOUR

As will be shown below, TSF foams in the bending mode depend upon the Young's modulus as integrated across the beam cross-section. In order to provide a suitable value for the modulus as a function of foam density, it is necessary to investigate the relationship between Young's modulus and uniform-density foams. Uniform-density foams have been extensively studied and modelled for years.[8] Although most models are stick-and-ball or strut-and-membrane types and are applicable only to low-density foams, some can be extended to the higher TSF density foams. Most of the models that appear applicable to the high-density thermoplastic foams have, in fact, been developed for either elastomers or rigid polyurethane foams (low-density rigid foams are considered in Chapter 2A). In a recent review,[16] it was shown that models that were constructed around the concept of an inclusion or void in a solid resin yield the most consistent values (Fig. 7.5). One of the most important models, particularly in composite work, is that of Halpin and co-workers.[17,18] Young's modulus, E_c, of a two-component composite, where E_i is the modulus of the inclusion and E_0 is that of the continuous phase, is given by:

$$E_c/E_0 = (1 + AB\psi_i)/(1 - B\psi_i) \tag{7.1}$$

where ψ_i is the volume fraction of the inclusion, A is a loading factor, and B is given as:

$$B = (E_i/E_0 - 1)/(E_i/E_0 + A) \tag{7.2}$$

They further show that $A = 2$ for beams in a bending mode, and:

$$A = (7 - 5v_0)/(8 - 10v_0) \tag{7.3}$$

for beams in shear. Nielsen modified the Halpin–Tsai equation to include a maximum packing value.[19] However, for most TSF foams, it appears that such a modification yields only slight improvements in the prediction of the Young's modulus as a function of inclusion volume.

Probably one of the simplest relationships is the power-law model. It was first proposed by Egli[20] as a design equation for structural foam in the form:

$$E_f/E_0 = (\rho_f/\rho_0)^n \tag{7.4}$$

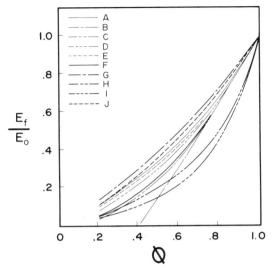

FIG. 7.5. Models for relationship between reduced TSF tensile modulus and reduced density, $\phi = \rho_f/\rho_0$. References given in original paper.[23]

A = Kerner–Rusch (13)
B = MacKenzie–Rusch (14)
C = Square relationship (21)
D = MacKenzie–Rusch II (16)
E = Baxter–Jones
F = Rusch (18)
G = Ogorkiewicz–Sayigh (20)
H = Okuno–Woodham (24)
I = Wilczynski
J = Nielsen (30)

where the subscript 'f' denotes the bulk-foam properties, with ρ being the material density. Subsequently, Throne[5] and Moore and co-workers[21,22] found that the appropriate value for n for uniform density foams was 2. This square-law model is seen to fit available uniform density TSF experimental data better than any theoretical model (Fig. 7.6). More important, it seems to fit available experimental data on uniform-density polyurethane and elastomer foams as well or better than theoretical models. From a pragmatic view, the square-law model is much more tractable in complex integrations (to follow), and thus throughout the rest of the discussion on the bending of TSF foams, we will consider the local or point value of the Young's modulus to be proportional to the square of the local or point foam density:

$$E_{local}/E_0 = (\rho_{local}/\rho_0)^2 \tag{7.5}$$

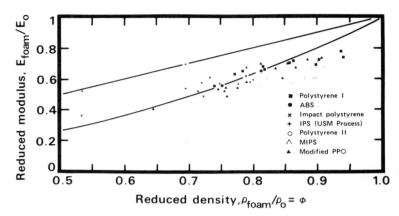

FIG. 7.6. Typical data on reduced flexural modulus of various styrenics as function of reduced density. Top curve is linear law; bottom curve is square law.

Please note that we have not resolved the appropriateness of using the 'o' value for either the Young's modulus or the density.

7.4 THE I-BEAM CONCEPT

In the early days of TSF design guides, great credence was given to the I-beam theory, designed specifically for laminated composites.[6] Specifically it was claimed that a TSF beam, in flexure, could be considered as a laminated composite, with the skin having one modulus and the foam core a lower modulus (Fig. 7.7). In this way, the relationship between load,

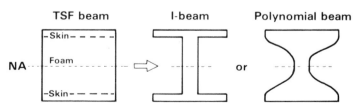

FIG. 7.7. Schematic of TSF beam in bending mode. Left, skin-core schematic. At centre, typical I-beam concept. Right, concept accounting for non-uniform density core.

P, and maximum deformation, δ_{max}, is given as:

$$\delta_{max} = (-PL^3/48EI) - (PL/4GbD) \tag{7.6}$$

where L is the span, E is the modulus of the skin, G is the shear modulus of the foam core, D is the panel cross-section, b is its width and I is the moment of inertia of the span, given in terms of the skin thickness, t, as:

$$I = bD^2t/2 \tag{7.7}$$

The basic equation assumes that the core supports only shear and therefore has no elastic modulus. This implies that at the skin–core interface, a shear plane exists. For sandwich panels, where the skins are high modulus solids such as aluminium or plywood and the core is a uniform density foam, this assumption is perhaps valid. Throne[5] showed that if the shear modulus is assumed to be a power-law function of foam density in the same way as Young's modulus:

$$G_f/G_0 = (\rho_f/\rho_0)^2 \tag{7.8}$$

then:

$$G_f/G_0 = (1+e)^2/[1/R' + e]^2 \tag{7.9}$$

Thus the maximum deformation under load becomes:

$$\delta_{max} = \frac{-PL^3}{48ED^3}\left(\frac{4}{e} + \frac{D^2}{L^2}\frac{12E}{G_0}\frac{[1/R' + e]^2}{(1+e)^2}\right) \tag{7.10}$$

He then proposed that for a single component TSF foam, the I-beam theory predicted deflections under load that were orders of magnitude greater than those measured experimentally. The casting out of the shear layer concept has apparently been successful. The problem now focuses on proper selection for a model to replace the I-beam concept. Several of these models will be discussed below. However, one bears mention at this time, since it is a form of the I-beam model in which the centre of the I-beam (the web) has been replaced by a beam having bending strength substantially less than that of the skin.[22] In this model, the moment of inertia given by eqn (7.7) has been replaced by the moment of inertia of a simple bean in bending:

$$I = bD^3/12 \tag{7.11}$$

The final beam stiffness is given quite simply as:

$$(EI)_f = S_f = (E_0 + E_c)I/2\varepsilon \tag{7.12}$$

where S_f is the stiffness of the beam and ε is the expansion ratio of the moulded part, which according to our definitions is (ρ_0/ρ_f). This simple equation unfortunately does not establish relationship between the foam density and the Young's modulus. If the square law is observed, the equation would become:

$$(EI)_f = S_f = E_0 I (1 + R'^2)\phi/2 \qquad (7.13)$$

The relationship between R' and ϕ must be obtained by measurement of the foam-core density and the skin thickness. Nevertheless, the Iremonger–Moore data correlate well with their equation. A weakness in their analysis will be pointed out shortly.

7.5 THE CORE-DENSITY PROFILE AND THE ROLE OF THE SKIN

7.5.1 Core-Density Profile

Early in TSF experiments, it became obvious that the process itself inhibited an abrupt skin–core interface of the type discussed above. In fact careful visual observation showed that there was a gradual change in bubble structure from the skin region to the core. This was discussed above when the cell morphology was considered. The first experiments on density as a function of distance from the surface were carried out by Throne on self-skinning thermoset polyester form.[5] The data showed a nearly monotonic decrease in density from the skin to the centre of the sample. He proposed that the density gradient at both the skin and the core approach zero:

$$d\rho/dy = 0 \quad \text{at } y = 0 \quad \text{(the surface)} \qquad (7.14a)$$
$$d\rho/dy = 0 \quad \text{at } y = Y \quad \text{(the centreline)} \qquad (7.14b)$$

In order to use a polynomial to describe the density across the sample:

$$\rho = a + by + cy^2 + dy^3 \ldots \qquad (7.15)$$

he set all other derivatives to zero at the surface and the centreline. This yielded a polynomial of the form:

$$\rho_f/\rho_o = R' + (1 - R')[(C + 1)Z^C - CZ^{C+1}] \qquad (7.16)$$

where C is the only polynomial coefficient to be determined and $Z = y/Y$.

The remarkable fact about this profile is that it offers great flexibility in shaping the density profile across the TSF foam sample and yet is sufficiently tractable to allow integration for beam stiffness.

Recently, Progelhof and Eilers[23] have examined dozens of structural foams using soft X-ray (Fig. 7.8). They found that they can fit the profile

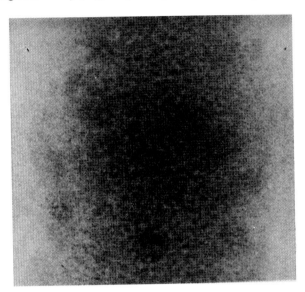

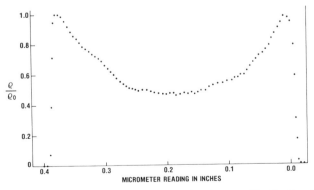

FIG. 7.8. Soft X-ray (top) and measured density profile (bottom) for TSF polystyrene foamed with chemical foaming agent.

to nearly every case. Progelhof has corrected eqn (7.16) to allow for a skin of finite thickness, yielding:

$$\rho_f/\rho_0 = R' + (1 + R') \left[\frac{(C+1)Z^C}{(1-e)^C} - \frac{CZ^{C+1}}{(1-e)^{C+1}} \right] \quad (1-e) > Z > 0$$

(7.17a)

$$\rho_f/\rho_0 = 1 \qquad\qquad\qquad\qquad 1 > Z > (1-e)$$

(7.17b)

The application of this density profile to the stiffness in bending of a TSF beam is the subject of a section following. However, note that both eqns (7.16) and (7.17) reduce to earlier forms when limits are taken. For example, when $e = 0$, eqn (7.17) reduces to eqn (7.16). When $C = 0$, there is no foam, and it can be shown, using L'Hopital's rule, that eqn (7.16) reduces to an identity, $\rho_f/\rho_0 = 1$. When $C = \infty$, the foam is of uniform density. Again using L'Hopital's rule, eqn (7.16) reduces to an identity, $\rho_f/\rho_0 = R'$. The accuracy of this polynomial density profile in predicting the bending strengths of TSF beams, using the ASTM D790 three-point bending method, is shown in Table 7.1, from Progelhof and Eilers,[23] for several foams.

7.5.2 The Skin

For a single component TSF, the skin or region of high density is apparently created by a combination of redissolution of foaming gas and inhibition of cellular growth during bulk expansion. Aiding these phenomena are the internal gas pressure pressing the densifying material against the mould surface for at least part of the cooling time and the contact with the cold mould itself. In order to determine the influence of the skin on the overall mechanical strength of TSF, it is necessary to determine the physical properties of unfoamed resin containing suitable foaming agents. It is well documented that certain chemical foaming agents are not compatible with certain resins. The most extreme case of resin-foaming agent compatibility is 1,1'- bisformamide (AZ) and polycarbonate. One of the foaming-agent residues is ammonia (NH_3), which reduces the otherwise tough resin to an extremely brittle consistency.[24] Other combinations are shown in Table 7.2. Recall now that these represent the relative effect of the foaming agent or its residue on the mechanical properties of unfoamed resins. As with any

TABLE 7.1
Comparison of Calculated Integrated Moduli with Experimental Data[23]

Material/Process	Density parameters[a]			Flexure		Tension/compression		Transverse compression	
	e	R'	C	Theory	Exp.	Theory	Exp.	Theory	Exp.
1. Single-component styrene[b] —chemical blowing agent	0	·48	5	·64	·75	·43	·41		
2. Single-component styrene —fluorocarbon blowing agent	·2	·44	3	·86	·89	·58	·46		
3. Two-component high-density polyethylene	·35	·38	5	·85	·85	·58	·53	·30	·21
4. Two-component polypropylene—expanding mould	·10	·45	∞	·42	·46	·28	·28	·19	·15
5. Single-component styrene[c] —chemical blowing agent	0	·48	5	·64	·40	·43	·25		

[a] e, R', C are skin thickness, core-to-skin density ratio, and polynomial shape factor, respectively. Experimental fits to soft X-ray data profiles.
[b] Apparent skin modulus equal to $2·5 \times 10^5$ psi.
[c] Skin moduli obtained from supplier literature.

TABLE 7.2
Effect of Chemical Blowing Agents on Unfoamed Properties of Several Resins (95% CL)[26]

	Amoco R-3 GPS	Amoco H-3 HIPS	Tenite 6P4DF PBT	Merlon PC
		Elongation, yield, %		
Control	None	1.24 (0.05)	3.40 (0.04)	5.65 (0.10)
+ 1% AZ	None	1.23 (0.03)	None	[e]
+ 1% 5PT	DNT	DNT	3.51 (0.12)	5.58 (0.55)
+ 1% Talc	None	1.11 (0.04)	3.35 (0.10)	5.21 (2.4)
		Elongation, ultimate, %		
Control	1.88 (0.15)	37.9 (3.5)	< 360[c]	93.8 (24.3)
+ 1% AZ	1.76 (0.18)	36.9 (7.6)	2.27 (0.96)[a]	[a]
+ 1% 5PT	DNT	DNT	> 92[a,b]	12.6 (2.5)[a]
+ 1% Talc	1.63 (0.34)	29.9 (3.9)	> 360[c]	36.3 (23.1)
		Tensile strength, yield, lb in.$^{-2}$		
Control	None	2898 (17)	6961 (30)	8078 (127)
+ 1% AZ	None	2597 (18)	None	[e]
+ 1% 5PT	DNT	DNT	6502 (143)	8688 (76)
+ 1% Talc	None	2884 (36)	6762 (120)	8480 (50)
		Tensile strength, ultimate, lb in.$^{-2}$		
Control	6582 (61)	2878 (91)	[c]	9013 (1166)
+ 1% AZ	5786 (73)[a]	2474 (55)[a]	5440 (1260)[a]	[e]
+ 1% 5PT	DNT	DNT	4557 (110)[a,b]	6862 (95)[a]
+ 1% Talc (3)	6471 (85)	2972 (48)	[c]	7690 (660)

	Tensile modulus, tangent, × 10^{-5}, lb in.$^{-2}$			
Control	4·43 (0·28)	3·12 (0·05)	3·43 (0·10)	3·33 (0·13)
+1% AZ	4·15 (0·10)[a]	2·59 (0·17)[a]	3·29 (0·10)	[e]
+1% 5PT	DNT	DNT	3·29 (0·07)	3·34 (0·21)
+1% Talc	4·55 (0·21)	3·08 (0·08)	3·30 (0·10)	3·38 (0·10)

	Flexural strength, yield, lb in.$^{-2}$[f]			
Control	12 991 (292)	5948 (63)	11 976 (56)	14 387 (167)
+1% AZ	11 681 (101)[a]	5577 (73)[a]	11 584 (76)	[e]
+1% 5PT	DNT	DNT	11 235 (171)	14 681[e] (116)
+1% Talc	13 913 (67)	6000 (36)	11 897 (102)	14 058 (91)

	Flexural modulus, tangent, × 10^{-5}, lb in.$^{-2}$			
Control	4·91 (0·05)	3·04 (0·14)	3·53 (0·02)	3·45 (0·10)
+1% AZ	4·69 (0·04)[a]	2·86 (0·01)[a]	3·43 (0·02)	[e]
+1% 5PT	DNT	DNT	3·37 (0·03)	3·71 (0·07)
+1% Talc	5·11 (0·08)	3·07 (0·05)	3·52 (0·01)	3·58 (0·14)

DNT = did not test.
[a] Significant deviation from control.
[b] 3 of 5 samples did not break at 360% elongation.
[c] None of 5 samples broke at 360% elongation.
[d] Except 0·5% talc for Tenite PBT.
[e] No whole samples were obtained.
[f] Flexural strength at break for GPS.

experiment, one must ensure that the resin is not degrading simply because it is being held too long at excessive temperatures. As a result, the resins chosen are typical of those that might be processed into TSF products. In order to determine if the chemical foaming agents acted as aggressive agents rather than simply as diluents, data were obtained using 1% (wt) talc as an inert diluent. The exception is with polybutylene terephthalate, where it has been shown that 1% concentrations of talc dramatically change the crystallisation rate of the resin, resulting in embrittlement.[25] For PBT, then, 0·5% talc was used. Flexural and tensile properties are given in Table 7.2, with those that appear to have been affected by the addition of the chemical foaming agent being denoted by[a].

For general-purpose polystyrene (GPS), there seems to be some decrease in ultimate tensile strength, flexural strength and their respective moduli with 1% (wt) Celogen AZ130 that cannot be explained by dilution. For rubber-modified polystyrene, the results are similar, indicating that the rubber is apparently not selectively attacked by the chemical foaming agent, its gas, or its residuals. 5-Phenyltetrazole (5PT) is a higher-temperature foaming agent, and thus could not be evaluated with the polystyrenes.

However, both AZ and 5PT apparently do affect the unfoamed properties of PBT. Note the absence of a yield point and a dramatic drop in the elongation at break, even though the measured flexural strength and the corresponding moduli are essentially unchanged from the virgin resin values. The stress–strain behaviour of the unfoamed resin (and thus the skin of a TSF part) may behave as predicted at low strains but there is evidence that at higher elongations, there may be a chemically induced shift in the stress–strain curve. Schematically, this shift might appear as a 'knee' as shown in Fig. 7.9.

As mentioned, polycarbonate cannot be moulded successfully with a chemical foaming agent that generates even minute amounts of ammonia or biurea. However, 5PT is the recommended foaming agent for the material. As seen in Table 7.2, however, the ultimate properties of the unfoamed resin are also affected by the presence of the foaming agent in much the same way as with PBT. Thus Fig. 7.9 might also apply for the PC/5PT combination.

It should also be noted that many high-performance resins that are candidates for TSF moulding are highly moisture sensitve. The list might include nylons (6 and 66), PBT, the new polyethylene terephthalate (PET) moulding resin and PC. Thus, not only is it necessary to ensure that the resins are thoroughly dried, but that the selected foaming agent does not

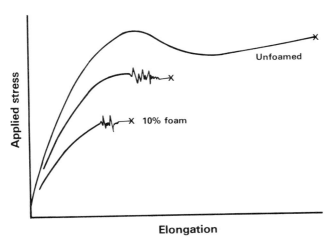

FIG. 7.9. Schematic of stress–strain behaviour of TSF at relatively low foaming levels, showing expected ductile–brittle transition.

liberate water as a by-product. The primary effect will be loss in ultimate properties in much the same way as shown in Fig. 7.9.

Perhaps it is necessary to emphasise the significance of the loss in ultimate properties (ultimate tensile strength, elongation at break, and so on) on the performance of the TSF moulded part. It would appear that nearly all the testing procedures for impact strength focus on the part performance at or near the ultimate strain level of the material. Reduction in ultimate properties would indicate that the parts, while performing satisfactorily at low strain levels (such as in static load bearing), might fail prematurely at higher levels. This topic is discussed in some detail below when impact and fatigue performances are considered.

Returning now to Table 7.1, of Progelhof and Eilers,[23] note that it was necessary to reduce the 'o' modulus for GPS and a chemical foaming agent in order to get agreement between theory and experiment. We have seen in Table 7.2 that the unfoamed properties are reduced by adding chemical foaming agent. What happens at very low foaming levels (of the order of 5% or so)? Progelhof and Eilers examined the modulus of PBT uniform density foam in this region and found that it behaved in accordance with the square law. If, in fact, the skin of a TSF foam had, say, only 95% of the density of the packed out solid resin, the modulus would be reduced by 10%.

This may help explain why the curves arbitrarily drawn through early experimental data[14] did not always extrapolate to the $E_f/E_0 = 1$, $\phi = 1$ point. The 40% reduction for GPS found by Progelhof and Eilers[23] can partially be explained by this fact, but one must also consider the very real plasticising capability of gas in a brittle polymer such as GPS. The extrapolation to the corner is referred to below as the '1:1' point.

Recent careful examination of TSF high-density polyethylene (HDPE) foam bars having foaming levels in the range of about 1% to 10% have shown that the presence of the bubbles, regardless of their level, dramatically affects the mechanical properties of the foam. In Fig. 7.10 the tensile modulus is presented as a function of reduced density, ϕ. Note that

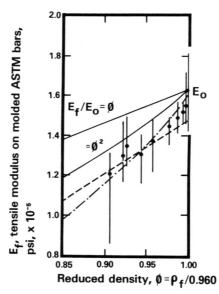

FIG. 7.10. Experimental data for tensile modulus for HDPE at near-solid foaming level. Dashed line, least-squares fit without solid resin modulus. Dash–dot line, least-squares fit with solid resin modulus.[24]

in this region, the data are considerably below the $E_f/E_0 = \phi^2$ curve. Similar results are shown for HDPE tensile strength at yield, as shown in Fig. 7.11. However, as seen in Fig. 7.12, this deviation is not as apparent when compared with older experimental data at higher foaming levels. Again, the reason for considering the load-bearing characteristics of un-

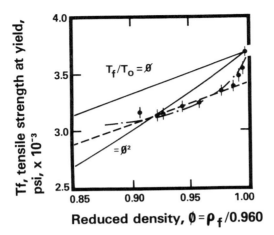

FIG. 7.11. Experimental data for TSF tensile strength at yield for HDPE at near-solid foaming level. Dashed line, least-squares fit through data without solid value. Dash–dot line, least-squares fit through data with solid value included.[24]

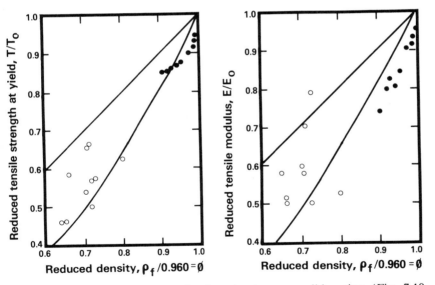

FIG. 7.12. Comparison of tensile data in the near-solid region (Fig. 7.10 and 7.11) with normal foaming level data for Amoco 650 HDPE. Solid circles: Ref. 24, chemical foaming agent nitrogen. Open circles: Ref. 63, physical foaming agent nitrogen.

foamed and near-solid resins is that the predictive models depend to a great degree upon the physical properties of the skin, e.g. the 1:1 point

Later, discussion will return to this point in an effort to understand the overall stress–strain behaviour and the role of the bubbles in the stress–strain behaviour of an integral-skin TSF foam.

7.5.3 Variable Characteristics in Foam Densities

In the early days of TSF foam testing, flexural and tensile bars were fabricated in various ways. As an example, Amoco Chemicals Corporation developed a test mould that yielded moulded rectangular and circular tensile and flex bars of various dimensions. Examples of these are shown in Fig. 7.13. Note in Fig. 7.14 that these bars had skins on all exterior surfaces. As a result, in the bending mode, for example, the side walls would add additional support to the beam. Similarly, the mechanical strengths of round beams could not be analysed satisfactorily using the conventional stiffness equations for rectangular beams. The data showed a significant amount of scatter and led to a more controlled programme

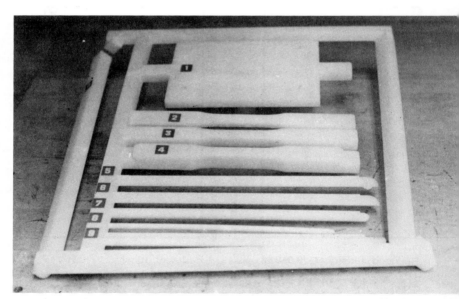

FIG.7.13 Old type of tensile-flex-impact ASTM test mould for structural foam. Preferred method now is to cut specimens from flat plaques.

STRUCTURAL FOAMS 285

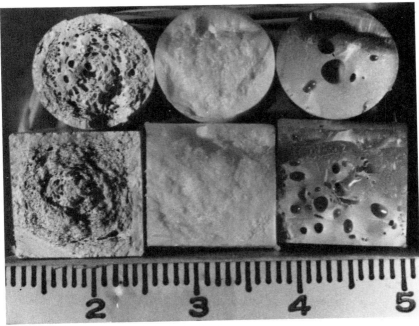

FIG. 7.14. Effect of foaming agent type on moulded flex bars of ABS. All bars foamed to $\phi = 0.75$.

for reporting testing results.[14] Another factor which has only recently been discussed[26] is that of variable properties along the length of a test beam. Recall that the majority of test results are given in terms of the bulk density of the beam. The standard test procedures have been simply extensions of test procedures for unfoamed injection-moulded specimens. It is well known that by its very nature, the TSF foam process does not yield uniform-density parts. It is usually the case that edges and thicker cross-sections have somewhat lower densities than thinner sections or areas near gates.[24] A simple example of this is seen in Fig. 7.15 for the foam moulding of a 40% glass reinforced nylon 6 baseball bat. The parameters are the speed of injection and the final weight of the bat. In nearly all cases, the thickest section of the bat had a lower measured density than the thinnest section or the runner. Although the arithmetic can be altered to consider end-to-end variation in density in the beam under flexure, in tension the beam will nearly always fail at the point of lowest density (at a constant cross-section). On occasion, this occurs at

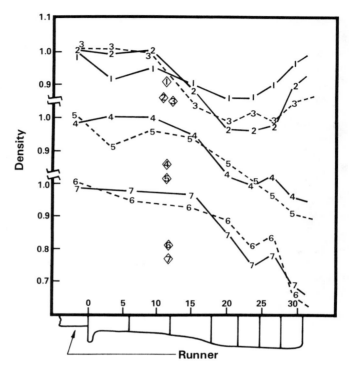

FIG. 7.15. Density profiles for TSF baseball bat. Solid resin density, 1·42 g cm^{-3}. Solid line, slow injection speed. Dashed line, high injection speed. Curves 1–7 represent effect of density reduction on material distribution along the bat length. Values in diamonds represent average bat density.

the transition to the shoulder region of the tensile bar. Furthermore, test specimens should be X-rayed before being tested and failed bars should be examined at the point of failure to ensure that the test specimen has not failed at a large void. Undoubtedly, some of the scatter in early data can be attributed to this type of failure.

Assume for the moment that the bar is of uniform bulk density from one end to the other. Can one be confident that the mechanical properties measured from testing this part are representative of the performance of the TSF foam? Apparently not. A recent examination of the interrelationship between the skin thickness, foam-core density, and the density profile across the foam core (assuming constant bulk density) indicates that under certain conditions thick skin and essentially zero foam-core density will yield parts having higher moduli than uniform density foams of the same bulk density.[27] This returns us to the discussion of the 'I-beam' concept. We consider this aspect of TSF foams shortly.

7.6 MODELS AND MECHANICAL PROPERTIES

7.6.1 Flexural Bending of Beams

As mentioned, the earliest models describing the bending of a TSF beam related the reduction in modulus to the bulk density of the foam (eqn 7.4). The inappropriateness of this equation and other models that relate the modulus of the foam to core density, polynomial shape factor, skin thickness and other aspects of the TSF foam is an *a priori* assumption that the modulus can be extracted from the stiffness equation.[23,26] The appropriate beam equation is given as:

$$M = \frac{PL}{4} - \frac{Px}{2} \tag{7.18}$$

where again P is the applied load, assumed concentrated at the centre of the weightless beam, L is the span, and M is the bending moment at any point x along the beam from the point of support. The appropriate form for M is:[28]

$$M = S(x)\frac{d^2\delta}{dx^2} = \frac{PL}{4} - \frac{Px}{2} \tag{7.19}$$

Now the beam stiffness, $S(x)$, is defined as:

$$S = 2\int_0^Y bEy^2 \, dy \tag{7.20}$$

where b is the beam width. Now assume that the modulus of the beam is a function of both spatial coordinates, x and y, e.g. $E = E(x, y)$. Further, assume that the modulus everywhere is proportional to the square of the foam density, e.g. $E = E_0(\rho_f/\rho_0)^2 = E_0\phi^2$. Then eqn (7.20) can be written as:

$$S(x) = \frac{bh^3}{4}E_0 \int_0^1 \phi^2(Z, x)Z^2 \, dZ = \frac{bh^3}{4}E_0 I(x) \tag{7.21}$$

where $h = Y$ and $I(x)$ is the integral (not the moment of inertia). This expression for stiffness is now substituted into eqn (7.19) and integrated for deflection as a function of load and span, subject to the conditions that $x = 0$, $d\delta/dx = 0$ and $\delta = 0$. The maximum deflection, δ_{max}, occurs directly under the concentrated load point, if and only if the beam is symmetric in any and all ways about this point. Assuming that to be the case:

$$\delta_{max} = \frac{PL^3}{48EI} = \frac{PL^3}{4bh^3}E_0 F(\phi_e) \tag{7.22}$$

where $F(\phi_e)$ is given by:

$$F(\phi_e) = \left\{ \frac{L^3}{4} \left[\int_0^{L/2} \int_0^x \frac{L\,dx}{I(x)}\,dx - 2\int_0^{L/2}\int_0^x \frac{x\,dx}{I(x)}\,dx \right]^{-1} \right\} \tag{7.23}$$

Equations (7.22) and (7.23) then define the degree of deformation of a TSF foam beam with symmetric properties about the centre concentrated load, assuming three-point bending. Several specific cases are worth considering here. If, for example, the foam has uniform properties everywhere, including uniform core density and no skin ($e = 0$), the stiffness eqn (7.21) becomes:

$$S_f = S_0 \phi^2 \tag{7.24}$$

Thus the maximum TSF foam beam deflection can be simply written as:

$$\delta_{max,f} = \delta_{max,0}/\phi^2 \tag{7.25}$$

If the core density is uniform but the foam has a finite skin thickness ($C = \infty$, $e \neq 0$), eqn (7.21) becomes:

$$S_f = S_0 \{(1-e)^3 R'^2 + [1-(1-e)^3]\} \tag{7.26}$$

The maximum TSF foam beam deflection will appear similar to that in eqn (7.25) with the term ϕ^2 replaced by the term in braces in eqn (7.26). A more complex but still tractable case is when the beam has uniform properties end-to-end, but where the density across the beam everywhere is given by eqn (7.17). Now eqn (7.21) becomes:

$$S_f = S_0 \left\{ 3 \left< (1-e)^3 \left\{ \frac{R'^2}{3} + 2R'(1-R')\left[\frac{(C+1)}{(C+3)} - \frac{C}{(C+4)}\right] \right. \right.\right.$$
$$\left.\left. + (1-R')^2 + \left[\frac{(C+1)^2}{(2C+3)} - \frac{2C(C+1)}{(2C+4)} + \frac{C^2}{(2C+5)}\right]\right]$$
$$\left.\left. + (1/3)[1-(1-e)^3] \right> \right\} \tag{7.27}$$

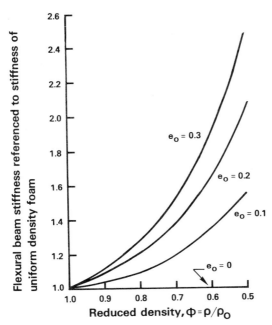

FIG. 7.16. Flexural bending of a typical I-beam. Effect of uniform skin thickness on relative stiffness of beam in flexure. All curves related to the bending of a beam of uniform density, where $S/S_0 = \phi^2$.

Again, the maximum TSF beam deflection is given by eqn (7.25) with the term ϕ^2 replaced by the term in braces in eqn (7.27).

In order to appreciate the importance of skin thickness and density profile on the stiffness of a beam in flexure, consider first Fig. 7.16. The effect of a uniform skin thickness on the relative stiffness of a foam beam is shown. The curves shown represent the ratio of eqns (7.26) and (7.24), where R' has been determined from:

$$R' = (\phi - e)/(1 - e) \qquad (7.28)$$

ϕ represents here the bulk density of the foam beam. It is apparent here why the concept of the 'I-beam' was initially proposed. In fact, it can be shown that for this model (where $C = \infty$), the sandwich model as given in eqn (7.10; where G_0 is replaced by E_0) yields identical results. Note, however, that significant deviations from the uniform density stiffness occur for rather high foaming levels (25% or more) and substantial skins (20% of the total foam thickness or more).

Another way of looking at the role of the skin is shown in Fig. 7.17, where for uniformly thick skin, uniform-density core stiffness is compared with the linear law of mixtures and the square law (assumed correct for skinless foams, where $e_0 = 0$). It is apparent that a considerable increase in stiffness occurs with rather thick skins on TSF foams.

What happens to stiffness when the polynomial density profile is incorporated into the model? This assumes that we are no longer dealing with an 'I-beam'. Shown in Fig. 7.18 is the stiffness of the polynomial beam referenced to the 'I-beam' model. In essence, the curves represent the ratio of eqns (7.27) and (7.26). Note that increased skin thickness leads

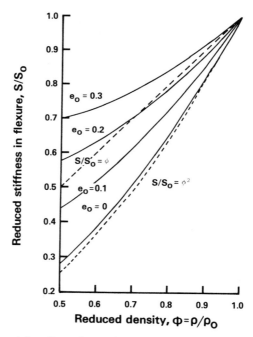

FIG. 7.17. Flexural bending of a typical I-beam. Effect of uniform skin thickness and uniform density core on beam stiffness in flexure. e_0 is fraction of cross-section that is skin. Dashed line is law of mixing, dotted line is lower limit square law.

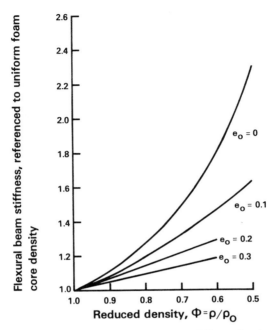

FIG. 7.18. Flexural bending of a typical I-beam. Effect of polynomial density profile on flexural beam stiffness. Reference is uniform density in foam core. Polynomial shape factor, $C = 3$. Uniform foam core shape factor, $C = 103$.

to a reduction in the apparent stiffness advantage of the polynomial or variable foam density profile. Undoubtedly part of this is because the core density, R', is rapidly approaching zero as the skin thickness increases. R' is given by:

$$R' = \frac{(2 + C)}{C} \frac{(\phi - e)}{(1 - e)} - \frac{2}{C} \quad (7.29)$$

for finite values of C.

Consider now the more complex but more practical case where the skin thickness, foam core density, and polynomial profile are not constant throughout the length of the beam.[26,27] Schematically, the foam beam may appear as shown in Fig. 7.19. Assume that the characteristics of the beam are symmetrical about the loading point, and that every-

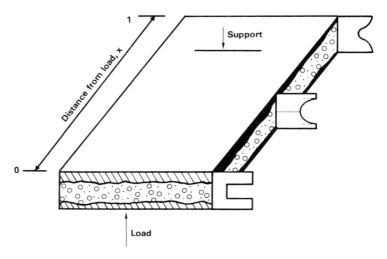

FIG. 7.19. Hypothetical variable skin-thickness and core-density profile for TSF beam in flexure.[26]

where along the beam, the foam has a constant density, ϕ. Now let $e = e_0 f(x)$ and $C = C_0 g(x)$. As an extreme example, we have considered the following conditions:

$$e = e_0(1 - x)^n \tag{7.30a}$$

$$C = C_1 + C_0 \exp(-Bx) \tag{7.30b}$$

where $C_1 = 3$ (considered to be the most gentle gradient expected in foams) and $C_0 = 100$ (assumed to yield conditions approaching $C = \infty$). For purposes of this example, $n = 0$ and $n = 1$ have been assumed. The latter yields a linear skin of zero thickness at the point of loading. The former is the uniform density skin. For 4 values of the base skin thickness, e_0, typical cases are shown in Fig. 7.20. The shaded area represents our best estimate of the accuracy of measuring stiffness of foam beams. The results, as before, are compared with the stiffness of a uniform skin, uniform core density beam, as given by eqn (7.26). For all cases, the value of R' was allowed to float, and was determined by eqn (7.29). It is apparent that for typical foaming levels of TSF foams the variabilities in skin thickness and density gradients in the foam core are considerably less significant to the stiffness of a TSF foam than is skin thickness. Again, this points

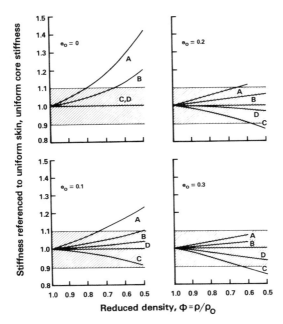

FIG. 7.20. Flexural bending of typical I-beam. Effect of position-dependent skin thickness, core profile on stiffness. Comparison with uniform skin. A: uniform skin, variable core $B = 5$; B: uniform skin variable core $B = 10$; C: linear skin, uniform core; D: linear skin, variable core $B = 5$.

up the importance of identifying any problems that might cause the strength of the skin to be less than that predicted by extrapolation to the 1:1 point.

The reader should be cautioned at this point that the models proposed above have been developed primarily as ways of explaining test results obtained for typical ASTM three-point bending of TSF beams. It is not apparent that the present single-component TSF foam parts can be fabricated in a manner that will yield a given polynomial profile or skin thickness. Thus, the square law, determined to be a satisfactory way of determining stiffness on skinless, uniform density foams, apparently represents a safe lower limit to TSF foams, purportedly with skins and polynomial density profiles, but subject to the inaccuracies of the processing methods.[14]

7.6.2 Tension and Compression

Models for other modes of stress–strain behaviour for TSF foams are considerably more rudimentary than those for the bending mode. In tension or compression one normally assumes that deformation of the entire section perpendicular to the direction of applied forces is uniform. Thus:

$$F = \int_A \sigma \, dA \tag{7.31}$$

where A is the cross-sectional area perpendicular to the application of force and σ is the local stress. According to Progelhof and Eilers,[23] there should be a Hookean relationship between the local stress and the deformation or strain:

$$\sigma = E_t \varepsilon \tag{7.32}$$

where ε is the local strain. Following their arguments for the moment, we can then write:

$$E'/E_0 = \int_A E_t \, dA/(AE_0) \tag{7.33}$$

where E_0 is the modulus of the unfoamed resin, and E' is an 'apparent zero strain tension–compression modulus'. Equation (7.33) is actually a definition of this modulus. The evaluation of eqn (7.33) for various geometries is rather straightforward, with the replacement of dA by the appropriate geometric factor and the replacement of E_t with the appropriate relationship to the local foam density. For example, if the beam is rectangular, $E_t = E_0 \phi^2$ is used locally, and the relationship between ϕ and y is given by eqns (7.17a and b). Progelhof and Eilers have worked out a similar relationship for a circular beam, as one might encounter in a chair leg under compression. The starting polynomial remains the same, except that $Z = y/Y$ should be replaced with $Z = r/R$, where R is the radius of the beam and r is the distance from the centre. Using their concept of apparent modulus, they find that the modulus is almost totally dependent upon the bulk foam density, with skin thickness and polynomial foam density profile of little importance. They show for example that for a typical tensile bar, for $\phi = 0.6$, $E'/E_0 = 0.40$ for $e = 0.2$, $C = \infty$, and $E'/E_0 = 0.41$ for $e = 0$, $C = 3$, these values representing the extremes in profile shape factors and skin thicknesses.

Although the Progelhof–Eilers approach does teach us about the effect of foam density on tensile modulus, it is perhaps not entirely correct in its basic assumption of local Hookean behaviour at zero strain. As we have seen earlier, there may be a knee in the stress–strain curve for foams in tension. The Progelhof–Eilers model does not allow for the occurrence of a non-Hookean behaviour in the foam. It is well documented that for low-density foams, the cell structure is characterised by a high degree of residual stresses and that additional strain results in early cell-wall rupture. An example of the elongation in the cells near the skin in a typical tensile failure of HDPE is shown in Fig. 7.21. Here the

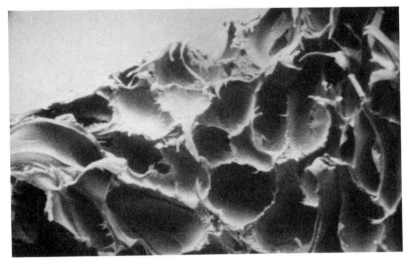

FIG. 7.21. Scanning micrograph of tensile-fractured polyolefin foam. Skin at lower left and upper right. Note difference in elongation in foam region. Cut tensile bar specimen. 60×.

skin on the right shows a characteristic high elongation at break, whereas the cell walls show rather early rupture. Recall now that the foam is anisotropic although the applied force is uniform. Thus an alternate approach to the Progelhof–Eilers model might be to return to the basic force-deformation eqn (7.31). Assume that we can separate the forces applied to the skin from those to the foam core. Assume further that we are dealing with a rectangular beam having skins on only two surfaces, and that the density of the foam core is uniform. Then the force balances for the skin and foam core would be:

$$F_s = 2bh \int_{1-e}^{1} \sigma \, dZ \qquad (7.34a)$$

$$F_c = 2bh \int_{0}^{1-e} \sigma \, dZ \qquad (7.34b)$$

where $Z = y/h$ and b is the width of the slab.

Obviously in the Hookean region, we can use a modified form of eqn (7.32) for the stress, but we must be careful here that we understand the role of the deformation or elongation to break, particularly with regard to the foam core.

$$F_s = 2bh \int_{1-e}^{1} E_0 \varepsilon \, dZ \qquad (7.35a)$$

$$F_c = 2bh \int_{0}^{1-e} E_f \varepsilon' \, dZ \qquad (7.35b)$$

The equation for the solid skin becomes:

$$F_s = 2bh E_0 e \varepsilon \qquad (7.36a)$$

Assuming the square-law relationship for the foam, and assuming that the density is uniform, we can write eqn (7.35b) as:

$$F_c = 2bh E_0 \phi^2 (1-e) \varepsilon' \qquad (7.36b)$$

Now the total force applied on the foam is the sum of F_s and F_c, which results in:

$$F = 2bh E_0 [\phi^2 (1-e) \varepsilon' + e\varepsilon] \qquad (7.37)$$

The measured deformation of the foam and the core must be identical at all times. Suppose for the moment that Hooke's law holds only to the point where $\varepsilon = \varepsilon_{max}$. If the foam also obeyed Hooke's law to the same strain level, $\varepsilon' = \varepsilon'_{max}$, and we should be able to recover the Progelhof–Eiler stress–strain curve. Suppose now that the foam core has some residual strain, such that it reached its maximum Hookean deformation at a much lower strain level, e.g. $\varepsilon' = \varepsilon'_{max} - \varepsilon_0$. For illustrative purposes, assume further that beyond this strain level, the foam no longer participated in load bearing, e.g. $F_c = 0$ beyond ε'_{max}. Examine for the moment the consequences. At maximum strain level in the foam, the stress–strain eqn (7.37) would appear as:

$$F = 2bh E_0 [\phi^2 (1-e) \varepsilon'_{max} + e\varepsilon] \qquad (7.38)$$

Beyond this point, the equation would become:

$$F = 2bhE_0 e\varepsilon \qquad (7.39)$$

until the skin material reached its maximum Hookean deformation. But note that for typical TSF structural foams, ϕ may be in the range of 0·75 and e may be as large as 0·2. This means that at the point of maximum Hookean strain in the foam (eqn (7.38)) typically 70% of the stress is being applied to material that is at maximum strain. Exceeding this, the constant applied stress rapidly elongates the skin material to failure. Certainly other models can be proposed which use more suitable constitutive equations of state for stress–strain behaviour, but this simplistic model seems to indicate that premature tensile failure may be caused by the nature of the prestressed state in the foam core. As the residual strain increases, the elongation to yield or break decreases at constant applied stress.

Incidentally, this implies that a highly instrumented tensile machine, in which the applied stress can be rapidly altered, might be able to determine the order of magnitude of residual strain in the cellular core.

7.6.3 Cantilever Bending

At times, the TSF foam part may be used in a cantilever bending way. Typical examples might be athletic equipment such as baseball bats, hockey sticks, tennis rackets, table tennis paddles and canoe paddles, and in utilitarian products such as axe handles. Normally in cantilever bending of uniform density and dimensioned beams, the maximum fibre stress occurs at the clamp, with the distortion of the rest of the beam being linear from the neutral axis. For irregularly shaped objects, however, the relationship between load and deflection is more complex. In one example, the static loading of a baseball bat, clamped at the handle end and loaded at the point of impact, the following approach was used. The moment of inertia of the cylindrical cross-section is given as:

$$I = \pi r^4/4 \qquad (7.40)$$

where:

$$r = a + bx + cx^2 + dx^3 \qquad (7.41)$$

Furthermore, the density of the bat as a function of distance down its length, x, was given by the polynomial:

$$\rho_f = A + Bx + Cx^2 + Dx^3 \qquad (7.42)$$

The coefficients for the density profile were determined from sectioning and weighing bats of several average densities. Again the basic elastic bending equation is given as:

$$EI\,d^2\delta/dx^2 = PL \qquad (7.43)$$

where P is the load concentrated at point L from the grip. The integration of this equation can be carried out using conventional techniques or the concept of finite elements.[29] Here the beam is segmented into discrete elements and the bending moment eqn (7.43) is applied to each element. The objective is to approximate the displacement in each element, knowing the moment of inertia and modulus of elasticity in that element. Thus the total deflection is given as the solution of a variation function:

$$\Pi = \int_0^N [\,1/2\ EI(x)(d\delta^2/dx^2)^2 - P\cdot\delta\,]\,dx \qquad (7.44)$$

Here the displacement for each element is approximated by:

$$\delta = ax^3 + bx^2 + cx + d \qquad (7.45)$$

Now:

$$\delta(x) = \begin{bmatrix} 2x^3 - 3x^2 - 1 & x^3 - 2x^2 + x & -2x^3 + 3x^2 & x^3 - x^2 \end{bmatrix} \cdot \begin{bmatrix} \delta_1 \\ \theta_1 \\ \delta_2 \\ \theta_2 \end{bmatrix} \qquad (7.46)$$

or in tensor notation:

$$\delta(x) = \mathbf{V}^t \cdot \mathbf{W} \qquad (7.47)$$

Incidentally, θ is the slope at the end of each element, $\theta = d\delta/dx$. The stiffness for each element is given in approximate terms as:

$$EI(x) = ax + b \qquad (7.48)$$

or:

$$EI(x) = \begin{bmatrix} 1 - x & x \end{bmatrix} \cdot \begin{bmatrix} EI \\ EI \end{bmatrix} = \mathbf{U}^t \cdot \mathbf{EI} \qquad (7.49)$$

Now for the ith element, having a length, l, we obtain:

$$\Pi_i = \int_0^l [1/2\ \mathbf{U}^t \cdot \mathbf{EI} \cdot \delta^t_{xx} \cdot \mathbf{V} \cdot \mathbf{V}^t \cdot \delta_{xx} + \mathbf{V}^t \cdot \delta]\,dx \qquad (7.50)$$

With appropriate substitution of the approximations, we then obtain:

$$\Pi_i = 1/2\delta^t \cdot \left[(EI_1/l^3) \cdot \begin{vmatrix} 6 & 4l & -6 & 2l \\ 4l & 3l^2 & -4 & l^2 \\ -6 & -4 & 6 & -2l \\ 2l & l^2 & -4l & l^2 \end{vmatrix} \right.$$

$$\left. + (EI_2/l^3) \cdot \begin{vmatrix} 6 & 2l & -6 & 4l \\ 2l & l & -2 & l^2 \\ -6 & -2 & 6 & -4l \\ 4l & l^2 & -4l & 3l^2 \end{vmatrix} \right] \cdot \delta - \mathbf{P} \cdot \delta \qquad (7.51)$$

This can be written as:

$$\Pi_i = 1/2\delta^t \cdot (\mathbf{X}_1 + \mathbf{X}_2) \cdot \delta - \mathbf{P} \cdot \delta \qquad (7.52)$$

When Π_i is minimised subject to the external constraints on the model, the following relationship between load and deflection obtains:

$$\mathbf{P} = (\mathbf{X}_1 + \mathbf{X}_2) \cdot \delta \qquad (7.53)$$

Note that this analysis requires computer matrix inversion. Nevertheless, the analysis is extremely powerful, allowing for determination of deflections of any type of beam of any arbitrary shape, any loading profile and any relationship between the Young's modulus, density and density distribution across and along the beam. As a test of the analytical tool, the bending of a baseball bat with various loads applied 21·5 in. from the handle, was considered. The shape of the bat was determined by measurement, in order to obtain the coefficients in eqn (7.41). The density profile down the bat was determined for a specific case by curve-fitting the data shown in Fig. 7.15. A typical plot of the deflection as a function of distance down the bat is shown in Fig. 7.22. Here two relationships between the local density and modulus were chosen: $E_f/E_0 = \phi$ and $E_f/E_0 = \phi^2$. The experimental data fall between the two theoretical curves for three different loadings. It should be noted that the density across the beam thickness was assumed to be constant. This would be equivalent to assuming that $C = \infty$, $e = 0$ in the polynomial models discussed above. Certainly had these models been coupled with the finite element model, the analytical results would have been more accurate in predicting cantilever beam bending. Finite-element analysis is potentially the most important method of analysis of elastic bending of complex TSF foam beam shapes that we have today. It deserves more attention.

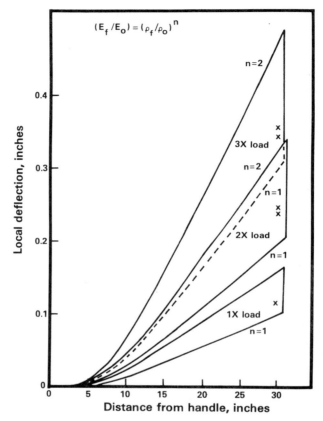

FIG. 7.22. Experimental confirmation of the application of finite element analysis to the bending of a structural foam beam (baseball bat). Crosses represent measured deflection.

7.6.4 Impact Strength

There are several tests proposed for the evaluation of the impact strength of TSF foams.[14] These include notched and unnotched Izod, Charpy, dart impact, Gardner impact and tensile impact. Consider the problem of notching TSF foams. If the notch is moulded in, the results will strongly depend upon the skin in the notch region. Thus one would expect results similar to the notched tests of the unfoamed resin. However, the concern here is whether these results truly represent the impact performance of foam having no notch. Furthermore, if the notched bar is moulded, there must be an accounting of the role played by the bar wall perpendicular to the notch. This point was discussed above when commenting on moulded flexural bars.

Unnotched Izod and tensile impact appear to represent flexural and tensile tests in which the speed of load application is extremely high. Although it has never been tested, one might expect a reasonable correlation between unnotched Izod values and flexural strength at break, and similar results for tensile impact and tensile strength at break.

In an early work, a correlation between unnotched Izod strength and bulk foam density was attempted.[14] It appeared that the unnotched Izod strength of the foam dropped dramatically with density and increased with foam slab thickness, T, as:

$$I_f/I_0 = \phi^{-n}(T/T_0)^m \tag{7.54}$$

where n was about 4 or more and m was about 2. Although these values were obtained on many foam samples, no attempt was made to determine skin thickness and the density profile through each foam sample. It has been noted that at the low moulding pressures, TSF foam is not in intimate contact with the mould surface. This is seen in a 1000 × optical micrograph of a TSF polystyrene surface (Fig. 7.23[30]). The shiny material in the microcrevice has not contacted the mould surface, and thus would be a point of crack initiation under reverse-side impact. Certainly for crystalline materials, the morphology and degree of crystallinity of the material at the surface will also influence the impact resistance of the material. The greater the amorphous content and the finer the crystallites, the greater the impact resistance (up to a point). As mentioned, the role of the skin had not been recognised as important in earlier work. In impact, as seen in Fig. 7.24, for several high-density polyethylenes, the impact at a given plaque thickness, T, and density, ϕ, is seen to be exponentially dependent on skin thickness, e, as:

$$I_f = B \exp(be) \tag{7.55}$$

where B is the impact strength of the foam core ($e = 0$), and b is the slope of the curve. This exponential relationship should probably be replaced with a power-law curve if one includes the high-density microvoid layer just beneath the skin, as shown in Fig. 7.25:

$$I_f = B'[1 + be']^n \tag{7.56}$$

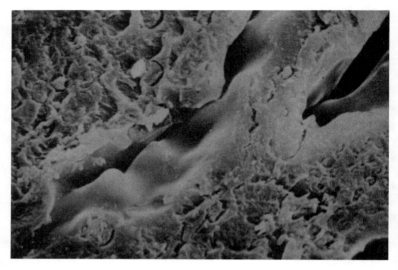

FIG. 7.23. Optical micrograph of TSF polystyrene surface, showing folded surface not in contact with mould surface. Flow is left to right. 1000×.

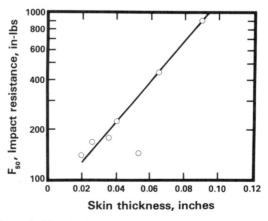

FIG. 7.24. Effect of skin thickness on impact resistance for several types of HDPE's. All bars 0·500 in. thick, tested on Gardner impacter with unsupported ring.

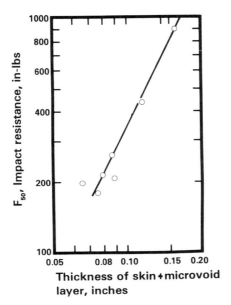

FIG. 7.25. Effect of skin thickness and microvoid layer on impact resistance of several types of HDPE's. All bars 0·500 in. thick, tested on Gardner impacter using unsupported ring.

where e' includes the microvoid layer thickness. The value for n, at least for HPDE's, is approximately 2. Although it might now be possible to correct eqn (7.54) to include the role of the skin, through application of the polynomial density profile and the above correlations, there does not seem to be sufficient experimental evidence to warrant this exercise at this time. This is particularly true if in fact the skin is bearing most of the load during impact, as seems to be the case with the exponential behaviour of eqn (7.55). Perhaps a more appropriate correlation would be to replace the average foam density in eqn (7.54) with a term that included the core density, R', and the skin thickness, e:

$$I_f/I_0 = R'^2(T/T_0)^2 \exp(be) \tag{7.57}$$

There are insufficient data to justify this equation, however. Ball-and-dart impact procedures have been touted over others because there is apparently less chance for failure at a surface blemish or a large void in the cell structure. However, the stairstep testing method, where the same foam sample is exposed to ever-increasing impact velocities, should not be used. There is evidence that failure may be occurring within the cellular structure at impact levels below that at which the first crack is observed on the foam surface. Thus the data may not be reproducible or may lead to artificially low impact values. Several types of impact failures are observed. The first is a characteristic star crack occurring at the maximum fibre stress directly under the tup. A second, as shown in Fig. 7.26, is a

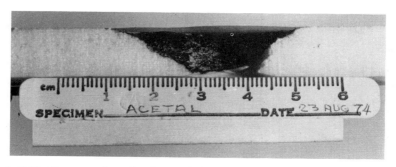

FIG. 7.26. Cross-section of polyacetal TSF panel impacted with 0·500 in. steel tup underside at 3·5 cm. Fractured area blacked with ink for visibility. Note 45° brittle fracture particularly on left of impact zone.

typical brittle impact failure. The impacted piece has been driven out of the TSF foam part, with the fractured zone showing the typical 45-degree fracture angle.[31] Crystalline materials such as polyacetal (polyoxymethylene) and polybutylene terephthalate exhibit this type of brittle failure. A third type of failure begins with a compressive failure in the cell structure. The cracks propagate from cell to cell through collapse and splitting of the cell walls parallel to the surface. This is seen for a ductile material in Fig. 7.27. Once the cells in a given area have collapsed, the fracture propagates to the layers below. This type of impact failure is best identified by a tup dent in the top surface but no appreciable crack pattern on the underside of the TSF foam sample.

7.6.5. Failure

Flexural fatigue failures in solid resins are frequently attributed to heat generation within the sample owing to internal friction.[32] Since plastics are well known thermal insulators, there is relatively little heat loss from

FIG. 7.27. Compressive impact damage on ABS structural foam. Note cellular collapse in centre section. Impact with 1·00 in. steel tup on underside at about 2·4 cm.

flexural samples, even at relatively low frequencies (100 Hz or so). The heating can be described in terms of the stress–strain hysteresis curve, where the energy loss between application of stress and relaxation of stress is given as ΔW. Dally and Broutman show that the specific energy loss is independent of stress level (for low levels of strain), thus yielding:

$$\Delta W = HW \tag{7.58}$$

where H is a material constant. Strain energy, W, is given as:

$$W = \sigma^2/2E \tag{7.59}$$

where σ is the applied stress and E is the modulus of the material. The amount of heat generated per unit volume and time, q''', is defined as the energy loss per cycle times the number of cycles per unit time:

$$q''' \equiv \Delta W f = H\sigma^2 f/2E \tag{7.60}$$

The equation that describes energy distribution within the sample under cyclic load is the one-dimensional transient heat-conduction equation with an internal energy generation term:[34]

$$\rho c_p \frac{\partial T}{\partial t} = \frac{k \partial^2 T}{\partial x^2} + q'''(x,t) \tag{7.61}$$

subject to appropriate boundary conditions of symmetry about the centre ($x = a$), insulation at the surface ($\partial T/\partial x = 0$ at $x = 0$), and a uniform initial temperature T_0 at $t = 0$. The boundary condition at the surface of the specimen can be modified by considering a convective heat transfer coefficient, but the correction is generally minor. Dally and Broutman[33] show that the solution to eqn (7.61) can be represented adequately by the first term of the analytical eigenvalue equation:

$$\Theta/4q'''a^2/k = 1/2[x/2a - (x/2a)^2] - \quad (7.62)$$
$$- (4/\pi^3)\sin(\pi x/2a)\exp[-(\pi/2a)^2\alpha t]$$

where q''' is assumed to be independent of x and t, α is the thermal diffusivity, $= k/\rho c_p$, and $\Theta = (T - T_0)$. The maximum temperature occurs at the centerline, where $x = a$:

$$\Theta_{max} = (q'''a^2/2k)[1 - \exp\{-(\pi/2a)^2\alpha t\}] \quad (7.63)$$

Note that the approach to maximum temperature is one of first-order exponential decay with time. This has been verified experimentally by many authors.[35,36] The upper limit of maximum temperature depends upon the orders of magnitude of frequency (from eqn (7.60)), Young's modulus, and the thermal conductivity of the resin. Assume for the moment that the loading frequency and stress level are constant but that the resin density changes. It has been shown that the thermal conductivity of foam decreases with foam density.[37] Although most of the models are quite complicated, a simple power-law model[38] can be used here for illustration:

$$k_f/k_0 = (\rho_f/\rho_0)^m = \phi^m \quad (7.64)$$

where m is of the order of 0·5 to 1·0. Assume further that we are considering a uniform-density foam, such that the modulus decreases in proportion to the square of the foam density, $E_f/E_0 = \phi^2$. Now the upper limit of maximum temperature can be written as:

$$\Theta_{max,f} = [H\sigma^2 fa^2/4]/[k_0\phi^m E_0\phi^2] \quad (7.65)$$

or:

$$\Theta_{max,f}/\Theta_{max,0} = \phi^{-(2+m)} \quad (7.66)$$

As an example, assume $\phi = 0.75$ and $m = 0.5$. Now the increase in the centreline temperature of the foam is twice that of the unfoamed sample of equivalent thickness, at the same loading conditions. Thus, if the unfoamed sample temperature increase reaches 50°C, the foamed sample temperature increase will reach 100°C. Since the primary mode of failure is thermal,[32] one would expect foam samples to fail earlier at equivalent loading conditions, or fail at the same time at lower loading conditions.

One factor that must be reconciled here is the strain level. Note that the temperature is proportional to the square of the stress level. Recall our earlier discussion regarding limitations in applying Hooke's law to a possibly prestrained foam sample. We might expect some limit to the applicability of eqn (7.60), particularly at high stress loadings. One way of approaching this analysis would be to include a position dependent form for q''' in eqn (7.61). Assume that the applied strain across the sample is linear. Thus $\varepsilon = \varepsilon(x)$, where $\varepsilon = \varepsilon_0$ at $x = 0$ and $\varepsilon = 0$ at $x = a$. Now since W is the area under the stress–strain curve, we can write:

$$W = \varepsilon\sigma/2 \qquad (7.67)$$

Now assume that between $x = 0$ and $x = e'$, the skin thickness, $\sigma = E_0\varepsilon$, and between $x = e'$ and $x = a$, in the foam core, $\sigma = E_f\varepsilon'$, where ε' may represent the prestrained condition. Since we have agreed that strain is linear across the sample, we can write:

$$W(x) = E_0\varepsilon_0(1 - x/a)/2 \qquad 0 < x < e' \qquad (7.68a)$$

$$W(x) = E_f\varepsilon'(x)/2 \qquad e' < x < a \qquad (7.68b)$$

Consider the conditions at the skin-core interface, where $e'/a = e_0$.

$$\varepsilon'(e_0) = [E_0/E_f]\varepsilon_0(1 - e_0) \qquad (7.69)$$

Consider again a foam having uniform density foam core with $\phi = 0.75$ and $E_f/E_0 = \phi^2$ and $e_0 = 0.2$. $\varepsilon' = 1.4\varepsilon_0$, or the strain in the foam is 40% greater than that in the skin. Thus, any prestraining of the foam would result in very early failure in the foam nearest the skin.

Thus, there are two factors working toward fatigue failure first in the foam core, then in the skin. The first is temperature build-up which is more severe with foam than with the unfoamed resin and the second is an abnormally higher strain level in the foam nearest the skin. Support of this hypothesis is sketchy. In one example of the flexural failure of an HDPE TSF foam rib, shown diagrammatically in Fig. 7.28 the initial failure was traced to a rather large void some distance from the point of maximum fibre stress. In Fig. 7.29, one can see the fatigue lines radiating toward the void and the arrest lines or beach markings in the skin region, indicating that failure had occurred at or near the void.[39] Scanning

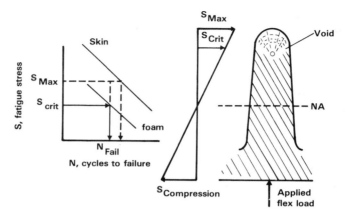

FIG. 7.28. Diagram illustrating how fatigue failure might occur in TSF structures at stress levels below maximum fibre stresses.

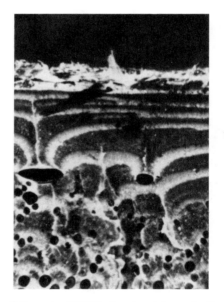

FIG. 7.29. Fatigue failure in HDPE structural foam in the skin region. Note that fatigue failure apparently begins in cellular portion, moves toward skin. Note arrest lines in skin area. 12×.

electron micrographs of another section of a fatigue-failed TSF part show fatigue failures that are seen primarily in glassy materials.[40] As seen in Fig. 7.30, about $\frac{1}{2}$ in. from the skin region, edges of some cells exhibit shiny spots beyond which radiate fatigue lines to the next cell or intersect with fatigue lines from other cells. Note the prominence of beach markings. Examination of the topology of the material in the skin region at low magnification (Fig. 7.31), reveals only radiating fatigue lines and beach markings, indicating that failure occurred in the foam and the failure radiated outward to the skin. At high magnification, the material in the beach markings for both the cellular area and the skin area (Fig. 7.32 and 7.33) shows remarkable similarity in the topology, indicating that the presence of the cells is the primary disrupting factor in the premature failure of TSF parts. One might anticipate that the region of prestraining may be restricted to a very thin shell around a cell, with the interstitial material having strain characteristics similar to the un-

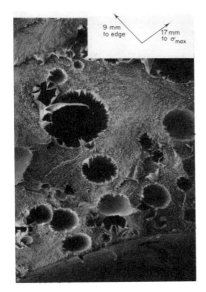

FIG. 7.30. Scanning micrograph of HDPE foam core near site of large void (bottom). Flexural fatigue failure; note mirror zones around some cells, drawn fibrils around others. 35×.

FIG. 7.31. Scanning micrograph of HDPE in skin layer in flexural fatigue failure. Note arrest markings radiating from foam core toward skin. 120×.

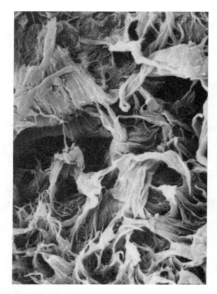 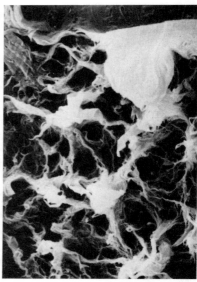

FIG. 7.32. Scanning micrograph of foam core of HDPE in flexural fatigue failure. Higher magnification than Fig. 7.30. Note fine microstructure and compare with Fig. 7.33. 1900×.

FIG. 7.33. Scanning micrograph of skin of HDPE in flexural fatigue. Higher magnification than Fig. 7.31. Note fine microstructure near arrest line and compare with Fig. 7.32. 1900×.

foamed material in the skin region. This, of course, is conjecture at this point, and requires careful isothermal flexural fatigue testing. One technique that might be employed is the soft X-ray technique mentioned earlier.

Hobbs has examined crack-induced fracture of TSF PC foams with 5% glass fibre reinforcement with a foam level of about 25%.[41] For thin samples of uniform density, the results closely follow the Griffith yield criterion:

$$\sigma_F = (2E\gamma/\pi c)^{1/2} \qquad (7.70)$$

where σ_F is the fracture stress, E is Young's modulus, c is the crack length and γ is the sum of surface and plastically dissipated energies. As Hobbs points out, this equation is frequently used to estimate the critical flaw size. For polyurethane foams, this size is of the same order of magnitude as the nomial cell dimension. For PC TSF foams, extrapolation yields a critical flaw size of about 0·09 cm or a factor of 4 or 5 greater than the nominal cell dimension. Hobbs attributes this discrepancy to cell-wall collapse and cell consolidation at low tensile loading. Careful examination of the crack region shows multiple and secondary cracking. This leads to an enhanced fracture toughness (particularly in GR foams). Hobbs believes that the crack extension force, given as $G = 2\gamma$ in eqn (7.70), is enhanced by local yielding in the interstitial resin between bubbles, although plane strain conditions are satisfied. Again, the role of the cell confounds application of general fracture principles applied to glassy polymers.

7.7 DUCTILE–BRITTLE TRANSITIONS AND SHEAR BANDS

As mentioned earlier, the stress–strain behaviour of resins can be dramatically influenced by addition of an aggressive foaming agent, by foaming the resins and by morphological alteration within the foam itself in the case of crystalline resins. Examine for the moment the low foaming level physical properties obtained for Eastman PETG, an amorphous polyethylene terephthalate copolymer, foamed with 0·5% 5PT, as shown in Table 7.3. Note the dramatic drop in elongation at break from 120% for the unfoamed resin to around 8% for a resin foamed to 9% level ($\phi = 0.91$) and to 5% at 13% ($\phi = 0.87$). This is contrasted to the expected gradual drop in moduli and strengths at yield owing to foaming. The tensile bars for the three foaming levels are shown in Fig. 7.34. The fracture zones at 5% and 10% foaming levels are shown in Figs. 7.35 and 7.36. Note the diagonal shear bands and the dislocations at the surface, shown diagrammatically in Fig. 7.37 for the resin foamed to the 5% level. In metals these are referred to as Luders bands, and are characterised by slip planes at about 30° to 45° from the perpendicular.[31] Ward[42] notes that such shear bands have been detected in polymers by Zankelies as early as 1961,[43] whereas Shultz[44] attributes the first accurate description of shear band yielding in homogeneous polymers to Whitney[45] in 1963. Sternstein and Ongchin[46] observed that shear bands in polystyrene were 58° to the tensile axis. As seen in Fig. 7.34 (and diagrammatically in Fig. 7.37), the shear bands observed in the foamed PETG were at an angle of about 60° to the tensile axis. Shultz further notes that shear band yielding is relatively

TABLE 7.3
Physical Properties of Foamed Eastman PETG (95% CL)[26]

Property	Control (unfoamed)		9% Foaming level		13% Foaming level	
	Absolute	Reduced	Absolute	Reduced	Absolute	Reduced
Density, g cm^{-3}	1·280	1·00	1·166	0·911	1·114	0·870
Flex strength, yield, 16 in.$^{-2}$	10 380 (46)	1·00	9 960 (254)	0·959	8 845 (34)	0·852
Flex modulus, tangent, 10^3 lb in.$^{-2}$	311·6 (3·35)	1·00	301·2 (5·27)	0·967	275·8 (4·95)	0·885
Tensile strength, yield, lb in.$^{-2}$	7 110 (27)	1·00	6 070 (45)	0·854	5 450 (148)	0·767
Tensile elongation, yield, %	4·35 (0·08)	1·00	4·04 (0·07)	0·929	3·52 (0·20)	0·809
Tensile modulus, tangent, 10^3 lb in.$^{-2}$	257·1 (9·6)	1·00	233·2 (11·3)	0·907	221·0 (7·0)	0·860
Tensile strength, break, lb in.$^{-2}$	3 780 (126)	1·00	4 360 (173)	1·153	4 720 (322)	1·249
Tensile elongation, break, %	124·3 (7·2)	1·00	7·68 (3·73)	0·062	4·86 (0·43)	0·039

FIG. 7.34. Photograph of unfoamed, 5% foamed and 10% foamed PETG, showing ductile–brittle transition at low foaming levels in amorphous materials.

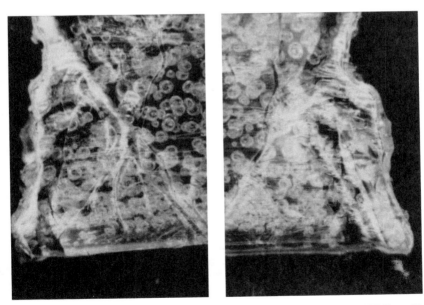

FIG. 7.35. Two optical micrographs of shear band yielding in PETG at 5% foaming level. 12×.

Fig. 7.36. Optical micrograph of the fracture zone in tension for 10% foamed PETG amorphous polymer. 12×.

close to the von Mises criterion, thus relating this type of yielding to a critical value of elastic shear strain energy. The appearance of shear bands in Fig. 7.34 indicates that foaming is influencing the elastic–plastic deformation behaviour of amorphous materials at very low foaming levels.

Most recently, Petrie et al.[47] have analysed the yielding behaviour of unfoamed polycarbonate in various external stress environments. Petrie et al. point out that when crazing and microcavitation are suppressed, ductile yielding is the preferred mode of failure (at least with PC and apparently with PETG in the unfoamed state). Further, they conclude that when crazing or microcavitation is present, ductile yielding begins and cavitation creates either crazes or microcracks. If the microcavitation is allowed to move unimpeded throughout the

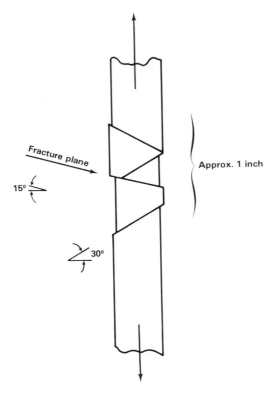

FIG. 7.37. Diagram of foam-induced shear bands in amorphous PET tensile test bar. The 30° angle corresponds well with typical values for yields obeying the von Mises criterion.

test sample, the result will be a shear-band ductile yielding. If the microcavitation is restrained, a flaw will develop resulting in brittle failure. In light of their analysis on PC, compare the failure areas of the 5% and 10% foam PETG specimens. It appears that at 5%, ductile yielding is proceeding, at least in the maximum strain area, although it does not produce the characteristic full ductile yielding seen in the unfoamed sample. The bubbles are apparently inhibiting microcavitation. In the case of the 10% foam specimen, the microcavitation is obviously suppressed, resulting in a brittle failure with very little signs of shear banding (except possibly at the surface). Schematically, this can be pictured as a rapidly increasing embrittlement of the specimen with foaming level. Recall now that the hypothesis is that the foam cells somehow inhibit microcavitation. A mechanism for this has not been put forth as yet, but one might conjecture that the excess surface energy around a foam cell and its possible anisotropic prestrained condition may resist crack propagation into, across, and beyond the foam cell. Recall the flexural fatigue scanning electron micrographs which show that the foam-cell fracture morphology is quite different from one side of the cell to the other. Frequently, a mirror zone is seen at one side of the cell, indicating initiation of the fracture, whereas the structure on the opposite side shows strong indication of highly elongated material before fracture. Another aspect of the yielding phenomenon with very low foaming levels can be seen by carefully examining the bubble shapes in Fig. 7.35. The normally spherical bubbles are elongated into teardrop shape in the yield zones, with the points of the teardrops protruding into the shear band. This indicates that additional energy must be supplied to generate the additional surface area in the bubbles, and that these points might act as stress concentration points for ultimate failure of the foam.

In crystalline polymers, such as HDPE, the problem is complicated by morphological changes in the material owing to the presence of the cellular core. First note that the cellular morphology is several orders of magnitude greater in dimension than the molecular morphology. We have seen this in the fatigue behaviour in the cellular region and in the foam skin. In Fig. 7.38 is shown the transcrystalline to spherulitic morphological transition typical of injection moulded crystalline materials.[48] In this case, the mould surface is to the left and the foam core of HDPE is to the far right of the photograph. The persistence of the transcrystalline region far into the HDPE TSF foam is apparently due to the processing conditions. TSF foams are made by injecting foamable resin into a cold mould. The only pressure available to press the skin to the mould surface is that of the dissolved gas, e.g. 200 lb in.$^{-2}$ or so. As the foam cools, there is strong indication that it shrinks from the mould surface, thus leaving a ragged part surface and a very prominent and irregular transcrystalline region. Careful examination of the right portion of the 318 × (on 3 × 4 in. film) polarised photograph shows a second band of transcrystalline morphology, and between is a faint region of flattened microvoids. At the bottom of the photograph is a near-spherical bubble, with a transcrystalline region parallel to and moving away from the mould surface. It has been shown that HDPE resins having large transcrystal-

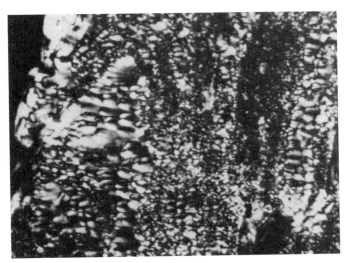

FIG. 7.38. Morphology of HDPE in the skin region of a foamed tensile specimen. Note the transcrystalline zone to the left at the moulded surface and the spherulitic zone appearing on the far right. Also note the bands of transcrystalline material around the flattened bubbles in the skin. Polarised optical micrograph.

line regions will experience poor impact strength, since any energy impacting the surface will penetrate to the interior due to a deficiency in tie molecules between the columnar crystallites. More important, the transmission of energy from cell to cell may occur along the columnar crystallites that connect the cells, as seen in this photograph.

How prevalent is the appearance of these transcrystalline regions in HDPE foams (and perhaps other crystalline foams as well)? Examine carefully Fig. 7.39, which shows HDPE foamed to about 10% level, as viewed through polarisers. The distinct white areas around the cells at 24 × are quite apparent, and the contrasting transcrystalline and spherulitic morphology is seen quite distinctly around a single cell at 240 ×. The change in morphology owing to the presence of the cells might have several causes. It has been hypothesised[49] that there is a strongly uneven temperature gradient around individual cells during the cooling phase of TSF foams. The resin between cells may be cooled at a much more rapid rate than that behind the cells, owing to the insulative nature of the gas in the cell. This might be the cause of spherulitic morphology directly

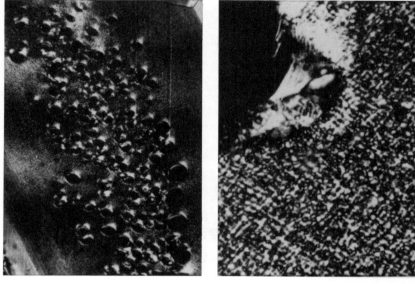

FIG. 7.39. Optical polarised micrographs of low foaming levels in HDPE, showing the complex interaction between transcrystalline morphology around the bubbles and spherulitic morphology between them. Left, 24 × ; right 240 × .

behind (and ahead of) the cell, and transcrystalline morphology at the edges parallel to the part surface (and perpendicular to the direction of heat transfer). Another factor might be shear-induced crystallisation owing to the rapid growth of the bubbles, but this effect would lead to nearly symmetrical morphology around the bubble.

In order to assess fully the importance of the changes in morphology with the formation of cellular growth, it is necessary carefully to examine impact failed samples using cross-polarisers. Heating or annealing should relieve stresses in the transcrystalline region. Thus, samples that have failed in fatigue owing to internal heating may be annealed to the extent that a morphological change has taken place around the cells and the failure mode may be more ductile than brittle. Furthermore, the crack propagation may not follow the transcrystalline grain boundaries as it might with impact failed samples. It is apparent that the crystallinity and possible morphological changes in the resin with process conditions may confound prediction of failure mechanisms in TSF foams.

7.8 TEST RESULTS

There have been many test results published over the last decade or so. Many of these have been summarised in an earlier work on mechanical properties of TSF foams.[14] As cautioned earlier, early results did not measure skin thickness, foam core density, or foam core density profile. In some cases, there was no assurance that the test bars were of uniform density end-to-end. Many early flexural tests used moulded rather than cut bars, thus yielding 'reduced flexural modulus' values significantly higher than data obtained from cut bars at the same apparent density. Probably the most important early study was that of Harry who carefully measured skin thickness and foam core density. His results show that at low foaming levels the flexural modulus followed that predicted by the models proposed in Section 7.6.1. As the foaming level increased, the modulus rapidly dropped, until at a reduced density approaching 0·5, the measured modulus was about that predicted by the square law. At this point, Harry noted that the skin thickness was rapidly approaching zero, confirming the role of the skin in the strength of a TSF beam in bending.

Specific testing procedures have been identified elsewhere,[14] particularly with regard to measurement of skin thickness, microvoid layer thickness, and predominance of large bubbles. To date, this author knows of no published data in which these factors are used as a means of qualifying the test results. Undoubtedly, the uncertainty of the respective roles of the skin and foam core, the residual strain that may occur around a foam cell, and the complex morphology of crystalline TSF foams may overshadow the importance of accurate testing and reporting at this point in time.

ACKNOWLEDGEMENT

The author wishes to acknowledge the decade of free idea exchange with Dr Richard C. Progelhof, Professor of Mechanical Engineering, New Jersey Institute of Technology, Newark, NJ, USA. Certainly Dr Progelhof's correct analysis of the stiffness of an integral skinned TSF beam and the experimental verification of the analytical solution helped put the bending of TSF beams into proper light. Most recently, his early work on impact strengths of TSF structures begins to illuminate the structure-property problems of gas bubble-induced ductile–brittle transition. The author looks forward to the next decade of co-operative problem solving.

REFERENCES

1. BEYER, C. E. et al. (to Dow Chemical), US Patent No. 3,058,161 (Oct. 16, 1962).
2. ANGELL, Jr., R. G. (to Union Carbide), US Patent No. 3,268,636 (Aug. 23, 1966).
3. ANGELL, Jr., R. G. (to Union Carbide), US Patent No. 3,436,446 (Apr. 1, 1969).
4. THRONE, J. L. (1976) *J. Cell. Plast.*, **12**, 264.
5. THRONE, J. L. (1972) *J. Cell. Plast.*, **8**, 208.
6. HARTSOCK, J. A. (1969) *Design of Foam-Filled Structures*, Technomic Publications, Westport, CT.
7. BENNING, C. J. (1969) *Plastic Foams*, 2 Vols., Wiley-Interscience, New York.
8. MEINECKE, E. A. and CLARK, R. C. (1973) *Mechanical Properties of Polymeric Foams*, Technomic Publications, Westport, CT.
9. Hewlett-Packard Faxitron Soft X-Ray Equipment, Palo Alto, California.
10. ESHBACH, O. W. and SOUDERS, M. (Eds) (1975), *Handbook of Engineering Fundamentals*, 3rd Edn. Wiley, Section 5.
11. KIESSLING, G. C. (1976) *Extrusion of Structural Foams*, in *Engineering Guide to Structural Foam*, Bruce C. Wendle (Ed), Technomic Publication, Westport, CT.
12. THRONE, J. L. (1979) *Principles of Thermoplastic Structural Foam Moulding: A Review*, Keynote Paper, 1977 International Conference on Polymer Processing Proceedings, MIT Press.
13. A typical industry instructional brochure is, *Structural Foam Resins*, General Electric Plastics Dept., Pittsfield MA, 01201, SFR-1(30M) 4-73, 1973.
14. THRONE, J. L. (1976) *Mechanical Properties of Thermoplastic Structural Foams*, in *Engineering Guide to Structural Foams*, Bruce C. Wendle (Ed), Technomic Publication, Westport, CT.
15. THRONE, J. L. (1976) *J. Cell. Plast.*, **12**, 161; Errata, *ibid.*, **12**, 284.
16. PROGELHOF, R. C. and THRONE, J. L. (1978) *Young's Modulus of Uniform Density Thermoplastic Foam*, SPE Technical Papers **24**, 678.
17. HALPIN, J. C. and TSAI, S. W. (1967) *Environmental Factors in Composite Materials Design*, AFML TR67-423.
18. ASHTON, J. E., HALPIN, J. C. and PETIT, P. H. (1969) *Primer on Composite Materials: Analysis*, Technomic Publication, Stamford, CT, 06901.
19. NIELSEN, L. E. (1970) *J. Appl. Phys.*, **41**, 4626.
20. EGLI, E. (1972) *J. Cell. Plast.*, **8**, 245.
21. MOORE, D. R., COUZENS, K. H. and IREMONGER, M. J. (1974) *J. Cell. Plast.*, **10**, 155.
22. IREMONGER, M. J. and MOORE, D. R. (1976) *PRI Materials Appl.*, **1**, 30.
23. PROGELHOF, R. C. and EILERS, K. (1977) *Apparent Modulus of a Structural Foam Member*, paper presented at SPE Engineering Properties and Structures DIVTEC III *Processing for Properties*, Proceedings, p. 1, Woburn, Mass.

24. THRONE, J. L. (1979) *Effect of Cellular Structure and Chemical Foaming Agents on Resin Properties in the Almost-Solid Region*, paper presented at SPE Annual Technical Conference, New Orleans.
25. DEMPSEY, R. E. et al. (to Standard Oil Company) US Patent No. 4,127,631 (Nov. 28, 1978).
26. THRONE, J. L. (1978) *J. Cell. Plast.*, **14**, 21.
27. THRONE, J. L. (1979) *Structure-Property Relationships in Structural Foams*, paper presented at SPE PACTEC, Feb. 2, Costa Mesa, CA.
28. SINGER, F. L. (1951) *Strength of Materials*, Harper and Brothers, New York, Chapter VI.
29. MARINACCIO, P. J. and DEAN, M. W. (1979) *Plast. Des. Forum*, **4** (2), 47.
30. THOMPSON, W. F. (1975) paper presented at Third Structural Foam Conference, University of Wisconsin, Madison, Nov.
31. HALL, E. O. (1970) *Yield Point Phenomena in Metals and Alloys*, Plenum Press, New York.
32. CESSNA, L. C., LEVENS, J. A. and THOMSON, J. B. (1969) *Flexural Fatigue of Glass-Reinforced Thermoplastics*, Section 1-C, 24th ANTEC, SPE/Reinforced Plastics/Composites Division, p. 1.
33. DALLY, J. W. and BROUTMAN, L. J. (1967) *J. Comp. Mat.*, **1**, 424.
34. THRONE, J. L. *Plastics Process Engineering*, Marcel Dekker, Inc., Chapter 6.
35. CARSLAW, H. S. and JAEGER, J. C. (1946) *Conduction of Heat in Solids*, Oxford University Press, Oxford.
36. RIDDELL, M. N., KOO, G. P. and O'TOOLE, J. L. (1966) *Polym. Eng. Sci.*, **6**, 363.
37. PROGELHOF, R. C. et al. (1975) *Methods of Predicting Thermal Conductivity of Composite Systems: A Review*, SPE Engineering Properties and Structure DIVTEC, October 7–8, Technical Papers, p. 221.
38. PROGELHOF, R. C. and THRONE, J. L. (1975) *J. Cellular Plastics*, **11**, 152.
39. HERTZBERG, R. W. (1976) *Deformation and Fracture Mechanics of Engineering Materials*, John Wiley and Sons, New York, Figure 7.15.
40. TAKAHASHI, K. (1974) *Cracking of PMMA Caused by Plane Stress Waves*, in P. H. Geil et al. (Eds), *The Solid of Polymers*, Marcel Dekker, Inc., 673.
41. HOBBS, S. Y. (1977) *J. Appl. Pys.*, **48**, 4052.
42. WARD, I. M. (1971) *Mechanical Properties of Solid Polymers*, Wiley-Interscience, New York, 296.
43. ZANKELIES, D. A. (1961) *J. Appl. Phys.*, **33**, 2797.
44. SCHULTZ, J. (1974) *Polymer Materials Science*, Prentice-Hall, Inc., p. 344.
45. WHITNEY, W. (1963) *J. Appl. Phys.*, **34**, 3633.
46. STERNSTEIN, S. S. and ONGCHIN, L. (1969) ACS Polymer Preprints.
47. PETRIE, S. P., DIBENEDETTO, A. T. and MILTZ, J. (1978) *Polym. Eng. Sci.*, **18**, 1200.
48. CLARK, E. S. (1967) *Soc. Plast. Eng. J.*, **13** (7), 46.
49. PROGELHOF, R. C. and THRONE, J. L. (1975) *Cooling of Crystalline Foams with Developing Density Profiles*, SPE Technical Papers, **21**, 455.
50. HOBBS, S. Y. (1975) *Polym. Eng. Sci.*, **15**, 854.
51. MEHTA, B. S. and COLOMBO, E. A. (1976) *J. Cellular Plastics*, **12**, 59.

52. BAXTER, S. and JONES, T. T. (1972) *Plast. Polym.*, **40**, 69.
53. OKUNO, K. and WOODHAMS, R. T. (1974) *J. Cell. Plast.*, **10**, 237.
54. GONZALEZ, Jr., H. (1976) *J. Cell. Plast.*, **12**, 49.
55. OKUNO, K. and WOODHAMS, R. T. (1974) *J. Cell. Plast.*, **10**, 295.
56. OGORKIEWICZ, R. M. and SAYIGH, A. A. M. (1972) *Plast. Polym.* **40**, 64.
57. SCHLEITH, O. (1975) *Kunststoffe,* **65**, 421.
58. NARKIS, M. (1978) *J. Appl. Polym. Sci.,* **22**, 2391.
59. SCHRAGER, M. (1978) *J. Appl. Polym. Sci.,* **22**, 2379.
60. WILCZYNSKI, A. P. (1977) *Int. J. Polym. Mat.*, **5**, 291.
61. HARRY, D. H. (1975) *Structural Foam Testing and Properties,* paper presented at Third Annual Conference on Structural Foam, University of Wisconsin, Madison, WI, Nov. 11.
62. WASSERSTRASS, I. D. and THRONE, J. L. (1975) *Flexure and Cantilever Bending of Structural Foam Beams,* SPE Technical Papers, **21**, 404.
63. PRAMUK, P. F. (1976) *Polym. Eng. Sci., ,* **16**, 559.

Chapter 8

REINFORCED FOAMS

J. M. METHVEN
Fillite Limited, Runcorn, Cheshire, UK
and
J. R. DAWSON
*Department of Metallurgy and Materials Science,
University of Liverpool, Liverpool, UK*

8.1 INTRODUCTION

8.1.1 General Considerations

The development of reinforced foamed materials has taken place over the past 20 years with a considerable upsurge in interest in such materials being shown over the past five years. Invariably the combination of foam and reinforcement has been directed towards specific goals and while many patents and articles have been published on this topic during this time, very little work of a systematic nature has appeared.

The main objective of this review is to present an adequately functional model of the reinforcement of a foamed matrix by fibres, and this is most easily achieved by considering polyurethanes and thermoplastic structural foams each containing short dispersed glass fibres. For this reason the major portion of this review is concerned with both the mechanical properties and fabrication routes of these materials. Such models are developed from more general composite relationships typically applied to reinforced and filled plastics.

The description of reinforced foam systems involving glass mats and other 'reinforcements' is presented together with such properties described in the original literature with the minimum of comment since invariably in such work the systems described are rather complex and frequently inadequately characterised especially from the microstructure viewpoint.

As with the more general development of composites, reinforced foams reflect a very specific desire to produce structures that are predominantly characterised by high specific properties together with other properties not readily achieved by any molecular architecture within the chemistry of the resin.

Although the desire to combine fibres and a gas together in a matrix has probably been present for many years it is only now that the processing difficulties encountered in such systems are being overcome. This has been particularly true of reinforced polyurethanes produced by the Reaction Injection Moulding (RIM) process described later. Indeed those reinforced foam systems available ten years ago were derived more or less entirely from existing technology taken from other resin processing routes (particularly unsaturated polyesters) taking the form of spray up or resin injection variations.

It is instructive at this stage to review such routes since they are not conducive to any systematic analysis.

8.1.2 Fabrication Routes

Closed-Mould Systems
Perhaps the most publicised of such systems is that of the Depotmat process develped in the late 1960s by Bayer.[1] This can be considered as a very sophisticated sandwich structure consisting of GRP skins between which is formed a core of reinforced polyurethane foam. The density of the foam *and* the level of reinforcement within the foam vary over the cross-section of the core from high at the interface with the skins to low in the centre. In this way high stress concentrations at the skin/core interface are essentially eliminated and the strength of any lamina is a continuous function of distance or depth into the section. The success of this system was due almost entirely to the appropriate selection of suitable reinforcement, this generally taking the form of the layers of glass mat, both woven and needled continuous filament mat. The properties of such a system reflect its excellent impact and fatigue properties and this is reflected in its use as vehicle chassis, boat hulls and military equipment elements.[2]

A more straightforward closed mould technique for reinforced polyurethane mouldings owes its development to resin injection of unsaturated polyesters.[3] This procedure involves placing the reinforcement within the mould cavity, closing the mould and injecting the foamable resin or resin froth into the cavity. This penetrates the reinforcement thoroughly, producing a reinforced foam with more or less uniformly distributed reinforcement throughout. The reinforcement typically used for such processes is usually continuous-filament glass mats to aid resin penetration, although recent developments have produced a needled very bulky mat with good resin penetration and with the added advantage of a three dimensional orientation of the reinforcement.[4]

Although of general applicability to any (thermoset) foam resin this process has been restricted in practice to polyurethanes. Various structures have been produced in this way and are characterised by excellent impact properties particularly at high strain rates[5] and under ballistic conditions.[6] As a consequence such materials have found applications predominantly military in nature.

An intermediate approach to reinforcing foams between these two systems uses preformed skins (usually of GRP) as the 'mould'. This is filled with loose, continuous filament mat and then injected with polyurethane to produce a more conventional sandwich structure than that produced by the Depotmat process. This is naturally much less expensive to produce than the latter material and currently is the basis of the design of the 'plastic' cab used on the Advanced Passenger Train produced by British Rail. Such structures can accommodate missile impacts at up to 350 km h^{-1} without complete penetration of the sandwich.[7]

Reinforced Foam Panels and Boards
Various modifications to those techniques described in the previous section have been developed to enable the production of boards of reinforced foams economically. Again the main interest has centred around polyurethanes but some relevant work has also been generated recently on unsaturated polyester resins.

The markets for such structures are similar to those previously described namely the production of a lightweight material of exceptional impact and fatigue resistance. Freight container panels have been the most actively pursued of the non-military outlets for such materials.

It is interesting to note that one significant bonus derived from such structures concerns their improved fire performance as a consequence of the reinforcement. This literally holds the structure together during exposure to flame and consequently retards flame penetration of the structure by reflecting heat back from the char attached to the reinforcing fibres.[8]

The standard panel process involving reinforced foams is attributed to the Xentex organisation who developed a process for the production of large (up to 40′ × 10′) panels based on glass mat and chopped strands dispersed in a polyurethane foam matrix. Hollow microspheres were also used in the matrix.[9] The uniqueness of this process lay in the fact that the reinforcement could be arranged within the matrix in a variety of ways depending upon the properties required of the final material.

Reinforced polyurethane sheets have also been produced using synthetic fibre reinforcement both in the form of woven mats[10] and chopped fibres.[11] Pre-tensioned fibres in the form of undirectional rovings have also been used to reinforce polyurethanes.[12]

One of the most interesting developments over the past few years has been the introduction of unsaturated polyester resin foams. Unsatured polyester resins are probably the most widely used thermosets for conventional GRP production. They are reasonably inexpensive (compared with a polyurethane resin for example) and can be tailored to have a variety of properties in terms of chemical resistance, electrical performance and mechanical performance. Since they became readily available many attempts to prepare foams have been tried and proved unsuccessful.

The curing characteristics of these materials did not permit the use of conventional fluorocarbon or hydrocarbon materials to function as blowing agents and led invariably to cracked structures as a result of the expansion occurring after the resin had gelled.

This failure led to the developing of other techniques to produce foamed structures, the most important of which were the incorporation of water in the resin as a fine stable emulsion followed by its subsequent removal by heat leaving fine cells throughout the structure. Both of these techniques had limitations not the least of which was the difficulty encountered in reproducing the procedure exactly. Coupled with the fact that the densities obtained were not low (above 600 kg m^{-3}) and the problem of high shrinkage exhibited by the WEP approach, this work was gradually discontinued.

In the early 1970s however a novel azo initiator with a very low decomposition temperature was developed for polyesters.[13,14] This produced nitrogen during the gelation period and enabled excellent foams both with and without reinforcement to be prepared at low densities (ca. 200 kg m^{-3}). Unfortunately this initiator required refrigerated transport and storage and this general handling limitation precluded its widespread use. It was also shown to be a health hazard and consequently undesirable.[15]

Following these efforts in the 1970s an alternative process was developed in Germany based on the use of certain carboxyester anhydrides as the blowing agent.[16] A typical recipe for this system was rather complex requiring a special resin, blowing agent, catalyst, water and a small amount of di-isocyanate.[17] Despite this, these foams could be produced at low densities (ca. 50 kg m^{-3}) in simple moulds at room temperature and in Germany are being used commercially to produce a lightweight resin concrete (using an expanded glass aggregate) of excellent structural quality and fire resistance as the main component of pre-fabricated building panels and even complete walls.[17,18]

The combination of this resin with continuous-filament glass reinforcement in a semi-continuous analogue to those processes described above for polyurethanes is presently being pursued as a wood replacement especially for large structures, such as container panels.[19]

Reinforced Sprayed Foams

The cell size obtained by spraying a polyurethane foam is finer than that obtained by pouring with consequent improvements in mechanical properties.[74] Thus from this viewpoint it would seem desirable to apply this to reinforced systems where possible. In practice, however, a spray-up process is rather wasteful when applied to small structures due to the difficulties encountered with overspray. It is therefore ideally suited to very large structures, such as buildings or prefabricated building elements. In such applications, however, the desire is invariably insulation, and consequently the presence of reinforcement is not desirable nor indeed necessary.

The greatest potential for reinforced sprayed foams at present is in the construction of LNG (liquefied natural gas) and LPG (liquefied petroleum gas) ships. These gases are stored at low temperatures ($-162°$ C) in the ship holds which are constructed from reinforced polyurethanes. Obviously thermal insulation is important in such applications, but equally important are the low-temperature mechanical properties of the foams, particularly their fatigue properties and fracture toughness.

A great deal of work is presently being carried out throughout the world on such investigations, most of which is unpublished for proprietary reasons. The results which are available do however show clearly the contribution the glass makes[20] and enables a mechanism for reinforcement to be developed.[21] This is discussed in greater detail elsewhere in this chapter.

8.2 REINFORCED REACTION INJECTION MOULDING (RRIM)

8.2.1 Background

Over the past six years a considerable interest has been manifest in polyurethanes produced by so-called Reaction Injection Moulding (RIM). This process is basically a closed-mould operation offering as a consequence mouldings with excellent surface appearance. At the same time the materials used in the process are extremely reactive and provide very fast cycle times.[67] Moulding pressures, or more exactly clamping pressures, involved in this process are extremely low (ca. 150 lb in.$^{-2}$) compared with, say, injection moulding (ca. 3000 lb in.$^{-2}$) or compression moulding (ca. 600–1000 lb in.$^{-2}$) and hence tooling costs are comparatively low.[22] All of these attributes make the RIM process singularly attractive to the rapid production of fairly large items in quantity, and as a consequence the largest motivator of study into RIM and RRIM is the automotive industry, particularly in the USA but also in Europe.[23]

The energy consumption of a RIM system is low both in absolute terms and relative to other plastics fabrication routes which are in themselves rather inefficient, requiring at least two melting steps (requiring energy) and shaping steps (losing energy) to produce a finished product. The particularly inefficient part in this cycle being that the first melting and shaping steps are required simply to produce the raw material (a granule) for the second steps. Although quantitative data on fabrication energies is difficult to generate precisely some information presently available indicates that the energy consumption of a RIM system is at least one order of magnitude lower than that of an analogous thermoplastics injection-moulding system.[24]

8.2.2 Chemistry of the RIM Process

As the name suggests, the RIM process involves the transference of reacting materials into a mould where after a greater or lesser time period they react to form a (usually) crosslinked polymer. The process hitherto has been applied commercially only to polyurethanes, since the reactivity of the polyurethane precursors is both fast enough and controllable enough to be commmercially acceptable. There is, however, a growing interest in the applicability of the process to any appropriate two (or more) component system. These include epoxies,[25] unsaturated polyesters,[26] polystyrene,[27,28] and nylon 6.[29]

With the exception of the nylon 6 system all the above materials can be used in combination with each other or with a polyurethane to produce so-called interpenetrating polymer networks (IPNs).[30] This family of materials have novel properties for a variety of applications and can be considered as 'chemically reinforced' systems according to the domain structure of the resultant material.[31]

As indicated above, the RIM process is exclusively applied to only polyurethanes on a commerical scale—to the extent of approximately 3×10^4 tonnes of material annually in the USA, as so-called high modulus elastomers, and this figure is expected to reach 10^5 tonnes by 1985. It is these materials that have seen the greatest interest from the reinforcement viewpoint, especially in the context of their use by the automotive industry. Some properties of RIM and RRIM systems are shown in Table 8.1. High modulus RIM elastomers are used in the automotive industry as the basis of the energy absorbing bumpers and 'soft front ends' required by Federal Legislation (FMVSS 214) to withstand impacts of up to 5 mph with no damage.

A detailed description of the chemistry of polyurethanes used in RIM is beyond the scope of this review. It is sufficient to indicate that apart from species such as catalysts, surfactants and cell stabilisers, there are three components in a typical RIM recipe for so-called high modulus elastomers. These are a difunctional isocyanate (MDI), a high molecular weight polyfunctional polyol and a low molecular weight polyfunctional polyol or chain extender. The functionality of all of these species can range between 2 and 6. The mechanical properties of the materials are dependent on the nature (functionality) and level of the three components. It is interesting to note that as the isocyanate's functionality is reduced to the theoretical level of 2 (corresponding to pure MDI) its compatibility with the polyol components is reduced.[32] This is of singular importance with regard to the mixing of the precursors discussed in Section 8.2.4.(The formulation and manufacture of low density flexible polyurethane foams is considered in more detail in Chapter 3.)

Although a variety of property changes can be effected by the chemistry of the system, there are certain properties which can only be controlled in this way at the expense of other equally desirable properties. Thus it becomes important to consider reinforcement of the basic system as an alternative to this molecular architecture approach.

TABLE 8.1
Properties of RIM and RRIM Polyurethanes

	System[a]						
	PBA 1501	PBA 1501	PBA 1478	PBA 1478	×	×	×
Glass (content)	0	WX6012(5)	WX6012(5)	WX6012(8)	0	WX6012(5)	P117B(20)
Density (kg m^{-3})	1200	1100	1140	1200	1100	1200	1100
Tensile strength (MPa)	31	21	21	24	30·2	32·2	27·6
Ultimate elongation (%)	140	20	45	40	80	25	40
Flexural modulus (MPa), −30°C	1100	3090	820	1160	1385	1572	2313
Flexural modulus (MPa), +20°C	620	970	520	820	616	935	1316
Flexural modulus (MPa), +70°C	410	820	290	460	422	838	664
Coefficient of thermal expansion (K^{-1} × 10^{-6})	92	66	58	42	100	67	88

[a] Imperial Chemical Industries Limited.
× Data described in refs. 67 and 69. See also refs 84, 24.

The motivation to examine reinforcement is specifically attributable to the following reasons:
1. To reduce the high coefficient of thermal expansion of the unfilled material to a value closer to that of aluminium, SMC or ideally, steel. This will enable RIM materials to be directly attached to such supporting members without distortion caused by temperature cycling.
2. To produce a flatter modulus temperature profile as well as a higher modulus. The flatter profile enables design data generated at a convenient temperature (say ambient) to be more generally applied to other temperatures without correction.
3. To reduce high-temperature creep. This is of specific interest to the automotive industry where large areas of unsupported material are painted at temperatures up to 170°C over a time period of up to 30–45 min. It is essential that the component does not sag during this process and the reinforcement rate is of fundamental importance.[81,82]

8.2.3 RIM Processing—Basics

Details of RIM processing equipment are reviewed in greater depth elsewhere,[22,23] and for the purposes of the present review it is sufficient to examine only the basic principles of the process in order to have an adequate perspective for the modifications required to process RRIM materials.

All RIM systems involve the transport of the polyurethane precursors (polyol blend and di-isocyanate) to a mixing chamber. Pumps must be accurately controlled and reproducible metering devices which can operate at the high pressures required by impingement mixing must be used. This has resulted in the widespread use of axial piston pumps which are capable of delivering up to 300 kg min.$^{-1}$ of polyurethane precursors at pressures up to 2500–3000 lb in.$^{-2}$ This can result in shot times of as low as 2–3 sec for some components so it is essential that the shot is delivered precisely with minimum interference from transients. Typical precision for high-modulus elastomeric RIM systems is ± 3% before noticeable changes in physical properties of the polymer are appreciable.

Axial piston pumps can be used successfully only with liquids of fairly low viscosity (ca. 5000 cP) and the use of a dispersed phase either as a particulate filler or a reinforcing fibre is precluded through the damage such a material would have on the mechanism of the pump.

8.2.4 RIM Processing—Impingement Mixing

For two fluids impinging from diametrically opposed injectors in a cylindrical mixing chamber, it has been shown that if the fluids be combined in a given ratio, the mass flow rates and momenta of each fluid must each be the same.[33-35] If the densities and mixing ratio are known, this enables the size of the injectors to be defined according to

$$\frac{A_1}{A_2} = \frac{\rho_2}{\rho_1} \cdot K^2 \tag{8.1}$$

where A is the area of the injector (1,2), ρ is the density of the two fluid streams (1,2) and K is the ratio (by weight) of the desired mix.

The rate of flow through each injector (V) can be from

$$V = Q/A \tag{8.2}$$

where Q is the volume of material to be delivered in unit time and hence the pressure drop across each injector is given by

$$\Delta P = \rho V^2 / 2 C_L^2 \tag{8.3}$$

where C_L is a constant defined by the geometry of the mixing chamber. From fluid mechanics, the Reynolds number of each stream is given by

$$\text{Re} = \rho V D / \mu \tag{8.4}$$

where D is the injector diameter and μ is the fluid viscosity, hence

$$\text{Re} = 2 \Delta P C_L^2 D / \mu V \tag{8.5}$$

for each component stream.

Studies of mixing phenomena using model systems involving titrations[36] and more realistic studies using polyurethane precursors and following the temperature rise during reaction[37] have each shown the desirability of having high Reynolds numbers for efficient mixing. In the first instance, a value of Re > 50 was shown desirable, whereas in the second case values closer to 200–250 showed best correlation with the temperature changes. This difference may well be attributable to the inherent incompatibility between typical polyurethane precursors compared with the materials used in the former study and, for interest, compared with, say, unsaturated polyesters or epoxies processed by impingement mixing.

It is in fact worth stressing that the newest high modulus RIM elastomer systems presently pursued in RRIM are singularly incompatible systems so much so that they must be catalysed sufficiently strongly that they may react chemically *before* they separate physically. This accounts for the very fast reaction times of such systems (< 5 sec gelation) and also accounts for the need for RIM equipment to have as high a throughput as possible (up to and beyond 300 kg min^{-1}) since only 1–3 secs shot time is 'available' with such systems and typical (automotive) parts may require 5–15 kg material.

Returning to the mixing studies the most important point to note from eqn (8.5) is that the 'goodness' of the mixing required a high driving pressure, and for a constant value of Re; this driving pressure must increase with an increase in viscosity of the polymer precursor. This is of fundamental importance to the success of filled or reinforced RIM systems where the filler or reinforcement is typically dispersed in one or both precursors as a slurry whose viscosity can be much higher than that of the parent unfilled precursor.

It is also interesting to note that with typical RIM components, eqns (8.2) and (8.3) together with the length dimension of a typical injector nozzle can produce shear rates of up to 10^5 sec^{-1} during injection. This compares with values measured during conventional injection moulding, and is at least an order of magnitude higher than that encountered in compression moulding.[38]

8.2.5 RRIM—Incorporation of the Reinforcement

Strictly to answer such a question begs an additional question: 'What is the reinforcement?' Let us assume that this must remain a choice as wide as possible and therefore may cover 'materials' ranging from particulate fillers to fibrous materials such as glass, carbon or Kevlar.

In principle with fibrous reinforcement there are three routes that may be used to incorporate the material in a RRIM system. The simplest involves placing the reinforcement in the form of a mat directly in the mould, equivalent to the process and materials described in Section 8.1.2. This has as yet seen no widespread use, since it is a rather slow process and as such defeats to some extent the attraction of RIM, namely its high rate of production. It is, however, reasonable to assume that this technique may be adopted with some large mouldings possibly requiring the use of carbon or Kevlar as reinforcement.

The second, and possibly most attractive, route involves the direct addition of fibrous or filler materials directly into the mixed component stream at a point between the mixing head and the after mixer. In this way no modification to the existing RIM equipment is necessary, since the reinforcement addition is external. The main difficulty encountered with this approach is the problem of design of a metering device for a solid material capable of delivering adequate throughput with adequate precision.

One successful application of this process involves the addition of 6 mm chopped strand glass in relatively large amounts to a rigid polyurethane by fluidising the glass with silica sand.[39] The delivery is simply by gravity into a modified runner of an otherwise conventional RIM set-up. The main disadvantage of this particular route is that the enhancement in property effected by the glass is offset somewhat by the relatively high level of sand required for adequate fluidisation. This applies particularly to the strength and toughness of the resultant composite material.

It is likely that a great deal of investigation of this general procedure is presently under way in various commercial and academic establishments, since, in principle at any rate, it offers one way by which a high loading of long glass can be incorporated in the system.

The third route, and the one which has been pursued commercially by equipment manufacturers, concerns the dispersion of the reinforcement in one or both of the polyurethane precursors and the consequent manipulation of these slurries. As indicated in Section 8.2.3. typical RIM pumps are unsuitable for such slurry processing and as a consequence most equipment manufacturers have adopted single-acting-piston dosing units as 'add-on' devices for existing RIM equipment. These units are operated by the pumps on the existing RIM equipment moving hydraulic oil instead of the polyurethane precursors.

Since the RRIM process is concerned more or less exclusively at present with the manipulation and mixing (by impingement) of both particulate and fibre-containing slurries, it is essential that some analysis of these procedures be presented in order that some apprettication of the limitations these place on the nature and level of dispersed phase incorporated in such slurries. This in turn is reflected in the mechanical properties of the resulting reinforced foams.

8.2.6. RRIM—Slurry Viscosities

For the purposes of this review a slurry is defined as an intimate mixture of a liquid which acts as a continuous phase and a particulate or acicular solid as a dispersed phase.

The simplest slurry consists of a mixture of small hard spheres dispersed in a Newtonian fluid. The relationship between slurry viscosity and volume fraction of spheres, ϕ, has been the subject of many investigations starting with that of Einstein[40] who showed theoretically

$$\eta_s = \eta_o(1 + 2\cdot 5\,\phi) \tag{8.6}$$

where η is the viscosity of the slurry (s) and the continuous phase (o).

This equation has been modified to include second and higher order volume fraction terms in the form of a virial expansion

$$\eta_s = \eta_o(1 + 2\cdot 5\,\phi + 14\cdot 1\,\phi^2) \tag{8.7}$$

This equation, due to Mooney,[41] has seen widespread use in that it has more general applicability than that of the Einstein equation above. This takes the form

$$\eta_r = \frac{\eta_s}{\eta_o} = \exp\left\{\frac{2\cdot 5\,\phi}{1 + S\phi}\right\} \tag{8.8}$$

where η_r is known as the relative viscosity and $S = P_m^{-1}$ where P_m is the maximum packing fraction of the dispersed phase. It has been shown that there is an intimate relationship between the mechanical properties of a composite and the viscosity of the corresponding slurry, which takes the form[42]

$$\frac{G_c}{G_m} \simeq \frac{\eta_s}{\eta_o} \tag{8.9}$$

where G is the shear modulus of the composite (c) and matrix (m).

This correlation lends itself to use composite analyses to interpret rheological behaviour of slurries, particularly when the dispersed phase is non-spherical.

These relationships are considered in more detail in the interpretation of mechanical properties of reinforced foams, but in the present context it is pertinent to consider one such relationship, due to Nielsen. This takes the form[43,44]

$$\frac{\eta_s}{\eta_o} = \frac{1 + (k - 1)B\phi}{1 - B\phi\psi} \tag{8.10}$$

where k is a generalised Einstein coefficient which takes into account the shape of the reinforcing element,[43] B

$$B = \frac{(\eta_d/\eta_o) - 1}{(\eta_d/\eta_o) + (k - 1)}$$

where η is the viscosity of the slurry (s), dispersed phase (d) and matrix (o) and

$$\psi = 1 + \left\{ \frac{1-\phi_m}{\phi_m^2} \right\} \phi$$

where ϕ_m is the maximum packing fraction of the dispersed phase (m) and ϕ the volume fraction of the dispersed phase.

Before commenting on the merits of these relationships, it is instructive to note that they all include a 'closeness of approach' parameter, and a parameter reflecting the geometry of the dispersed phase (k or A).

The maximum packing fraction has been studied for both spheres[45] and fibres[46,47] as well as mixtures thereof.[48] It can be measured directly from the sedimentation volume of a suspension either by gravity or centrifugation or it can be measured less precisely from the relative bulk volume of a dry solid material. For hammer-milled glass fibres, the relationship between ϕ_m and their average aspect ratio has been measured and it has been suggested[49] that the value of ϕ_m is independent of the aspect ratio distribution and depends only on the average value of this parameter. Which average value to take (mean or medium) is however debatable.

Various experimental results have been published on the viscosity of fibre-containing slurries, most of which over the last two years have been concerned with milled glass fibres in polyurethane precursors.[50–54] These latter studies have shown that the method of measurement (concentric cylinder, cone-and-plate, extrusion) influence greatly the values of measured relative viscosity. These values can range over two or more orders of magnitude between a simple rotating spindle viscometer (Brookfield) and an extrusion rheometer of the Instron type (constant shear rate). These discrepancies are undoubtedly in the main due to the differences in shear rates applicable to the different instruments and, in order to provide meaningful results in the context of RRIM, shear rates of the same order as those encountered in the processing equipment should ideally be used. Indeed, the most appropriate measuring device could be the machine itself, provided suitable instrumentation can be assembled.

This shear-rate dependence of the slurry viscosity (pseudoplasticity) arises through orientation of the dispersed phase, the extent of which will determine the value of the maximum packing fraction of the dispersed phase.

Figure 8.1 shows the theoretical relationships derived for a milled fibre slurry. The value of P_m for the equivalent aspect ratio measured

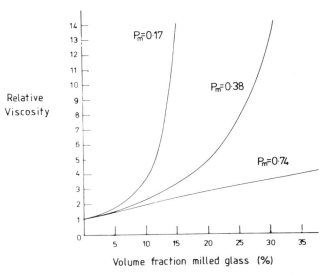

FIG. 8.1 Relative viscosities of slurries containing milled fibres assuming different maximum packing fractions for the fibres. The Einstein coefficient has been kept constant (11·5) throughout.

by sedimentation is 0·17, and the equivalent P_m for aligned fibres is 0·74. The results derived from these values of P_m are also included.

For typical milled-fibre loadings ($\phi = 0.3$) the relative viscosity of the slurry can thus be between 2 and 13 depending on the packing fraction of the dispersed fibres and hence the applied shear rate. Equation (8.10) also shows a marked dependence of slurry viscosity on fibre aspect ratio. In consequence, the impingement mixing of such slurries, the 'goodness' of which is derived by eqn (8.5), is more difficult unless the driving pressure is increased and by implication the capacity of the machine increased. In practice it is found that values of $Re \simeq 30$ provide adequate mixing, but there is at least an awareness that such values are approaching a lower limit. These low values do however enable existing RIM equipment to be tolerably efficient in mixing RRIM precursors provided a dosing unit such as that described above is used.

One further implication of eqn (8.9) and the results of Fig. 8.1 is that the expected mechanical property increase derived from the use of such reinforcement will be about half of the size of the relative viscosity (since the slurry is diluted 50:50 by the second precursor) giving $\phi = 0{\cdot}15$, namely between 1 and 7, probably closer to the value derived at the higher shear rates reflecting the planar orientation of the reinforcement. This sort of order compares reasonably with other GRP materials where modulus ratios of composite to matrix range between 2 and 10.

8.3 REINFORCED THERMOPLASTICS STRUCTURAL FOAMS

Thermoplastics structural foams (TSF) have been commercially available for many years and have presently widespread applicability to the machine housing sector and to the automotive market.

The structure of a TSF is analogous to that of an I-beam with hard, dense skins covering a low-density core. This gives the material an excellent stiffness-to-weight ratio, and hence makes it attractive to the manufacture of rather large unsupported sections otherwise precluded to more conventional plastics (more correctly thermoplastics). The mechanical properties of TSF and the relevance of the I-beam concept are discussed in Chapter 7.

Incorporation of reinforcement into TSF follows the route adopted by thermoplastics in general, namely the use of a masterbatch of reinforced polymer prepared usually by a compounding extruder or a crosshead extruder. The reinforcement is usually 3–6 mm chopped glass fibres treated with an appropriate coupling agent suitable for the polymer matrix.

Properties of reinforced TSF reflect in general those attributes required of RRIM and described in the previous section. Thermal expansion coefficients are reduced, temperature of deflection under load (so-called heat-distortion temperature) is increased, creep is reduced, the modulus is increased and the impact properties improved over those of the unreinforced foam.

Applications for reinforced TSF are growing, although at present restricted to very large mouldings or moulding operations at high temperatures. An example of the latter is a washing machine tub currently in commercial use.[55] Reinforced TSF was used in this case to provide stiffness with good high temperature creep resistance. Some comparative properties of TSF with and without reinforcement are shown in Table 8.2.

TABLE 8.2
Effect of Reinforcement on the Properties of Polypropylene Structural Foam[83]

Glass content (%)	Density ($\times 10^3$ kg m^{-3})	Flexural strength (MPa)	Flexural modulus (MPa)	Tensile strength (MPa)	Ultimate elongation (%)
0[a]	0·905[a]	48·7[a]	1·73[a]	34·6[a]	200–700[a]
20[a]	1·04[a]	70·2[a]	4·16[a]	63·0[a]	2–3[a]
20[a]	0·92[a]	61·0[a]	3·11[a]	52·6[a]	2–3[a]
20	0·73	43·6	2·77	22·9	2–3

[a] Solid polymers.

8.4 MODEL SYSTEMS FOR REINFORCED FOAMS

8.4.1 General Considerations

The derivation of a model to describe the mechanical properties and thermal properties of a reinforced foam provides a basis for the interpretation of structure/property relationships which in turn may be used to derive and predict design data for these materials.

Any model system must begin with an intimate knowledge of the spatial relationships of the components of the model as well as a knowledge of the properties of these components. In practice, this detailed knowledge derives from extremely careful microstructure characterisation before, during and after the application of a mechanical or thermal perturbation of the system. Such detailed microstructure is beyond the scope of this review and is considered in greater detail in Chapter 1. It is important that reference be made directly to this to confirm otherwise apparently unsubstantiated premises upon which some of the models described are based. Further microstructure details, for example in the determination of the role of struts, windows and nodes in a low-density foam during the propagation of a crack may be found in the original references cited (see also Chapter 2A).

For convenience, this section is divided into two parts, the first dealing with high-density foams reinforced with milled glass fibres and chopped strands. This discrimination is due entirely to the basic differences in microstructure leading to, in the first case, a composite with reinforcement and gas bubbles distributed throughout the matrix, and, in the second case, to reinforcement dispersed throughout and in some cases within the constituent skeleton of the foam itself.

The only phenomenon common to both types of structure is the effect the reinforcement has on the size and shape of the cells within the foam. In both cases, the heat-sink effect of the fibre causes it to be encased in predominantly solid matrix and, as a consequence no fibre penetrates a cell wall. In general this leads to much finer cell sizes in a reinforced foam compared with the equivalent unreinforced material, and as a consequence an improved 'matrix' property within the composite. More direct experimental evidence for this is discussed further in the original references.

8.4.2 High-Density Reinforced Foams—RRIM and TSF

Densities of typical RRIM systems range between 900 kg m^{-3} and 1200 kg m^{-3} and those of reinforced TSF range between 800 and 1000 kg m^{-3}. In each case the materials are prepared in closed moulds and, as a consequence, there exists a density distribution throughout any section analogous to that described for their unreinforced counterparts.[56] The lowest density is encountered in the middle of a section and, in each case, particularly in RRIM, is greater than about 500 kg m^{-3}. As a consequence of this, the microstructure of such reinforced materials is composed of discrete bubbles of gas distributed throughout a glass-reinforced matrix. These gas bubbles have, at least in typical systems, a spherical structure with no preponderance to the dodecahedral structure applicable to low-density foams, shown in Section 1.3. There is thus no discrimination necessary between cell struts and windows and these are simply described as cell walls.

These materials can therefore be described as three-phase composites either as foamed reinforced systems or reinforced foamed systems, depending on the notional order with which the final composite is considered assembled. The description of the mechanical properties of such a composite is therefore in its simplest terms a combination of the appropriate relationships which describe suitable permutations of two-phase composites (gas–solid and reinforcement–solid) taking account of the relative volume fractions of each of the constituents. This procedure has previously been adopted for gas–filler–resin systems[57] and for filler–fibre–resin systems.[58] Relationships describing the mechanical properties of gas–solid systems are reviewed elsewhere[59,60] and those concerned with fibrous reinforced solid systems have also been reviewed.[61]

Before proceeding with this analysis, some relationships pertinent to the structure of a reinforced foam should be noted in order to preserve consistent nomenclature throughout this section. These can be found as follows.

Consider a reinforced foam of total volume V_t, consisting of matrix, V_m, reinforcement, V_r, and gas V_g. Let the mass of each of the components be written as M_x where x can be m, r or g.

The following relationships can be defined.

$$\rho_{rf} = \frac{M_m + M_r}{V_t} = \rho_m \phi_m + \rho_r \phi_r \quad (8.11)$$

$$\phi_r = \frac{V_r}{V_t}; \quad \phi_m = \frac{V_m}{V_t}; \phi_g = \frac{V_g}{V_t}$$

where ϕ is the volume fraction of each of the constituents. Define now the following parameters

$$\rho_f^* = \frac{M_m}{V_m + V_g}$$

$$\rho_g^* = \frac{V_g}{V_m + V_g} \quad (8.12)$$

$$\rho_r^* = \frac{V_r}{V_m + V_g}$$

where the asterisk denotes the analogous property of a two-phase system (composite or foam), but these are measured from the relationships applicable to the three-phase system, namely the reinforced foam.

Using eqns (8.11) and (8.12), further relationships can be derived by simple algebra. These are

$$\rho_f^* = \frac{\rho_m(\rho_{rf} - \rho_r \phi_g)}{\rho_{rf} - \rho_r \phi_r + \rho_m \phi_g} \quad (8.13)$$

and

$$\phi_r^* = \frac{\phi_r \rho_m}{\phi_r \rho_m + \rho_{rf} - \rho_r \phi_r} \quad (8.14)$$

Consider now the series of relationships due to Kerner[62] and Halpin and co-workers.[63,64] These can be written for a unidirectional, discontinuous fibre-reinforced composite in the form

$$E_{c_{11}} = \frac{1 + AB\phi}{1 - B\phi} \quad (8.15)$$

where

$E_{c_{11}}$ is the composite modulus,
E_m is the matrix modulus, A is twice the aspect ratio of the reinforcement and

$$B = \frac{(E_r/E_m) - 1}{(E_r/E_m) + A}$$

The analogous transverse modulus $E_{c_{22}}$ and that due to a spherical reinforcing material can be written in the same form as eqn (8.15) with $A = 2$. The randomly reinforced composite properties can be found from a linear combination of the two equations and expressed as[64]

$$E_c = \tfrac{3}{8} E_{c_{11}} + \tfrac{5}{8} E_{c_{22}} \qquad (8.16)$$

This can be considered via $E_{c_{22}}$ above with $E_r = 0$ and $A = 2$ hence if the modulus of the resulting foam is E_f

$$\frac{E_f}{E_m} = \frac{1 - \phi_g^*}{1 + \phi_g^*/2} \qquad (8.17)$$

where ϕ_g^* is the volume fraction of gas in the two-phase composite.

$$\frac{E_{c_{11}}}{E_m} = \frac{1 + AB\phi_g^*}{1 - B\phi_g^*} \qquad (8.18)$$

$$\frac{E_{rf}}{E_{c_{11}}} = \frac{1 - \phi_g}{1 + \phi_g/2} \qquad (8.19)$$

where E_{rf} is the modulus of the reinforced foam. Combining eqns (8.17)–(8.19) gives

$$\frac{E_{rf}}{E_f} = \frac{1 + AB\phi_g^*}{1 - B\phi_g^*} \cdot \frac{\left(\dfrac{1 - \phi_g}{1 + \phi_g/2}\right)}{\left(\dfrac{1 - \phi_g^*}{1 + \phi_g^*/2}\right)} \qquad (8.20)$$

when $\phi_g = \phi_g^*$, i.e. when the unreinforced foam has the same density as the 'matrix' in the reinforced foam the second term on the right-hand side of eqn (8.20) disappears and the relationship is directly analogous to the original form of the Halpin–Tsai equation. The volume fraction of glass however, is not that measured directly from the composition of the reinforced foam but given by eqn (8.14). E_{rf} expressed by eqn (8.20) measures the modulus of the foam in the injection direction, but there is no reason to ignore the more complex analysis involving the equivalent transverse modulus to give a random figure.

Note that the value of B in eqn (8.15) contains the ratio of reinforcement modulus to *solid* matrix modulus.

An alternative approach to this can be based on a simpler analysis of the unreinforced foam properties—for example using the reduced density relationship.[65]

$$\frac{E_f}{E_m} = \left(\frac{\rho_f}{\rho_m}\right)^2 \tag{8.21}$$

this foam can then be reinforced by the addition of material (E_g, ϕ_g) to give

$$\frac{E_{rf}}{E_f} = \frac{1 + AB_1\phi_g}{1 - B_1\phi_g}$$

and hence

$$E_{rf} = E_m \left(\frac{\rho_f}{\rho_m}\right)^2 \left(\frac{1 + AB_1\phi_g}{1 - B_1\phi_g}\right) \tag{8.22}$$

where B_1 contains the ratio of the reinforcement modulus to the modulus of the unreinforced foam.

Equation (8.20) has been used to analyse RRIM systems based on a high modulus elastomer matrix,[66] the results of which are shown in Fig. 8.2.

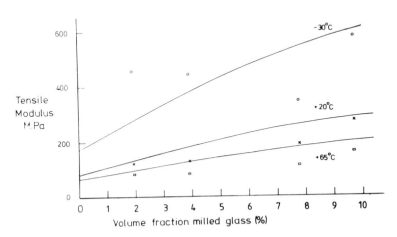

FIG. 8.2 Tensile moduli for reinforced foams containing $\frac{1}{16}$ in. hammer-milled glass.[24] Symbols show experimental points and the continuous curves are derived from eqn (8.20).

Agreement can be described as reasonable with some room for further investigation, particularly in the area of the interface between the reinforcement and matrix.

The constitutive relationships used apply to circumstances of perfect adhesion at the interface. This, as seen from Fig. 8.3, is dependent on the nature of the coupling agent present on the reinforcement, and a great deal of study is presently concerned with the optimising of this bond.[67] One such system has been developed for use with short-chopped-strand glass reinforcement to exploit further the higher aspect ratio of this material compared to that of the milled glass fibres referred to directly in the original analysis.

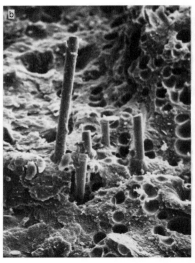

FIG. 8.3 Effect of different coupling agents on the bonding of hammer-milled glass fibres in the same RIM matrix. (a) 25% $\frac{1}{16}$ in. fibres with treatment P731. (b) 20% $\frac{1}{16}$ in. fibres with treatment P117B.

Fracture toughness and failure mechanisms applicable to RRIM systems have not been sufficiently investigated to warrant further comment in this review. There is however some evidence that the analysis developed below for low-density foams is sufficiently general to be applicable to these systems. This is in spite of the difference in matrix microstructure involving in the RRIM case no formal discrimination between cell struts and windows and the low-density foam core; a more geometrically precise (and hence analysable) structure. Evidence for this is predominantly from fracture surface microscopy and an example of this is shown in Fig. 8.5(b).

The benefit of higher aspect ratio of reinforcement on modulus is negated somewhat by the problems of incorporation of the material into the system as a slurry in one or both precursors. In practice, it is found that the same modulus enhancement is obtained using low volume fractions of 'long' glass as is obtained with higher volume fractions of short glass. This in itself suggests an irrelevance of which system is used in practice, but with the longer glass there are additional advantages in processing, such as less material to dry before the preparation of the slurry (hence less energy used) and, depending on the relative material costs less inventory depreciation.

8.4.3. Low-Density Reinforced Foams

Short Chopped Fibres
Short chopped fibres usually 3, 6 or 12 mm in length, can be used to reinforce low-density foams. These can be incorporated in the foam matrix either by blending with one or both precursors before mixing, or by chopping directly into the mixed component stream in a spray up operation. In the former case, the disadvantages described for RRIM (high slurry viscosity, difficult mixing, etc.) apply although the overall gel time of these low-density foams is generally longer than that associated with RIM systems, and hence mixing times may be extended if required.

Chopped fibres usually exist in the form of strands of a given overall tex (linear density), and these strands may be divided into smaller bundles according to how the strand was assembled in manufacture. The chemical size on the strand determines the rate of conversion of strand to bundles, and bundles to individual filaments, according to its solubility in the resin matrix and the rate of shear experienced by the material in the mixing operation. This, when incorporated in the urethane matrix is coated with resin and following expansion of the matrix, becomes in essence a 'composite strut'. These struts bridge across several cells with the individual cell struts radiating outwards, as shown in Fig. 8.4. There may be a local variation in cell size around the fibres, particularly if the fibre bundle is large and can hence act as a significant heat sink during

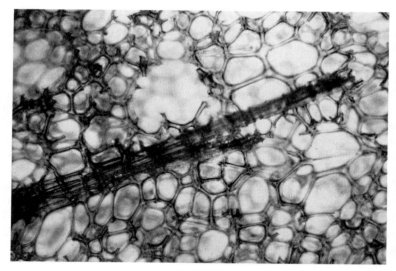

FIG. 8.4 Morphology of a low-density foam reinforced with 3 mm chopped glass.

the foaming process, thus reducing the expansion of the cells adjacent to the fibre bundles.

Table 8.3 shows the properties of a series of sprayed foams of density 80 kg m^{-3}, reinforced with various lengths of chopped fibres.[19,20,70] The properties of these foams exhibit a strong dependence on fibre length, with maximum strengths and moduli occurring with fibres of length 12 mm. This phenomenon can be explained by examining the behaviour of the fibres during tensile testing of the foams.

TABLE 8.3
Effect of Fibre Lengths on the Mechanical Properties of a Low-density (80 kg m^{-3}) Rigid Sprayed Polyurethane Foam

Fibre length (cm)	Tensile strength (kPa)	Tensile modulus (kPa)	$K_{IC}(kN^{-3/2})$	Compressive strength (kPa)^a	Compressive modulus (kPa)^a
0	1035	22.4	106.9	641.9 (705.3)	25.1 (22.5)
5	1082	27.9	129.5	667.9 (784.8)	26.8 (26.2)
12	1095	34.0	146.4	636.1 (797.3)	23.2 (25.6)
20	960	35.4	129.6	592.1 (708.1)	24.7 (25.3)
40	966	30.3	115.6	561.4 (627.1)	20.3 (24.3)

^{a} Properties measured parallel and (perpendicular) to foam rise.

A crack propagating through a foam under tensile load, towards a bundle of short fibres, is arrested as it reaches the bundle. As the load is increased, the crack is diverted parallel to the fibre bundle, fracturing the radial foam struts as it propagates towards the end of the bundle. This process results in the creation of a pull-out fragment on the fracture surface. Single glass filaments of length 5 mm and 12 mm, act in a similar manner; cracks are arrested and diverted, and pull-out fragments created. A typical pull-out fragment is illustrated in Fig. 8.5(a). The larger fibres of 20 mm and 40 mm, however, fractured in the plane of the matrix, crack without any crack diversion and subsequent pull-out. This is an illustration of the concept of critical fibre length, which is the length above which a fibre will fracture, and below which it will pull out from the matrix.

The critical length for fibres in a solid matrix can be calculated[71] and a similar analysis can be applied to the case of the fibres in foam.[72] This analysis may also be applied to RRIM systems (Fig. 8.5b).

When a tensile stress is applied to a reinforced foam, stress is transferred from the matrix to the fibres via both the resin sheath around the fibres and the foam struts which radiate from this. A tensile stress builds

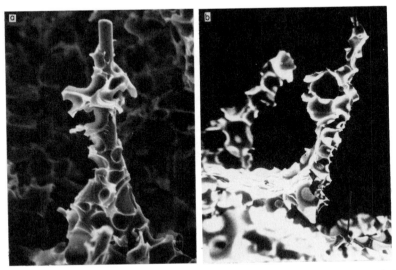

FIG. 8.5 Pull-out fragment in the fracture surface of (a) RRIM system containing $\frac{1}{16}$ in. hammer-milled glass; (b) low-density foam reinforced with 3 mm chopped glass.

up from the ends of the fibres to a maximum at their centre. If this exceeds the ultimate tensile stress of the fibre, then the fibre will fracture. Also, because of the difference in modulus between the fibres and the foam matrix, a shear stress is set up between them along the length of the fibres. If this shear stress exceeds the shear strength of the matrix, then the matrix will fail close to the fibre producing a pull-out fragment.

Consider a single fibre in its resin sheath, aligned parallel to the tensile axis. The simple rule of mixtures gives the stress as

$$\sigma_c = \sigma_f \phi_f + \sigma_s \phi_s \tag{8.23}$$

Where σ is the stress and ϕ the volume fraction and c refers to the composite (fibre + resin sheath), f the fibre and s the sheath. Expressing the volume fractions in terms of the radii of the different material elements gives

$$\sigma_c = \sigma_f \pi r_f^2 + \sigma_s \pi (r_s^2 - r_f^2)$$

Since $\sigma_f \gg \sigma_s$ and $r_s < 2r_f$ the contribution of the resin sheath can be neglected so

$$\sigma_c = \sigma_f \pi r_f^2 \tag{8.24}$$

The shear stress acting on the cylinder of foam surrounding the fibre is

$$\tau_c = \tau_m \pi \, dl \tag{8.25}$$

where τ_m refers to the foam matrix, d is the diameter of the shear zone, and l the fibre length. When $\sigma_c = \tau_c$

$$\sigma_f \pi r_f^2 = \tau_m \pi \, dl$$

This can be rearranged as

$$l = \frac{\sigma_f r_f^2}{\tau_m d}$$

Since the above applies to half the fibre length, then the critical length is

$$l_c = \frac{2\sigma_f r_f^2}{\tau_m d} \tag{8.26}$$

The following values are used to calculate the critical length

$\sigma_f = 3 \cdot 65$ GPa.
$r_f = 6 \cdot 5 \, \mu$m.
$\tau_m = 0 \cdot 51$ MPa.

The diameter of the shear zone is taken as the filament diameter plus one quarter of the cell size, where the cell size = $142\,\mu m$ so that $d = 48\cdot5\,\mu m$.[74] Using these figures, the critical length for the foams in Table 8.3 is $l_c = 12\cdot5$ mm. This figure is supported by the experimental observations that fibres of 5 mm and 12 mm pulled out, whereas those of 20 mm and 40 mm fractured in the plane of the matrix crack.

The value of critical length varies with foam density, since this in turn determines τ_m the shear strength of the foam. The cell size, and hence shear-zone diameter, is also density dependent. The expression for critical length, eqn (8.26), can be generalised by substituting expressions for the dependence of shear strength and cell size on foam density[74,75]

$$\tau_m = 5\cdot71\,\rho^{1\cdot1} \text{ (kPa)}$$
$$\text{cell size} = 15\cdot3 \times 10^{-3}\rho^{-1\cdot1} \text{ (m)}$$

Thus
$$d = 3\cdot825 \times 10^{-3}\rho^{-1\cdot1} + 2r_f$$

Substituting these into eqn (8.26) gives

$$l_c(\text{m}) = \frac{2\sigma_f r_f^2}{5\cdot71 \times 10^{-3}\rho^{1\cdot1}(3\cdot825 \times 10^{-3}\rho^{-1\cdot1} + 2r_f)}$$

This can be further generalised by considering σ_f and r_f as constants. Then

$$l_c = \frac{14\cdot2 \times 10^{-3}}{1 + 3\cdot4 \times 10^{-3}\rho^{1\cdot1}} \tag{8.27}$$

As seen from the results in Table 8.3, fibre length has an effect on fracture toughness and critical strain-energy release rate, these having maxima for fibres of length close to the critical length. Values of G_{IC} versus fibre length are plotted in Fig. 8.6. Fibres of length close to l_c produce the longest pull-out fragments and hence require the most energy to pull-out.

The form of the curve of G_{IC} in Fig. 8.6 can be explained by considering the work of fracture, since this is twice G_{IC}. The work of fracture due to fibre pull-out, is derived as follows:[76] all fibres crossing the plane of fracture, with ends within a distance $l_c/2$ from the plane of the matrix crack, will pull out during fracture. The fraction of fibres pulled out, is l_c/l. The work of fracture done in pulling out a fragment whose end is a distance x from the matrix crack is

$$W_f = \pi \left(\frac{d_s}{2}\right)^2 \int_0^x \sigma\,dx \tag{8.28}$$

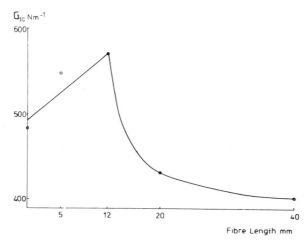

Fig. 8.6. Dependence of critical strain energy release rate (G_{IC}) on the fibre length of the reinforcement in a low-density (80 kg m^{-3}) polyurethane foam.

where d_s is the diameter of the shear zone. From eqn (8.26)

$$\sigma = \frac{\tau_m d_s x}{2 r_f^2}$$

Substituting in eqn (8.28), the total work done in withdrawing all of the pull-out fragments is (from eqn (8.26))

$$W = \frac{V_f}{\pi \left(\frac{d_s}{2}\right)^2} \frac{l_c}{l} \int_0^{l/2} \frac{W_f \, dx}{l_c/l}$$

(8.29)

$$\text{i.e.} \quad W = \left(\frac{V_f l_c}{24 \, l}\right) \cdot \sigma_f l_c$$

For fibres of length $l > l_c$ the work of fracture due to pull-out, is proportional to $1/l$ which is the shape of the curve of G_{IC} for fibre lengths above the critical length of 12·5 mm.

For fibres of length $l < l_c$ the work of fracture is found by setting $l_c = l$ in eqn (8.29), then

$$W = \left(\frac{V_f \tau_m d_s}{48 \, r_f^2}\right) \cdot l^2$$

Thus for fibres shorter than the critical length, the work of fracture, and hence G_{IC}, should be proportional to l^2. In practice this relationship is not strictly obeyed by typical foam systems, particularly sprayed foams, because of a more dominant effect played by the orientation of the fibres. Sprayed foams are built up in layers by repeated passes of the spray gun until the required thickness of material is attained. The layers of foam are separated by so-called weld lines consisting of thin regular layers of higher-density foam. The fibres are constrained within the layers and do not cross the weld lines. With the 80 kg m^{-3} foams reported, the layer thickness is typically 5–8 mm, and hence fibres longer than this cannot become aligned in the rise direction. Fibres of 5 mm, however, can be aligned in the rise direction in these foams. A maximum work of fracture is obtained with fibre length $\approx$ 12 mm, as seen in Fig. 8.6. Presumably this phenomenon could be exploited to optimise the work of fracture by increasing the layer thickness so that fibres of critical length can be oriented in the rise direction.

This orientation effect can also be seen in the compressive properties of foams tested parallel to the rise direction, where the foams with 5 mm fibres had the highest compressive strength. When foams are tested in compression, perpendicular to the rise direction, however, where there is no restriction to the alignment of fibres, then the highest strength is exhibited by the foams containing 12 mm fibres, these being closest to the calculated critical length.

The effect of fibre bundles on the compressive behaviour of a low density polyurethane foam is illustrated in Fig. 8.7, where a compression band propagating through the specimen, has been arrested by a fibre bundle of less than the critical length. A second compression band has then initiated at the other end of the fibre bundle. With fibres longer than the critical length, the failure mode in compression is for fibres at an angle to the compression axis to try to rotate within the foam, causing compressive failure in the radial foam struts. This initiates a compression band which then propagates through the specimen.

Hammer-milled Fibres

The structure of low-density rigid foams consists of a three-dimensional network of struts which form the edges of pentagonal dodecahedral cells (Fig. 2.3). Restrictions to the expansion of the cells during foaming generally causes the cells to be elongated in the rise direction, as explained in Section 2A.3 and Section 1.3. The struts are roughly triangular in cross-section, and may or may not be covered by thin

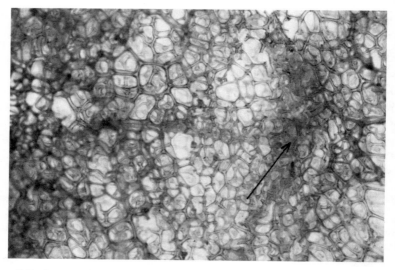

Fig. 8.7. Arrest of compression band (arrowed) by a fibre bundle in a low-density polyurethane foam reinforced with 3 mm chopped glass fibres.

membranous windows, depending on whether the foam is closed cell or open cell. The majority of the polymer is concentrated in the struts, which are the main load-bearing elements of the structure.

Under compressive loading, the struts support the load until a point is reached where their critical buckling load is exceeded. The struts then buckle, causing a collapse of the cellular structure and yield of the foam.

One method of strengthening low-density foams is to incorporate very short glass fibres into the struts, producing composite struts with a higher critical buckling load than unreinforced struts.[77-79] Short glass fibres are available commercially as hammer-milled glass fibres with a typical maximum length being 400 μm. These are easily added to low-density polyurethane foams by preblending with the polyol before mixing with the isocyanate. The viscosity of the polyol–glass slurry is not as high as with the longer, chopped fibres and fibre loadings of up to 30% by weight are possible. That the fibres do align within the struts is illustrated in Fig. 8.8, which shows the fracture surface of a low-density foam with a fibre pulled out from a broken strut. The fibres are not visible by optical microscopy, but since they cannot be seen, they must lie 'hidden' within the polymer structure.

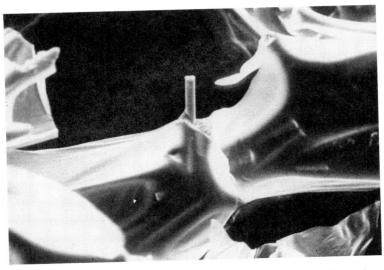

FIG. 8.8. Low-density foam containing hammer-milled glass fibres showing the alignment of the fibres within the foam struts.

The values of compressive modulus and strength of a series of low-density rigid polyurethane foams of density 80 kg m^{-3} presented in Table 8.4 show an increase in strength and modulus with increasing fibre content. The compressive strength of a low-density rigid foam can be rationalised by considering the critical buckling load of a strut under axial loading as described in detail in Chapter 2A. This is given

TABLE 8.4
Effect of Hammer-milled Glass-fibre Reinforcement on the Properties of Low-density (80 kg m^{-3}) Rigid Polyurethane Foam

Fibre loading (% by weight)	Compressive strength (MPa)	Compressive modulus (MPa)
0	2·11	48·4
5	2·21	50·6
9	2·26	62·6
15	2·54	67·9
25	3·11	89·3

by the Euler formula[80]

$$P_{cr} = \frac{\pi^2 EI}{(kl)^2}$$

Where E is Young's modulus, I is the second moment of area of the strut, l is the strut length and k is a factor that depends on the restraints on the ends of the strut. If the strut is considered to contain reinforcement, then the rule of mixtures can be used to predict the modulus.

$$E_r = \phi_f E_f + \phi_m E_m$$

where ϕ is the volume fraction and the subscript r refers to the reinforcement strut, f the fibre and m the solid polymer. The volume fraction of fibre within the strut is observed to be about 0·5. The modulus of glass fibres is 76 GPa and the modulus of the solid polyurethane is taken as about one tenth of this hence

$$E_r \simeq 41 \cdot 8 \text{ GPa}$$

The increase in modulus compared to an unreinforced strut is

$$\frac{E_r}{E_m} \simeq \frac{41 \cdot 8}{7 \cdot 6} \simeq 5 \cdot 5$$

Since the strut is approximately triangular in cross-section, see for example Fig. 1.3, its second moment of area is

$$I = \frac{bh^3}{36}$$

where b is the length of the base of the triangle and h its height. If adding a fibre to a strut doubles its cross-sectional area, then,

$$I_r = 4I_m$$

Thus adding a fibre to a strut increases its modulus by a factor of 5·5 and doubles its area, so that I is increased by a factor of four, and hence the critical buckling load is increased by a factor of 22. Not every strut is reinforced, however, and the fraction of struts containing fibres can be estimated from the cell size and the volume fraction of fibres. A foam of 80 kg m^{-3} with 25% fibres has about 3% of the struts reinforced. The rule of mixtures can be used to predict the overall increase in compressive strength since this is directly proportional to the critical buckling load of the individual struts.

$$P_{rf} = N_m P_m + N_r P_r$$

where rf refers to the reinforced foam and N is the number fraction of struts. Since

$$P_{rf} = 0.97 P_m + 0.03 \times 22 P_r = 1.63 P_m$$

Thus the compressive strength of an 80 kg m^{-3} foam containing 25% hammer-milled fibres should be 1·63 times stronger than the equivalent unreinforced foam. The results presented give a value lower than this, namely 1·47. The reason for this is the presence of curved struts in the structure, one of which is shown in Fig. 8.9(a). These curved struts are

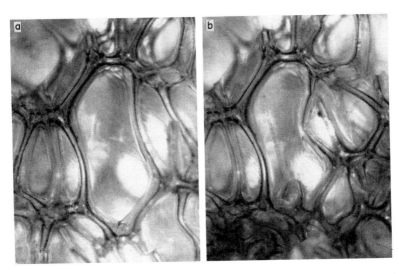

FIG. 8.9. Low-density foam containing hammer-milled glass fibres showing the presence of curved struts. (a) Before failure; (b) after failure.

produced during foaming by the internal pressure within the expanding cells. Struts containing fibres offer more resistance to expansion, hence the unreinforced struts tend to bow out under the pressure. A curved strut has a lower critical buckling load than an equivalent straight one, and a microscopic examination of foams under compressive load shows that the curved struts tend to fail at a lower stress than the straight struts (Fig. 8.9). An estimation of the critical buckling load of a curved strut is a fraction of that of an equivalent straight strut then,

$$P_{rf} = N_r P_r + N_m P_m + N_c P_c \qquad (8.30)$$

where c refers to the curved strut. Observations indicate that the fraction of curved struts is about one in five, and since $P_{rf} = 1 \cdot 47\, P_m$

$$1 \cdot 47 P_m = (0 \cdot 03 \times 22 P_m) + 0 \cdot 77 P_m + 0 \cdot 2 n P_m$$

This expression gives $n = 0 \cdot 3$, i.e. the critical buckling load of a curved strut in the foam, is about one fifth that of a straight strut. In order to further increase the strength of a reinforced low-density foam, the term N_r in eqn (8.30) must be increased. This involves increasing the volume fraction of fibres and exercising greater control over the fibre-length distribution, since with hammer-milled fibres, many of the individual fibre are too short to have any reinforcement value and act merely as fillers. It is also conceivable that the number of curved struts could be reduced by the selective use of specific surfactants.

REFERENCES

1. HOPPE, P. (to Bayer A. G.) UK Patent 1191902 (1970) and 1233910 (1971).
2. HOPPE, P. (to Bayer A. G.) German Patent 2253323 (1974).
3. PENN, L. (1977) *Reinforced Plastics*, **21** (9), 279.
4. 'Cermat' Product Data Sheet, T.B.A. Industries Ltd.
5. HOPPE, P. *et al.* (1972) *Kunststoffe*, **62**, 735.
6. US Patent 3916060 (1975); see also ref. 7.
7. GOTCH, T. M. (1979) *Plast. Rubb. Int.,* **4** (3), 119.
8. ANON (1974) *Mod. Plast. Int.*, **4** (6), 14.
9. MURPHY, J. (1976) *Du Pont Mag.* **70** (3), 19; Anon (1975), *Plastics Rubber Weekly*, 574, 16.
10. KLEINHOLZ, R. and NEUMANN, A. (1974) *Kunststoffe*, **64**(12), 742.
11. CHEREPANEV, V. P., SHAMOV, I. V. and TARAKANOV, O. G., (1974) *Plastichevski Massay*, **5**, 73.
12. MODIGLIANA, P. (1970) 25th SPIR/P Conf., 8-C.
13. GALLACHER, R. B. *et al.* (1976) 31st SPIR/P Conf., 6.
14. ANON (1978) *Mod. Plastics Int.*, **8**(6), 37.
15. WOOD, A. S. (1980) *Mod. Plast. Int.*, Dec, 28.
16. KLÖCKER, W. (1978) BPF R/P Conference, Brighton, 271.
17. LANGLIE, C. *et al.* (1972) *Kunststoffe*, **62** (12), 790.
18. Mitex Technical Data, Cellmico Ltd., 1978.
19. KLÖCKER, W. (1978) *Eur. J. Cell. Plast.,* Oct. 186.
20. COTGREAVE, T. C. and SHORTALL, J. B. (1978) *J. Mater. Sci*, **13**, 722.
21. COTGREAVE, T. C. and SHORTALL, J. B. (1977) *J. Mater. Sci.*, **12**, 708.
22. SCHULTE, K. W., BODEN, H., SEEL, K. and WEBER, C. (1979) *Ind. Prod. Eng.*, **3**, 13.
23. METZGER, S. G. (1980) *Reaction Injection Moulding*, Van Nostrand, New York.
24. METHVEN, J. M. and SHORTALL, J. B. (1978) *Eur. J. Cell. Plast.,* Jan, 27.

25. WADDILL, H. G. (1980) 35th SPI R/P Conference, 22-B.
26. CHARLESWORTH, D. (1980) *Plast. Rubb. Int.*, Sept; also ref. 28.
27. KUBIAK, R. S. and HARPER, R. C. (1976) 34th ANTEC, 481.
28. KUBIAK, R. S. and HARPER, R. C. (1976) 34th ANTEC, 59.
29. KIRCHER, K. and MENGES, G. (1976) *Plastics Engineering*, Oct, 37.
30. KIRCHER, K. and MENGES, G. (1976) 34th ANTEC, 271.
31. MASON, J. A. and SPERLING, L. H. (1976) *Polymer Blends and Composites*, Plenun Press, New York, Chap. 8.
32. MCBRAYER, R. L. (1977) Design Forum and Equipment Seminar, Oct. 10–14.
33. MALGUARNERA, S. C. and SUH, N. P. (1977) *Polym. Eng. Sci.*, **17**(2), 111.
34. *Idem, ibid*, p. 116.
35. LEE, L. J. and MACOSKO, C. W. (1978) SPE ANTEC Washington.
36. TUCKER, C. L. and MACOSKE, N. P. *ibid*, p. 158.
37. LEE, L. J. and MACOSKE, C. W. *ibid*, p. 151.
38. BROWN, J. (1978) *Injection Moulding of Plastics Components*, McGraw-Hill, p. 113 ff.
39. US Patent 5005035 (Technics Inc.)
40. EINSTEIN, A. (1906) *Ann. Phys.*, **4**, 19.
41. MOONEY, M. (1951) *J. Colloid Sci.*, **6**, 162.
42. NIELSEN, L. E. (1968) *J. Comp. Mat.*, **2**, 120.
43. NIELSEN, L. E. (1970) *J. Appl. Phys.*, **41**, 4626.
44. NIELSEN, L. E. (1969) *Mechanical Properties of Polymers and Composites*, Marcel Dekker, New York; also ADAMS, D. F. and TSAI, S. W. (1969) *J. Comp. Mat.*, **3**, 368.
45. VAN KAO, S. and NEILSEN, L. E. (1974) Monsanto/Washington University ONR/ARPA Association HPC-74-166.
46. MASCHMEYER, R. O. and HILL, C. T. (1974) *idem* HPC-74-170: also BAYLY, M. B. (1969) *J. Comp. Mat.*, **3**, 705.
47. MILEWSKI, J. V. (1973) 28th Am. SPI R/P Conference, Section 3F(1).
48. MILEWSKI, J. V. (1974) 29th Am. SPI R/P Conference, Section 10-B(1).
49. ISHAM, A. B. (1977) Int. Conference Polymer Processing, MIT Cambridge, Mass.
50. SCHULTE, K. W., BODEN, H., SEEL, K. and WEBER, G. (1978) *Kunststoffe*, **69**(9), 510.
51. LEIDTKE, M. W. (1978) *J. Cell. Plast.*, **14**(2), 102.
52. ZLOCHOWER, I. A., DAS, B. and HUDSON, H. J. (1979) SPI RP/C 34th ANTEC, 11-C.
53. HARPER, A. C. and REBER, H. W. (1978) SPE ANTEC April, 151.
54. HARPER, A. C. and REBER, H. W. (1979) SPE ANTEC November, 673.
55. NORGEN, M. (1979) *Plastics Today*, **6**, 18.
56. THORNE, J. L. (1976) in *Engineering Guide to Structural Foams*, B.C. Wendle (Ed), Technomic, Westport, Conn., Chap. 6.
57. COHEN, L. J. and ISHAI, O. (1967) *J. Comp. Mat.*, **1**, 390.
58. DICKIE, R. A. (1973) *J. Appl. Polym. Sci.*, **17**, 2509.
59. PROGELHOF, R. C. and THRONE, J. L. (1979) *Polym. Eng. Sci.*, **19**, 493.
60. e.g. MEINECKE, E. A. and CLARK, R. C. (1973) *Mechanical Properties of Polymeric Foams*, Technical Publishers, Westport, Conn.

61. NICOLAIS, L. (1975) *ibid.*, **15**, 137.
62. KERNER, E. H. (1956) *Proc. Phys. Soc.,* **698**, 808.
63. HALPIN, J. C. and KANDOS, J. L. (1976) *Polym. Eng. Sci.*, **18**,344.
64. ASHTON, J. E., HALPIN, J. C. and PETIT, P. H. (1969) *Primer on Composite Materials*, Technomic, Westport, Conn.
65. MOORE, D. R. and IREMONGER, M. J. (1974) *J. Cell Plast.*, **10**, 1.
66. METHVEN, J. M. and SHORTALL, J. B. (1979) *Eur. J. Cell Plast.*, **2**(2), 88.
67. CHISNALL, B. C. and THORPE, D. (1980) 35th SPIR/P Conf. February, 22-A.
68. PINDDEMANN, E. P. (1974) 30th SPIR/P Conf.
69. CHAPMAN J. F. and FORSTER J. M. W. (1980) *Reinforced Plastics*, **24**(1), 14.
70. DAWSON, J. R. Unpublished Data.
71. KELLY, A. (1966) *Strong Solids*, Clarendon Press, Oxford.
72. COTGREAVE, T. C. and SHORTALL, J. B. (1978) *J. Cell Plast.*, July, 137.
73. QUINN, J. (1978) *Design Data for Fibreglass Composites*, Fibreglass Ltd.
74. DAWSON, J. R. and SHORTALL, J. B. (1981) *J. Mat. Sci.,* In Press.
75. TRAEGER, R. K. (1967) *J. Cell Plast.*, **3**, 405.
76. HOLISTER, G. S. and THOMAS, C. (1967) *Fibre Reinforced Materials*, Applied Science Publishers Ltd, London.
77. CHEREPANOV, V. P., SHAMOV, I. V., TARAKANOV, O. G., MEL'NIKOV, E. N. and DOBROSKOKIN, N. V. (1973) *Soviet Plastics*, **12**, 61.
78. CHEREPANOV, V. P., SHAMOV, I. V., TARAKANOV, O. G. and SAMSONOV. V. D. (1975) *Int. Polym. Sci. Tech.*, **2**, 4.
79. CHEREPANOV, V. P., SHAMOV, I. V., TARAKANOV, O. G., DOBROSKOKIN, N. V., MEL'NIKOV, E. N. and IVANOV, YU. V. (1975) *Int. Polym. Sci. Tech.*, **2**, 9.
80. TIMOSHENKO, S. P. and GERE, J. M. (1961) *Theory of Elastic Stability*, McGraw-Hill, New York.
81. DE SILVA, A. R. T. (1968) *J. Mech. Phys. Solids*, **16**, 169.
82. GOTTENBURG, W. G., LO, K. H. and ALLEN, R. C. (1980) 35th Annual SPIR/P Conf., New Orleans.
83. TITOW, W. V. and LANHAM, B. J. (1975) *Reinforced Thermoplastics*, Applied Science Publishers Ltd., London, Chap. 10.
84. MCBRAYER, R. L. (1980) *Elastomerics*, **112**(7), 33.

Chapter 9

SYNTACTIC FOAMS

A. R. LUXMOORE and D. R. J. OWEN
*Department of Civil Engineering, University of Wales,
Swansea, UK*

9.1 INTRODUCTION

Syntactic foams are usually manufactured by including hollow microspheres into resin matrices. The adjective 'syntactic' is used to indicate a 'constructed' foam, as opposed to the normal foaming process of introducing a random arrangement of cavities into a single material. Hence this term can also be used to cover foams where the cavities are introduced in a regularly ordered arrangement, which can lead to a far more efficient packing of the cavities without reducing the desirable mechanical properties. To avoid confusion, the latter range of materials will be referred to as 'ordered foams', and are discussed at the end of this chapter.

The main objective of syntactic foam manufacture is to produce a lightweight material with reasonable compressive and shear strengths. The motivation for the development of this material came from the design of deep submergence vehicles, which required buoyancy aids that were not easily damaged, e.g. by the penetration of a single hollow container. Foamed polymers provided an attractive possibility, but, in their simplest state, they did not possess sufficient strength to prevent either ingress of water or implosion at any significant depth of water, and so hollow inclusions were used to provide extra strength. Eventually, these materials proved so successful, and relatively cheap, that they are currently used in other applications, such as the central filling for sandwich plates and shells.

The mechanics of syntactic foams can be studied at two levels: the micromechanics of individual microsphere behaviour, taken alone and in combination with neighbouring microspheres; and the macromechanics of the foam, where the material is treated as homogeneous and isotropic, due to the statistically random dispersion of the microspheres in the matrix.

Syntactic foams are three-phase materials, consisting of a matrix, the sphere wall and air. The basic structure is a random arrangement of strong hollow microspheres held together by a resin matrix in a volume ratio of approximately 2:1 respectively (see Fig. 1.11). A wide variety of materials have been used for the two main components, but on the basis of cost alone, the most popular microsphere material is glass, as this can be obtained from the pulverised fuel ash (PFA) which forms the main waste product from coal-burning power stations. In the UK these are known as Cenospheres, and are composed of mixed metal silicates, believed to be formed at about 1400°C by carbon dioxide generated when ferric oxide is reduced to a lower oxide. The hollow molten spheres are rapidly cooled in the boiler flue, freezing to form spheres in the size range 20–200 μm before significant contraction or collapse can occur. The microspheres, which make up to 5% by weight of the PFA, are readily separated from the other waste products by flotation, keeping the cost very low.

Typical densities are around 500 kg m^{-3}, and the spheres fail under hydrostatic pressure by implosion, with typically 60% of spheres exceeding a pressure of 10MN m^{-2}.

The strongest foams are made by mixing the spheres with resin under a partial vacuum, thus preventing the entrainment of free-air voids in the resin during the mixing process. However the addition of extra air voids reduces the density of the foam, albeit with a corresponding loss in strength, and this additional density reduction can prove useful in some applications.

9.2 MECHANICAL BEHAVIOUR OF RANDOM SYNTACTIC FOAMS

9.2.1 General Considerations

Providing the foam is made under controlled conditions which prevent segregation of the microspheres into different size ranges due to gravity, etc., the foams are essentially statistically isotropic and homogeneous. This is true even though different batches of foam will show variations in their mean mechanical properties because of variations in the mean properties of their constituent components.

A typical stress–strain curve for a foam subjected to uniaxial tension and compression is shown in Fig. 9.1. The compressive section of this diagram shows two curves, corresponding to two types of failure: the low-density foam specimen (containing phenolic spheres) barrelled outwards during plastic collapse, producing a constant collapse load for increasing strain; the high-density foam (containing Cenospheres) failed by the formation of a 45° shear plane, and the reduction in the collapse load is due to the transition of the specimen from a continuum into two separate units sliding relative to each other along the shear plane (cf. change from static to dynamic friction).

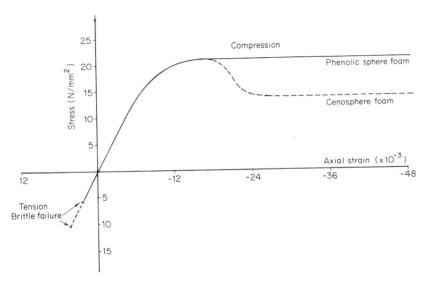

FIG. 9.1. Typical uniaxial stress–strain curves for syntactic foams.

In both uniaxial tension and compression the failure of the foam is connected primarily with the failure of the resin matrix. In tension, the addition of both stiff spheres and/or air voids reduces the ductility of the matrix, so that it fails in a brittle manner, with a crack propagating catastrophically across a plane perpendicular to the tensile axis. The crack usually initiates from an oversize void in the foam. These brittle failures will always occur under uniaxial tension, even though the matrix resin shows a reasonable ductility in its 'unfoamed' form. This reduction in ductility will be discussed later.

In compression, the shear failure of the high-density foam is again related to the brittle behaviour of the matrix when containing stiff microspheres, and can be likened to the shear failure that commonly occurs in uniaxial compressive tests on cementitious materials, which epitomise brittle behaviour. The low-density phenolic foams behave in a more ductile manner with the matrix material undergoing a ductile stretching in the transverse direction as the phenolic spheres, which are very flexible, become flattened into narrow ellipsoids by the load.

Different combinations of microsphere and resin can produce other uniaxial compressive behaviour. DeRuntz[1] reports that the presence of microspheres can cause tensile microstresses which produce longitudinal cracks in the specimen, and these propagate through the specimen producing catastrophic failure. Alternatively, the microspheres can fail along localised shear bands, and failure is apparently ductile.

A further useful insight to the behaviour of the foams is given by Fig. 9.2, where the hydrostatic compression versus dilatation is given for a typical Cenosphere foam. The curve shows a linear region up to a 'yield' point, where the microspheres start to collapse, and the collapse pressure remains reasonably constant until the large majority of the spheres have failed, and the foam becomes much stiffer fairly quickly, as the matrix alone withstands the hydrostatic pressure, although the effective density of the foam has increased at this stage by approximately 40%. In practice, once the collapse pressure has been reached, the foam would be considered as having failed.

In torsion, the foams again show classic brittle behaviour.[1] Torsional specimens fail by cracks forming on the 45° tensile principal stress plane,

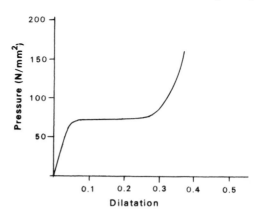

FIG. 9.2 Typical hydrostatic pressure–dilation curve for a Cenosphere foam.

i.e. perpendicular to the tensile principal stress. These failures are sometimes accompanied by secondary cracking along the compressive principal stress plane, and this failure is associated with the crushing of the microspheres along these planes.

The tests reported to date have only considered the effects of increasing load. Most resins show considerable viscoelastic and viscoplastic behaviour and syntactic foams are no exception. However, in most engineering applications, time-dependent effects are ignored wherever possible, and the analysis and discussion in this chapter will be similarly limited. The justification is that the inclusion of an inert material in a viscoelastic material helps to reduce the time-dependent effects, and the essential behaviour of syntactic foams can be explained without their use.

9.2.2 Elastic Moduli

The random arrangement of microspheres indicates that the 'rule of mixtures' should be adequate in predicting the elastic constants of the composite. For example,

$$E_c = \frac{E_1}{V_1} + \frac{E_2}{V_2} \tag{9.1}$$

where E represents the Young's modulus (or elastic modulus) and V the fractional volumes of the two components. Strictly, formulae of this type provide upper and lower bounds for the moduli, but they still provide close estimates for most practical purposes.

Unfortunately, the use of hollow microspheres prevents the direct measurement of their elastic properties, apart from the bulk modulus, and even this is difficult to measure accurately with such small specimens. DeRuntz[1] has studied this problem, and devised a method of estimation which seems to work quite well.

The bulk modulus of a thin-walled sphere, of wall thickness t and radius R is given by

$$K_s = \frac{2E_g}{3(1-v_g)} \frac{t}{R} \tag{9.2}$$

where E_g and v_g are Young's modulus and Poisson's ratio for the sphere material. Because of the spherical symmetry, this single value could equally well represent the hydrostatic behaviour of a solid sphere which has a bulk modulus $K_g = E_g/3(1-2v_g)$. As will be shown later, this fact can be used to optimise the performance of the composite. For the present, we need at least one other material constant for the sphere in order to deduce the necessary two elastic constants for the statistically isotropic, homogeneous composite. This may be done by considering a solid sphere subjected to a biaxial stress state of $-\tau$ and $+\tau$ along two orthogonal axes. This produces a stress distribution within the sphere which is independent of the radial coordinate, and with a strain energy density which is constant throughout the sphere and equal to $\tau^2/2\mu_s$, where μ_s is the shear modulus for the sphere. Hence the total strain energy for a sphere of unit radius must be

$$U = 2\pi\tau^2/3\mu_s$$

and the loading may be considered to be equivalent to pure shear.

For a thin-walled sphere, the same loading situation will produce a total strain energy given approximately by

$$U = \frac{4\pi\tau^2(7 + 5v_g)R}{15E_g t} \tag{9.3}$$

and an equivalent shear modulus for the hollow sphere may be defined simply by equating these two energies, giving

$$\mu_s = \frac{5E_g}{2(7 + 5v_g)} \frac{t}{R} \tag{9.4}$$

These two constants, K_s and μ_s, may be associated with a fictitious solid sphere which effectively acts in the same manner elastically as the hollow sphere. Using the usual relations between elastic constants of an isotropic material, we may thus define an equivalent Young's modulus, E_s, and Poisson's ratio, v_s, for the sphere by

$$E_s = \frac{10E_g}{(11 + 5v_g)} \frac{t}{R} \tag{9.5}$$

$$v_s = \frac{3 + 5v_g}{11 + 5v_g} \tag{9.6}$$

As the elastic properties of the sphere are linearly proportional to t/R (except for Poisson's ratio), we can calculate the mean elastic properties for spheres of varying size and wall thickness by defining a mean sphere density $\bar{\rho}_s$. For a thin walled sphere, the density ρ_s is given by

$$\rho_s \simeq 3\rho_g(t/R)$$

where ρ_g is the density of the wall material. The mean sphere density can be defined as the total mass of N spheres divided by the total volume, i.e.

$$\bar{\rho}_s = \sum_{}^{N} v\rho_s / \sum_{}^{N} v$$

$$= 3\rho_g \sum_{}^{N} v(t/R) / \sum_{}^{N} v \qquad (9.7)$$

For the mean bulk modulus of the spheres, K_s, we can write

$$\bar{K}_s = K_s \left(\sum_{}^{N} vt/R / \sum_{}^{N} v \right)$$

$$= \frac{\bar{\rho}_s}{3\rho_g} K_s$$

$$= \frac{2E_g}{3(1-v_g)} \left(\frac{\bar{\rho}_s}{3\rho_g} \right) \qquad (9.8)$$

Similarly

$$\bar{\mu}_s = \frac{5E_g}{2(7+5v_g)} \left(\frac{\bar{\rho}_s}{3\rho_g} \right) \qquad (9.9)$$

and

$$\bar{E}_s = \frac{10E_g}{(11+5v_g)} \frac{\bar{\rho}_s}{3\rho_g} \qquad (9.10)$$

The Poisson's ratio is independent of variations in (t/R).

These mean values can be used in conjunction with the elastic constants of the matrix to obtain upper and lower bounds of the composite moduli. For example, we can use the limits proposed by Haskin and Shtrikman[2] for the bulk and shear moduli:

$$\bar{K}_s + \frac{v_m(K_m - \bar{K}_s)}{1 + 3v_s(K_m - \bar{K}_s)/3\bar{K}_s + 4\bar{\mu}_s} \leq K_c \leq K_m \frac{v_s(\bar{K}_s - K_m)}{1 + 3v_m(\bar{K}_s - K_m)/3K_m + 4\mu_m} \qquad (9.11)$$

$$\bar{\mu}_s + \frac{v_m(\mu_m - \bar{\mu}_s)}{1 + 6v_s(\mu_m - \bar{\mu}_s)(\bar{K}_s + 2\bar{\mu}_s)/5\bar{\mu}_s(3\bar{K}_s + 4\bar{\mu}_s)} \leq \mu_c$$

$$\leq \mu_m + \frac{v_s(\bar{\mu}_s - \mu_m)}{1 + 6v_m(\bar{\mu}_s - \mu_m)(K_m + 2\mu_m)/5\mu_m(3K_m + 4\mu_m)} \qquad (9.12)$$

Using these bounds, DeRuntz has shown that reasonably close estimates of the measured composite moduli can be obtained. Where significant deviations between predicted and experimental values occurred, these were ascribed to the presence of air voids, which lowered the composite values.

9.2.3 Stress Distribution

As mentioned in the previous section, the hollow sphere can be considered equivalent to a solid sphere of elastic material. This equivalence is only exact when the sphere is subjected to hydrostatic loading, and in this case the effective bulk modulus of the thin walled sphere can be equated to the modulus of the matrix material, i.e.

$$K_m = \frac{2E_g}{3(1 - v_g)} \frac{t}{R} \tag{9.13}$$

By choosing the appropriate values of E_g, v_g, t and R, the thin-walled sphere can exactly duplicate the hydrostatic load–displacement behaviour of the solid sphere of matrix that it replaces, and so the hollow spheres will produce no stress-raising effects, and should not weaken the matrix when loaded hydrostatically.

This effect has been noted before. Lee and Westerman,[3] in extending Hashin's[4] work on the bounds of matrices containing small hollow spherical inclusions, commented that if the wall thickness of the inclusions was suitably chosen then the inclusions have no effect, the bulk modulus of the composite being identical to that of the matrix. They did not explain that the bulk modulus of the spheres must equal the matrix modulus. A more general application of the same principle is Mansfield's 'neutral hole'.[5] He was solely interested in aircraft structures, i.e. plates, and postulated the shape of the hole and the necessary stiffener section and properties, to replace effectively, under specific loading, the material removed from the plate. His solution for a circular stiffener in a plate under equal biaxial stress is easily transformed into three dimensions, i.e. a hollow sphere under hydrostatic pressure.

The presence of a stiff liner, which counteracts the removal of the matrix material, is the main reason for the high strength of syntactic foams compared with foamed plastics with no microspheres. Under hydrostatic loading, the matching of material properties can be exact, but even under other loading conditions, e.g. uniaxial compression, the microspheres will be very significant in balancing the internal microstresses produced by holes in the matrix. Exact mechanical matching in these other loading cases is usually impossible, because a hollow sphere does *not* deform elastically in the same way that a solid sphere does for generalised loading. This can be understood by comparing the expression for the effective bulk modulus of a sphere (eqn (9.2)) with the bulk modulus of a solid sphere, in terms of Young's modulus and Poisson's ratio. The former involves a term of $(1 - v)$, as well as the wall thickness to radius ratio t/R, whereas the latter involves a term of $(1 - 2v)$.

The gain in strength–weight ratio of the syntactic foam is at the expense of high stresses in the liner. The tangential stress σ_t in a thin-walled sphere under hydrostatic pressure P is given by

$$\sigma_t = \frac{PR}{2t} \tag{9.14}$$

and for a microsphere matched to the matrix bulk modulus, the pressure P will equal the hydrostatic loading. Hence, σ_t will be increased beyond the mean stress in the matrix by a factor $(R/2t)$ which can vary typically between 5 and 30 for Cenospheres.

It is obviously inefficient to match the microspheres exactly, as the matrix can sustain some shear stress, and the critical wall thickness for any microsphere can be reduced further (hence increasing the microspere stresses). However, the weight savings are very small compared to the increases in matrix and microsphere stresses, especially when the ratio of microsphere material to matrix Young's modulus is high (typically 20 for Cenosphere/resin mixtures). Thus deliberate undermatching of the microsphere bulk modulus is unlikely to be useful in practice. A numerical example of this is given in Section 9.4.

Although the counterbalance of material properties provided by the microspheres explains the improved compression properties of syntactic foams compared with unreinforced foams, both types of foam still fail in a brittle manner under tension, and, as explained previously, the use of a stiff microsphere can produce a 'brittle' type failure under uniaxial compression. In non-syntactic foams, brittle tensile behaviour can be explained in terms of microscopic triaxial tensile stresses set up between adjacent voids (comparable with the triaxial tensile stresses set up in a necked tensile specimen) which reduce the ability of the material to yield prior to fracture, even though the matrix material is reasonably ductile. The presence of a stiff microsphere should reduce the microscopic tensile stresses, and, in fact, the writers have observed that foams using stiff microspheres can have a much higher tensile strength than similar foams formed with flexible microspheres. However, the stiff microspheres are much weaker under loading conditions other than hydrostatic compression, and it is believed that, when subjected to shear forces and diametral compression (as might arise in the transverse direction of a tensile specimen), the microspheres fail at relatively low loads, leading to localised failure planes in the uniaxial specimens.

9.2.4 Failure Envelope

DeRuntz and Hoffman[6] carried out a systematic investigation of failure in a typical syntactic foam (resin–glass microsphere combination) containing a minimum of air voids. They used biaxial specimens (cylindrical specimens subjected to combined uniaxial and torsional loading), hydrostatic compression of cubes and spheres, and triaxial testing (combined hydrostatic and uniaxial compression). They ignored

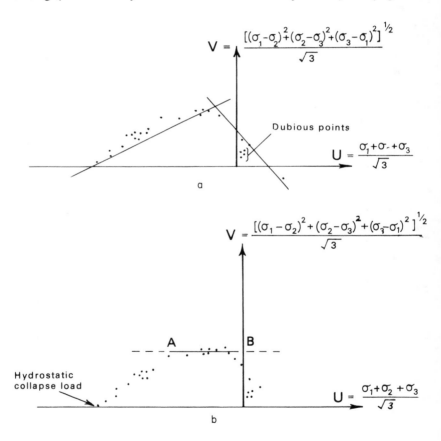

FIG. 9.3. Failure envelope for syntactic foams. (a) General failure envelope for syntactic foam suggested by De Runtz;[1] (b) failure envelope corresponding to von Mises criterion. Courtesy of J. A. DeRuntz.

the effect of the third stress invariant, and plotted their results as in Fig. 9.3, using variables u and v related to the first and second stress invariants by

$$u = (\sigma_1 + \sigma_2 + \sigma_3)/\sqrt{3}$$
$$v = [(\sigma_1 - \sigma_2)^2 + (\sigma_2 - \sigma_3)^2 + (\sigma_3 - \sigma_1)^2]^{1/2}\sqrt{3} \qquad (9.15)$$

where σ_1, σ_2 and σ_3 are the principal stresses.

The two full lines represent an approximate fit to the data, the left-hand line corresponding to 'compressive' failure, and the right-hand line to specimens demonstrating 'tensile' failure, as described in Section 9.2.1. In a subsequent paper, DeRuntz confirmed that the compressive failure line was too conservative, and a more accurate result is obtained by following more closely the obvious curve formed by the experimental points.

For specimens subjected to compressive loads only (which is the only loading suitable for syntactic foams), Yeo[7] pointed out that a horizontal line V = constant, could be used to fit the data, provided the following condition is satisfied:

$$1 \cdot 5 \, P_c < (\sigma_1 + \sigma_2 + \sigma_3) < 0 \qquad (9.16)$$

where P_c is the hydrostatic collapse load of the syntactic foam. This failure criterion is equivalent to the von Mises yield criterion, i.e.

$$(\sigma_1 - \sigma_2)^2 + (\sigma_2 - \sigma_3)^2 + (\sigma_3 - \sigma_1)^2 = \text{constant}$$

but the full acceptance of von Mises does require an investigation into the effect of the third stress invariant.

Design of structures using the foam in tension requires the application of statistical theories of brittle failures, and detailed analyses of syntactic foams under tension have not been reported.

9.3 NUMERICAL ANALYSIS OF ORDERED FOAMS

9.3.1 General Considerations

The approach of Section 9.2.3. can, at best, only provide an approximate result for the distribution of stress in a syntactic foam. A detailed analysis of displacements and stresses in such composites can only be undertaken by use of numerical techniques such as finite difference, finite element or boundary integral methods. Of prime interest are the stresses produced throughout the matrix for various void arrangements and relative geometric dimensions. For the case of lined voids a knowledge of the stress distribution in both matrix and liner is essential for producing a balanced design. A finite element analysis of various configurations will now be described.

The finite-element method is a very powerful numerical technique which can be employed to determine the displacements, strains and stresses in solids of arbitrary geometry when subjected to arbitrary load and boundary conditions.[8-10] The material behaviour can be elastic, or nonlinear effects such as those due to elastoplastic or viscoplastic deformation can be included.[11] The basic steps of the finite element process are summarised below:

1. The region to be analysed is subdivided into a series of distinct non-overlapping regions known as *elements*.
2. These elements are connected at a discrete number of points along their periphery known as *nodal points*.
3. The displacement variation within each element is postulated in terms of the nodal values by means of *shape functions*.
4. For each element the stiffness matrix and applied load vector are calculated.
5. The contributions of each element are assembled to give the global stiffness matrix and global load vector, and the resulting system of simultaneous equations solved for the unknown nodal displacements.
6. Finally the stress components within each element are evaluated.

The elements employed in this analysis take the form of the 20-node isoparametric element illustrated in Fig. 9.4. This hexahedral element has curved sides defined by the Cartesian coordinates of its nodal points and, at best, a parabolic variation along each edge can be modelled. The shape functions for the element are described with respect to the rectangular parent element shown in Fig. 9.4(a) with transformation to the global element of Fig. 9.4(b) being performed by a conformal mapping. An important feature of these elements is that relatively coarse mesh subdivisions provide accurate solutions.

In order to restrict the computation costs, the number of inclusion shapes and arrangements were limited. After studying the results of an extensive experimental programme by the US Naval Applied Science

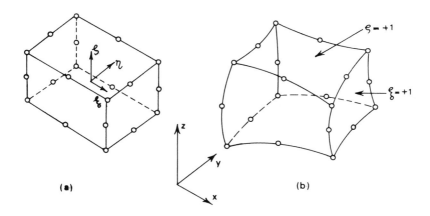

FIG. 9.4 Three dimensional 20-node isoparametric element employed in numerical analysis, showing (a) parent element and local axes (b) global curved side hexahedral element.

Laboratory,[12] it was considered sufficient to take cubes and spheres as the two extremes of a number of inclusion shapes. This decision also limited the number of regular distributions that could produce maximum packing, and some consideration was given to possible manufacturing difficulties. It was decided to consider only cubes stacked face to face, and spheres in simple cubic and face-centred cubic distributions. The symmetry inherent in these distributions allowed three basic modules (or unit cells) to be considered for analysis, so that, *except at the surface*, the displacements and stresses within any one module would be representative of the entire composite.

The modules were loaded by applying uniform displacements to each plane of symmetry, to represent a hydrostatic load (the main load to be sustained by the foam), and then calculating the total reactions on these planes. Other displacements were also applied to ascertain the elastic constants (the composites under consideration are anisotropic[7]) but it was found that accurate moduli could be obtained from approximate analytical methods.[13]

The symmetry inherent in the loading enabled the basic modules to be reduced even further, introducing considerable savings in computation. These reduced modules, occupying only one-eighth of the volume of the full modules are illustrated in Figs. 9.5–9.7, showing their subdivision into elements.

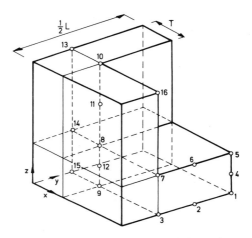

FIG. 9.5. Division of the cubic module into elements.

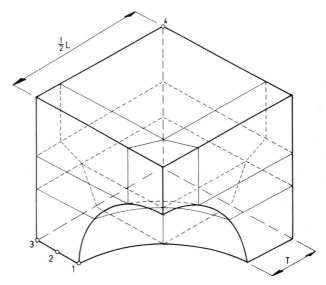

FIG. 9.6. Division of the cubically arranged spherical void module into elements.

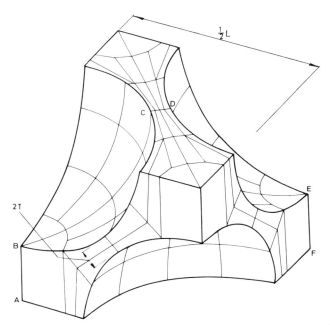

FIG. 9.7. Division of face-centred cubic module into elements showing plane of diagonal symmetry ABCDEF.

9.3.2 Results for Unlined Voids

In the initial computation, any liners used to form the voids were assumed to have negligible strength. The assumption of elastic behaviour allows the stresses to be plotted independently of the matrix elastic modulus, but not of its Poisson's ratio. Experimental tests[14] on different matrix materials, i.e. resins filled with microspheres and voids, gave a mean value for the Poisson's ratio of 0·30, and this was used in the computations. Further computations were made with a Poisson's ratio of 0·34, but the maximum shear stresses were changed by less then 1%, and so these results are not included here.

The spacing of the voids is expressed in dimensionless terms as the ratio T/L, where $2T$ is the minimum matrix thickness between two adjacent voids, and L is a measure of the void spacings (these parameters are illustrated in Figs. 9.5–9.7). This definition allows a sensible comparison between the three geometries, and the densities can be calculated readily from this ratio.

9.3.3 Module with Cubic Void

The maximum principal stress occurred at node 1, and the highest shear stress at nodes 1, 2, 4, 5 and 6 and the corresponding symmetrical nodes. The variation of these stresses with T/L is shown in Fig. 9.8. This was

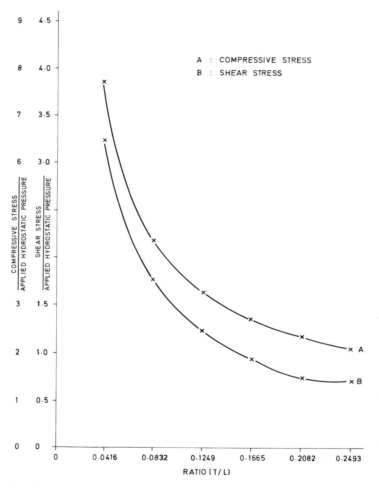

FIG. 9.8. Stresses occurring in cubic module when subjected to hydrostatic pressure showing the maximum matrix density to form a 320 kg m^{-3} module.

entirely unexpected, as there were no stress concentrations at nodes 7 and 8, and so the problem was recomputed with line supports along the three axes only. That is, only the nodes on the intersections of faces retained their constraints. The constraints of all other nodes were removed. With these supports, stress concentrations did occur at nodes 7 and 8. It must therefore be assumed that the constraint of one module on an adjacent one decreases the stress concentrations at the internal corners due to the interfacial pressures and associated bending moments.

The mesh used for this module was relatively coarse. Although locations of high stress concentration were apparent, the value of the stress at these points must be used with discretion. The element shape functions permit only parabolic strain variations through the element. This does not allow the large stress concentrations to be accurately represented in magnitude. However, in practice, no module would be constructed with sharp corners, as a fillet radius would automatically be added to relieve any stress concentrations. It is therefore suggested that the stresses, given by the programme at the corners, were not likely to be significantly different from those in the same location in the actual module constructed with rounded corners.

To check this supposition, a two-dimensional plane-strain analysis was performed on the two meshes shown in Fig. 9.9. The coarser mesh was consistent with that used for the three-dimensional analysis of this section, while the finer mesh allowed a fillet radius to be included at the corner. Displacement loads were applied to faces AB and CD as in the three-dimensional analysis. Nodes on faces AO and DO were constrained to move only in these faces.

The stresses (in MN m^{-2}) at node P for the applied displacements, from the two analyses were:

	σ_x	σ_y	σ_z	τ_{xy}
Coarse mesh	−12·18	−12·18	−7·31	5·33
Fine mesh	−11·37	−11·37	−6·82	4·68

The maximum difference of any of the stresses was 6·5% with the coarser mesh giving the higher stresses. The two-dimensional analysis therefore justifies the supposition, and indicates that the stresses at the corners, obtained from the three-dimensional analysis, are likely to be higher than those in the actual module manufactured with rounded corners, leading to a conservative estimate.

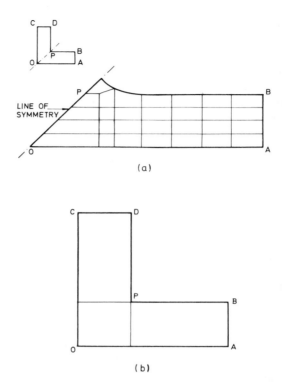

FIG. 9.9 Meshes employed in stress concentration examination.

9.3.4 Spherical Void in Simple Cubic Distribution

Figure 9.10 shows, for various T/L ratios, the maximum shear and compressive stresses in the module. The location of the highest shear and compressive stress were the pinch points between adjacent voids, i.e. nodes 1, 2, 3 (see Fig. 9.5). The lowest stressed areas were those furthest from the pinch points, i.e. the corners of the module—node 4. The major principal stress magnitude at the corners was approximately half that at the pinch points. As can be expected from the loading and geometric symmetry at the corners of the module, the shear stress at node 4 was zero, and the principal stress directions were along the axis of the module.

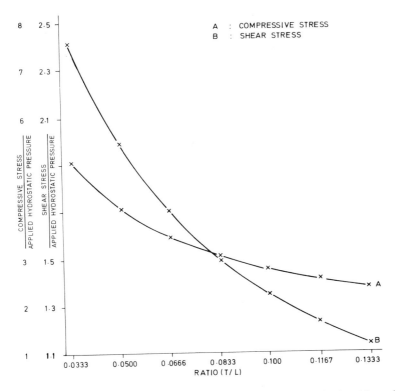

Fig. 9.10 Stresses occurring in the cubically arranged spherical void module when subjected to hydrostatic pressure showing the maximum matrix density to form a 320 kg m^{-3} module.

9.3.5 Spherical Void in Face-centred Cubic Distribution

The computer core size required to analyse module 3, the face-centred cubic distribution of voids, was greater than that available to the authors, and it was necessary to reduce the size of the problem. Hence use was made of the plane of symmetry ABCDEF, Fig. 9.7, which was applicable to hydrostatic loading. Using this plane of symmetry, only the structure on one side of this plane needed to be analysed. This reduced the mesh size by half and reduced the computing time by an estimated 65% of that needed to analyse the whole mesh.

The method of loading was similar to that previously described except that the nodes on the plane of symmetry ABCDEF were constrained to move only in that plane.

Figure 9.11 gives identical information for this mode as Fig. 9.10 gave for the spherical voids in cubic distribution.

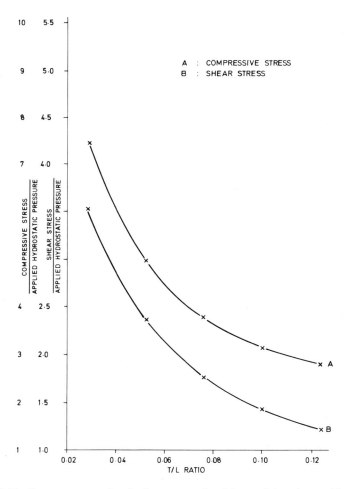

FIG. 9.11. Stresses occurring in face-centred cubic module when subjected to hydrostatic pressure showing the maximum matrix density to form a 320 kg m^{-3} module.

The maximum shear and compressive stresses occurred at the pinch points between adjacent nodes. However, the stress magnitude did not vary greatly throughout the module, but it was a minimum at the corners of the module.

9.3.6 Comparison of the Three Modules

The relative efficiencies of the three modules are compared in Fig. 9.12, where the maximum principal stress and maximum shear stress are plotted against the dimensionless density (i.e. the density of a module with a matrix density of unity), which is, of course, related to the T/L ratio defined in Figs. 9.5–9.7. The cubic void, due to its efficient packing

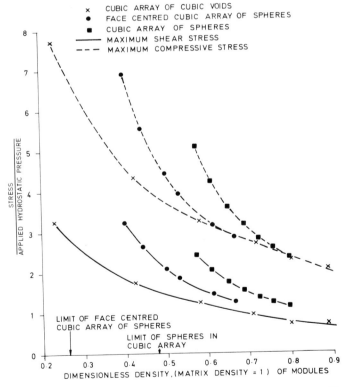

FIG. 9.12. Variation of maximum shear and compressive stresses with dimensionless density.

arrangement, has the best stress–weight ratio, as expected, and the face-centred cubic distribution of spheres is the next most satisfactory. The difference is quite pronounced at low densities, i.e. when T/L is small, diminishing as the densities approach unity (when the three modules are identical, with zero void content).

Because the loading was a hydrostatic compression, the maximum principal tensile stresses were always small compared with the compressive stresses (always less than 10%). However, this analysis has so far ignored the behaviour of modules near a surface, where considerable tensile stresses would be caused by bending, especially in a cubic void. A major advantage of syntactic foams is their ability to remain buoyant if the surface covering is damaged, which infers that if water enters any one void, the walls of the void must not collapse. In a pressurised cubic void, the tensile stresses produced in this fashion would be considerable and, as described previously, microscopically filled materials (which would be used as the matrix) usually have very low tensile strengths.

Buckling could conceivably be another source of failure in the cubic voids, but thin plate theory suggests that this is a possibility only at T/L ratios of less than 0·02, an unlikely value from practical considerations.

9.3.7 Analysis of Lined Spheres

Though cubic voids can be introduced into a matrix by the use of removable formwork, spherical voids can only be achieved by using an internal former which is permanently positioned. This can be made of a material of negligible density and stiffness, e.g. expanded polystyrene, but, as with the random foams, a stiffer former can assist in reinforcing the void.

The effect of liner stiffness on the composite was investigated using the simple cubic distribution of spheres only, as it was the most inexpensive to analyse.

When the bulk modulus of the liner matches that of the matrix, the pressure on the liner at the matrix/liner interface is uniform and equal to the applied pressure. Using the equation for the radial displacement of an externally pressurised sphere,[15] we can equate the displacements of the liner and the solid sphere of matrix it replaced, viz:

$$-\frac{1}{(3\lambda_L - 2\mu_L)} \cdot \frac{PR_0^4}{(R_0^3 - R_i^3)} - \frac{1}{4\mu_L} \frac{PR_0^3 R_i^3}{(R_0^3 - R_i^3)} = -\frac{PR_0}{(3\lambda_m + 2\mu_m)} \quad (9.17)$$

where λ and μ are Lamé constants, suffixes represent liner and matrix materials, and R_0 and R_1 are the external and internal radii of the liner, respectively. Rearranging, and replacing Lamé's constants by the elastic modulus, E, and Poisson's ratio, v, we obtain for the critical ratio R_0/R_1:

$$\left[\frac{R_0}{R_1}\right]^3 = \frac{E_m(1+v_L) + 2E_L(1-2v_m)}{2E_L(1-2v_m) + 2E_m(1-2v_L)} \qquad (9.18)$$

For this critical ratio, the stresses in the matrix are purely hydrostatic (and equal to the applied pressure). The stresses at any radius R within the liner are given by:

$$\text{radial stress} = -P\frac{R_0^3}{R^3} \cdot \frac{(R^3 - R_1^3)}{(R_0^3 - R_1^3)} \quad \text{(compressive)} \qquad (9.19)$$

$$\text{tangential stress} = \frac{P}{2} \cdot \frac{R_0^3}{R^3} \cdot \frac{(2R^3 + R_1^3)}{(R_0^3 - R_1^3)} \qquad (9.20)$$

A simple empirical model was employed to estimate the maximum matrix shear stress for liner thicknesses which were not optimum. It consisted of a liner surrounded by a spherical shell of matrix material with an external radius of $L/2$. For this model the largest shear stress occurred at the liner–matrix interface (in the numerical model, the largest shear stress occurred in the vicinity of the pinch points between two adjacent voids).

The numerical shear stress values are compared with the simple analytical results in Table 9.1 for different values of T/L, (liner thickness) $/L$, and E_m/E_L. The Poisson's ratio of both matrix and liner was taken as 0.3. Only maximum shear-stress values are compared, and for liner thicknesses up to the optimum value there is good agreement between numerical and analytical values. The rows marked with an asterisk are based upon an optimum liner thickness as calculated from eqn (9.18).

There is little stress variation across the minimum section between two adjacent spheres. Therefore the mean stresses in this region are plotted in Fig. 9.13 for various (liner thickness)/L ratios with $T/L = 0.0666$. The stresses in the liner are, of course, given by eqns (9.19) and (9.20). With increasing liner stiffness a predominantly uniaxial stress state σ_x in the direction of the line joining the centres of the voids is developed. This indicates a change from a matrix load-bearing mechanism to transfer of load from sphere to sphere across the pinch point regions with increasing liner stiffness.

TABLE 9.1
Analysis of the Cubically Arranged Spherical Void Module with Aluminium Liner Subjected to $6 \cdot 89$ MN m^{-2} External Pressure

Test no.	Elastic modulus of matrix[a] ($GN\ m^{-2}$)	T/L ratio	Liner thickness/ L ratio	Shear stress τ_{max} numerical model (kNm^{-2})	Shear stress τ_{THEOR} empirical model ($kN\ m^{-2}$)
1[b]	2·07	0·0333	0·0113	85·5	0
2[b]	2·76	0·0333	0·0153	58·6	0
3	2·07	0·0666	0·0066	1165·2	1103·2
4	2·07	0·0666	0·0083	613·6	537·8
5[b]	2·07	0·0666	0·0106	41·4	10·3
6	2·07	0·0666	0·0133	503·3	427·5
7	2·07	0·0666	0·020	1358·3	1013·5
8	2·07	0·0666	0·0267	2040·9	1316·9
9[b]	2·07	0·1000	0·01	64·8	0
10	2·07	0·1333	0·008	227·5	239·2
11[b]	2·07	0·1333	0·009	41·4	2·07
12	2·07	0·1333	0·010	234·4	210·3
13[b]	1·38	0·1333	0·006	40·7	0

[a] Elastic modulus of all liners is $72 \cdot 4$ GN m^{-2}.
[b] Indicates optimum liner thickness.

9.4 DESIGN OF BUOYANT COMPOSITES WITH STIFF LINERS

The advantages of a structurally strong liner arise from the use of stiff, low-density materials, such as aluminium and glass (both of which possess very similar stiffnesses and densities). These materials are 20 times stiffer than a typical syntactic foam, but their densities are only three times greater. Hence liners of these materials provide a considerable saving in the composite weight. From eqn (9.18) writing

$$R_0 - R_1 = t \ll R_0, v_m = v_L = 0 \cdot 3, \text{ and } E_L/E_m = 20$$

we obtain

$$t/R_0 = 0 \cdot 01$$

for an optimum liner thickness.

The density of the lined spherical void, compared to a similar sphere of solid matrix material, is

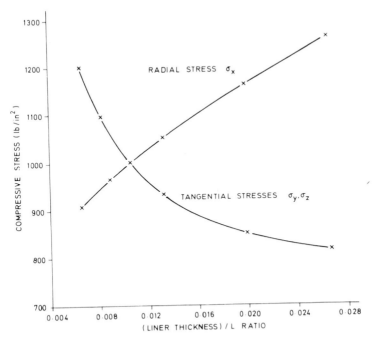

FIG. 9.13. Stresses in matrix in vicinity of the pinch point between the voids in the cubically arranged spherical void module with $T/L = 0.0666$ when subjected to 6.89 MN m^{-2} external pressure.

$$\frac{4\pi R_0^2 t}{\tfrac{4}{3}\pi R_0^3}\left[\frac{\rho_{\text{liner}}}{\rho_{\text{matrix}}}\right] = \frac{3t}{R_0} \times (3) = 0.09$$

Using this result in conjunction with the face-centred cubic array, and taking the ratio $T/L = 0.03$, the density ratio becomes 0.47. Therefore use of a stiff liner can halve the density with no significant change in the stress distribution of the matrix. This ratio compares with a value of 0.42 for the unlined face-centred cubic array with the same T/L value, but with large shear stresses occurring in the matrix for the unlined case.

This advantage is gained at the expense of high stresses in the liner. Applying the previous approximation to eqn (9.20), the expression for the tangential stress becomes the same as eqn (9.11), i.e.

$$\sigma_t = \frac{P}{2}\left[\frac{R_0}{t}\right]$$

For the numerical example considered, $\sigma_t = 20P$. A high-strength material must be used for the liner, and if manufacturing difficulties lead to non-uniform loading of the liner by the matrix, localised failures may occur.

Furthermore, weight savings are obtained if the liner-wall thickness is less than the optimum, resulting in some small shear stresses in the matrix (Table 9.1). However, this also leads to significantly higher stresses in the liner for relatively small savings in density, as shown in Fig. 9.14. In the case examined above, decreasing the liner thickness by half will reduce the composite density (in dimensionless terms) from 0·47 to 0·435, but will increase the liner stresses by more than a factor of two. When the strength of the liner is a limiting factor, it may be necessary to use a liner thickness greater than the optimum.

With the use of liners, the problem of the ingress of water is removed. For example, with an optimum liner, the stress in the composite is not

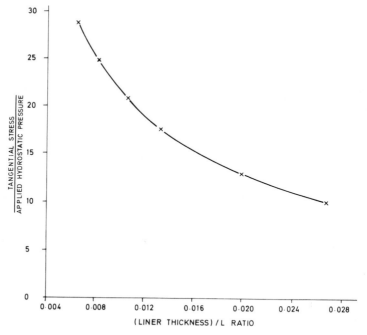

FIG. 9.14. Tangential stresses in stiff liner for cubic arrangement of spherical voids with $T/L = 0.0666$ when subjected to external pressure.

altered from the hydrostatic distribution if water entered a cell. Additionally, the water is unlikely to breach the liner and thus the only water entering the structure will be a small volume forming a lens between the liner and the matrix. The resulting decrease in buoyancy will therefore be negligible.

These principles apply equally well to random arrangements of microscopic spheres, and can be used to optimise the basic syntactic matrix.

9.5 ULTIMATE LOAD BEHAVIOUR OF ORDERED FOAMS

In the previous section a design process for composites with stiff liners was discussed assuming elastic behaviour of the constituent materials. Frequently a knowledge of the collapse behaviour of such foams is important and in this section a series of such failure studies based on the finite element method are described. Of prime importance for the calculation of safety factors are the applied pressure values required to cause initial yielding of the liner and matrix materials, respectively, and, of course, the ultimate collapse pressure. Since elastoplastic finite element analyses are computationally expensive, attention was restricted to the cubic distribution of spheres indicated in Fig. 9.6 with T/L being chosen as 0.1. After initial yielding both the liner and matrix material were assumed to behave elastic–perfectly plastically and the material properties employed are included in the appropriate figures.

The liner properties were kept constant; being chosen to represent those of commercially available spheres.[7] The first series of numerical analyses maintained the elastic modulus of the matrix at the value required for a balanced design, while the uniaxial yield stress was varied between 0.25 and 1.0 times the liner yield stress. The first two curves in Fig. 9.15 are horizontal indicating that, for the yield stress ratios used in analysis, the matrix does not yield until the liner has become fully plastic. Curves C and D are linear, allowing a significant conclusion to be drawn. When the liner becomes fully plastic it deforms under constant load, and this load is not significantly altered by the fact that the deformation of the liner is not purely hydrostatic; as happens after initial yield of the liner when the design becomes unbalanced. Therefore after the liner becomes fully plastic its effect could be replaced by an equivalent distributed surface load that is independent of the applied pressure. This allows the principle of superposition to be applied to the module after the liner becomes fully plastic and the module behaves in a linear manner.

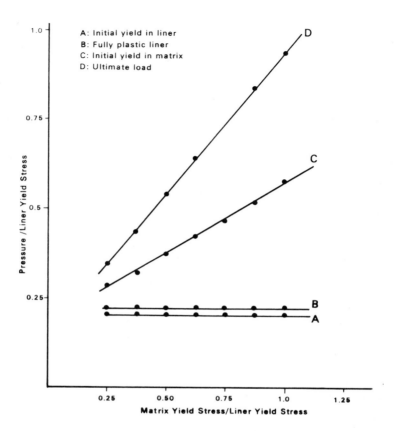

FIG. 9.15. Pressure causing yield in a cubically arranged spherical void module with a balanced design, for various ratios of matrix to liner yield stress (matrix elastic modulus = 390 N mm^{-2} [0·564 × 10^5 lb in^{-2}], liner elastic modulus = 5210 N mm^{-2} [0·755 × 10^6 lb in^{-2}], liner yield stress = 7·25 N mm^{-2} [1050 lb in^{-2}]).

In the second series of computations the elastic modulus of the matrix was varied while the yield stress was maintained constant. The results of the analysis are shown in Fig. 9.16. The curves for first yield and full plastification of the liner were obtained assuming elastic behaviour of the matrix at all times and are therefore only applicable under this condition. Provided that the liner is fully plastic the load causing first yield in the matrix is independent of the modular ratio. The curve denoting first

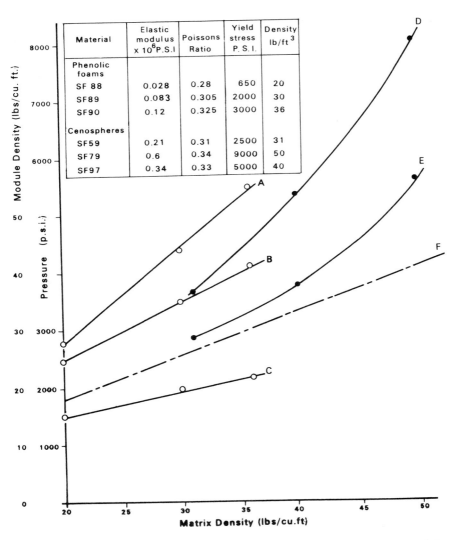

FIG. 9.16. Pressure causing yield in a cubically arranged spherical void module using commercial buoyancy spheres and matrices. *Phenolic foams*: A, pressure at which liner becomes fully plastic; B, pressure causing first yield in liner; C, pressure causing first yield in matrix (matrix yield stress = $0.875\ Y_L$). *Cenosphere foams*: D, pressure causing first yield in matrix (matrix yield stress = $0.625\ Y_L$); E, pressure causing first yield in matrix (matrix yield stress = $0.50\ Y_L$); F, pressure causing first yield in matrix (matrix yield stress = $0.375\ Y_L$).

yield of the matrix falls where it intersects the complete liner plastification curve, eventually levelling off when the liner stiffness becomes insignificant in relation to the matrix stiffness and yielding occurs in the matrix with only a small stress transfer across the liner–matrix interface.

A third series of analyses were undertaken using for matrix properties those obtained experimentally from existing foams. The T/L ratio was maintained constant at 0·1 as before. The basic aim is to indicate the pressure which could be resisted by composites utilising foams presently available. The results are shown in Fig. 9.17. An ultimate load curve is included for phenolic foams since these can sustain load over a large strain range after the onset of first yield. Also shown in Fig. 9.17 is the module density for various foam densities.

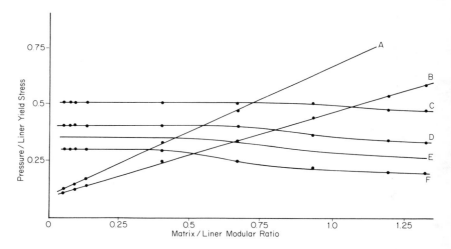

FIG. 9.17. Pressure causing yield in a cubically arranged spherical void module for various modular ratios (liner elastic modulus = 5210 N mm^{-2} [0·755 × 10^6 lb in^{-2}], liner uniaxial yield stress, Y_L = 7·25 N mm^{-2} [1050 lb in^{-2}]).

For an elastic design, the highest factors of safety are obtained by the intersection point of the matrix and liner first yield curves in Fig. 9.16. For matrices with different uniaxial yield values to those analysed, the matrix initial yield curve can be found by simple interpolation between curves A,B and C in Fig. 9.16. From Fig. 9.17 it can be seen that if plastification of the liner is permitted an increase of approximately 50% in the working load can be obtained for both types of foam.

The results of these elastoplastic studies cannot be directly extrapolated to the face-centred cubic module of Fig. 9.7 although it is expected that the results for this configuration would be qualitatively similar. Finally in Fig. 9.18 the development of plastic zones with increasing load is indicated for a modular ratio of 1·320. The matrix–liner yield stress ratio was 0·375 and the zone distribution to either side of the diagonal line is symmetric. Initial yield in the matrix occurs at the pinch points between the voids. With increasing load the plastic zones spread until they join so that the liner is then completely surrounded by the yielded matrix.

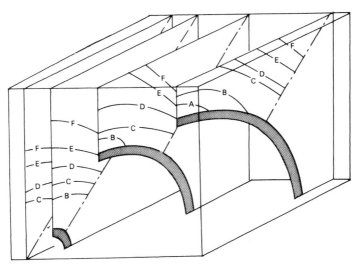

FIG. 9.18. Spread of plastic zones for a matrix/liner modular ratio of 1·320. Matrix/liner yield stress = $0.375\, Y_L$. The liner is completely elastic for the range of pressures shown. Pressure/liner yield stress: A, 0·214; B, 0·254; C, 0·298; D, 0·332; E, 0·362; F, 0·388; Collapse, 0·43.

9.6 SOME CONCLUDING REMARKS ON ORDERED FOAMS

The analyses performed indicate that only cubic void arrays and spherical voids in a face-centred cubic arrangement are of practical use. Although the cubic void module is the most efficient from a strength/weight viewpoint, it is potentially the most unstable if pressurised water leaks into one (or some) of the voids. Manufacturing difficulties associated with forming cubic voids with fillet radii also may make its use unattractive. On the other hand a face-centred cubic arrangement of spherical voids presents fewer difficulties.

The benefits of employing a structurally strong liner have also been demonstrated. Provided that the effective bulk modulus of the spherical liners is matched to that of the matrix foam, shear stresses in the matrix can be eliminated when the composite is subjected to a hydrostatic loading. This is wasteful of the shear strength of the matrix foam, but selecting a liner thickness less than the optimum increases the liner stresses quite considerably. Secondary causes of failure, such as debonding along sphere–matrix interfaces, have not been considered.

The main conclusion to be drawn from the elastoplastic computations is that a balanced design between the liner and matrix stiffnesses does not provide the highest factors of safety against yielding of the matrix. Provided that the liner does not exhibit strain hardening behaviour and has become fully plastic before the matrix begins to yield then the load required to cause matrix yielding is independent of the modular ratio.

REFERENCES

1. DeRuntz, J. A. (1971) *Structure, Solid Mechanics and Engineering Design*, M. Te'eni (Ed), Wiley-Interscience, London, p. 405.
2. Haskin, Z. and Shtrikman, S. (1961) *Franklin Inst. J.*, **271**, 336–341.
3. Lee, K. J. and Westerman, R. A. (1970) *J. Comp. Mat.*, **4**.
4. Hashin, Z. (1962) *J. App. Mech. ASME*, **29**.
5. Mansfield, H. E. (1953) *J. Mech. App. Maths*, **6**.
6. DeRuntz, J. A. and Hoffman, O. (1969) *J. App. Mech. ASME*, **36**, 551-557.
7. Yeo, M. F. (1973) *The design of syntactic foam structures by finite element analysis*, Ph.D. Thesis, University of Wales.
8. Zienkiewicz, O. C. (1977) *The Finite Element Method*, McGraw-Hill, London.
9. Hinton, E. and Owen, D. R. J. (1979) *An Introduction to Finite Element Computations*, Pineridge Press, Swansea, UK.
10. Gallagher, R. H. (1975) *Finite Element Analysis—Fundamentals*, Prentice-Hall, New Jersey.
11. Owen, D. R. J. and Hinton, E. (1980) *Finite Elements in Plasticity: Theory and Practice*, Pineridge Press, Swansea, UK.
12. Stechler, B. J. and Poneros, G. J., *Parametric Analysis of Optimum*

Buoyancy Module Designs, NASL Project 9300–57 (FR86) Progress Report No. 2.
13. COHEN, L. J. and ISHAI, O. (1967) *J. Comp. Mat.*, **1**, 393–403.
14. LONG, R. F. and MITCHELL, E. C., *Syntactic Foams from Cenospheres*, Admiralty Materials Laboratory Report No. 1/73.
15. LOVE, A. E. H. (1944) *The Mathematical Theory of Elasticity*, Dover Publications, Article 98, p.142.

INDEX

ABS, 266, 285
Acceleration efficiency, 182, 201–2
Acid concentration effects, 253
Adiprene systems, 210–11, 221
Advanced Passenger Train, 325
Air-filled systems, 164, 189, 195, 197
Allophonate prepolymers, 115–16
Anisotropy effects, 7–8, 13, 15, 22, 91–6, 186
Aspect ratio, 345
Azo initiator, 326

Ball-and-dart impact procedures, 303
Ball-and-strut model, 19, 23
Ball-rebound test, 105
Bending, 265, 266, 270, 271, 276, 287–93, 297–9
Bending moment, 287, 298
Bending strength, 58
Bending tests, 58
Biuret prepolymers, 115–16
Blowing agents, 4, 123, 251, 254, 259, 278, 327
Breathability, 103–4, 226, 227, 255, 257, 260
Brittle failure, 304, 315, 367
BRL systems, 255–6, 260
Bubble
 growth, 5
 nucleation, 5

Buckling, 190, 191, 212, 352, 354–6, 380
Buckling deformation, 87
Buckling strain, 84
Bulk
 density, 265
 modulus, 363, 390
 reduced density, 265
Buoyant composites, 382–5

Cantilever bending, 297–9
Carbonyl group, 126
Carboxyester anhydrides, 327
Catalysts, 112–13, 139, 234–6
'Catalytic window', 119
Cavitation, 314
Cell
 geometry effects, 212–17
 membrane distribution, 93
 morphology, 7
Cell-size distributions, 93, 96
Cell-wall tackiness, 107
Cellular growth, 318
Cellular matrix, 190–1
Cellular morphology, 266–70
Cellular polymeric foams, 99–100
Cellular structures, 5–7, 9–20, 91, 92, 228, 266
 dimensional description of, 13–15
 morphology, 9–13
Celogen AZ130, 280

Celuka process, 266
Cenospheres, 360, 362, 367, 387
Chemical blowing, 4
Circular structures, 217
Closed-cell foams, 9, 10, 18, 42–7
Closed-cell structures, 1, 2, 15, 19, 25, 28, 30
Closed-mould operation, 324–5, 328
Collapse behaviour, 385
Collapse load, 361, 369
Columnar struts, 212
Comfort aspects. *See* Cushioning
Complex modulus, 169
Compression, 190, 294–7
Compression–deflection characteristics, 194, 207–8, 212, 216–21, 225, 244–8, 256
Compression modulus, 81, 83, 87, 90, 185, 353
Compression moulding, 328
Compression set, 226
Compression–tension tests, 61
Compression testing, 51, 62
Compressive loading, 35, 60, 61, 66, 80–7, 92, 352
Compressive strain, 156, 159
Compressive strength, 38, 51, 259
Compressive stress, 45, 154, 376
Core–density profile, 274–6
Crack-induced fracture, 310
Crack propagation, 347
Crazes, 314
Creep
  modulus curves, 70–1
  properties, 69–70
    high-temperature, 331
Crosslinking, 4, 115, 118, 120, 126, 328
Cubic plate model, 18, 25
Cubic strut model, 18, 23
Cushioning, 99–142
  cell geometry, 106
  cell-wall thickness, 107
  closed-cell content, 107
  comfort
    characteristics, 100–5
    comparisons, 117
    evolution, 117

Cushioning—*contd.*
  design requirements, 140–1
  development and growth, 99
  fatigue, 128–41
    memory, 131
    optimisation, 133–40
    test methods, 131–2
    urethane foam, 128–32
  'fight-back', 100, 103, 106, 107, 112, 121, 122, 125, 126
  force–deflection behaviour, 100–1
  formulation effects, 134–40
  general considerations, 106
  hardness, 120
  material requirements, 102–5
  mechanics, 106–8
  openness of cells, 125
  polymer design, 114
  prediction of fatigue service performance, 132
  processability, 118
  structural consideration of reactants, 110–12
  ultimate design, 126–7
  water effects, 122
Cyclic tests, 61

Damage energy, 63, 64
Damping measurement apparatus, 210
Damping quality factor, 210
Damping tests, 207
Deflection, 227, 298
Deflectometer, 208
Deformation behaviour, 33, 65, 68, 74, 83, 191, 197, 216, 294, 296
Deformation limits, 65
Deformation measurement, 54
Deformation rate, 187
Deformation–time curves, 69–70, 195
Density
  distribution, 49, 299
  effects, 22, 50, 64–8, 76, 95, 122, 141, 234, 235, 240, 244, 248–50, 259, 260, 265, 270, 282, 284–7, 292–3, 295, 306, 319, 332, 349

Density—contd.
  profile, 298, 299, 301
  ratio, 22, 24
Density–load bearing relationship
  envelopes, 230–2
Density–property relationships, 21–3
Density–structure relationships, 23–5
Di-isocyanate, 108–15
Dilatometric determinations, 227
Dimethylsiloxane-polyoxypropylene, 113
Displacement ratio, 171–2
Distortion, 216
Dodecahedral voids, 16
Dodecahedron structure, 31
Double truncated cone model, 43
Ductile–brittle transitions, 311–18
Dynamic mechanical behaviour, 143–77
  comparison of predicted and measured behaviour, 161–9
  fluid-flow effects, 150–69
  polymer matrix, 145–50
  prediction of, 152–60
Dynamic modulus, 150

Eastman PETG, 311–15
Einstein coefficient, 335
Einstein equations, 335
Elastic behaviour, 18, 46, 52, 60, 61
Elastic constants, 363
Elastic modulus, 21, 39–41, 54, 55, 65, 91, 93, 144, 191, 363–5, 381, 386
  see also Young's modulus
Elastic wave, 191
Elastoplasticity, 370
Elements, 370
Elongation, 226, 295, 296, 311
Energy-absorbing capacity, 181, 191, 200–3
Energy-absorbing efficiency, 185
Energy consumption, 328
Epoxy resins, 329
Equation of motion, 180
Ethylene glycol, 238–9

Euler formula, 354
Expansion process, 3–5, 7

Failure
  criterion, 369
  envelope, 368–9
  mechanisms, 33, 304–11, 318, 344, 362–3, 380
Fatigue, 127, 129
  cumulative, 130
  cushioning, 128–41
  failures, 304, 308–10
  memory, 131
  prediction of service performance, 132
  resistance, 325
    optimisation, 133–40
  tests, 130–2, 228
Fibres
  hammer-milled, 351–6
  short chopped, 345–51
Fillers, 123
Finite element process, 370
Flame penetration, 325
Flexible polymer foams, 3, 20, 73–97
  closed-cell structures, 87–91
  fluid-flow effects, 150–69
  impact loading, 186–204
  mechanical anisotropy, 91–6
  mechanical properties, 83–6
  structural models, 86–7, 93–6
  structural theories, 76–9
  see also Polyurethane flexible foams
Flexural failure, 307
Flow
  orientation, 48
  velocities, 48
Fluid-flow effects, 150–69, 195
  compressible fluids, 159–60
  incompressible fluids, 152–9
  vibration isolation, 171–5
Fluidic effects, 189–90, 194–204
  behaviour prediction, 196
  model representation, 196
Foam
  core density, 319
  density, 21

Foaming agents, 280, 285, 311
Force–deflection characteristics, 179
Force–deformation diagram, 182, 190
Force–time curves, 195
Form factor, 172
Fracture, 310, 349–51
  morphology, 316
  surface, 12
  toughness, 344
'Fragility' requirement, 182
Frequency effects, 145, 154, 163, 164, 166–70, 174

Gas bubbles, 4
Gas-filled foams, 195
Gas–solid systems, 340
Gauge length, 55
Gent–Rusch model, 163, 198, 202, 204
Gent–Rusch theory, 166, 196
Gent–Thomas analysis, 95
Gent–Thomas equation, 78
Gent–Thomas theory, 78, 79
Glass transition temperature, 149, 150, 167, 227
Graft polyols, 121
Griffith yield criterion, 310
Gurley densometer, 104

Halpin–Tsai equation, 92, 270, 342
Hardness, 120
Health hazards, 326
Hennecke UBT machine, 228
Hexagonal structures, 217
Hooke's law, 296, 307
HR (high-resilience) foam, 117–20, 127
Hydrogen bonding, 21, 125–6, 129, 130, 133, 136, 149, 233
Hydrolytic stability, 226
Hydrostatic behaviour, 363
Hydrostatic pressure, 374, 377, 378, 380
Hydroxyl groups, 119
Hysteresis, 3, 102

I-beam concept, 272–4, 289–90
Impact
  failures, 304
  loading, 179
  flexible foams, 186–204
  test procedure, 189–90
  resistance, 302, 303, 325
  strength, 300–4, 317
Indentation force deflection (IFD), 101, 103, 107, 121–6, 130–2, 136–9
Injection moulding, 263, 266, 328
Instron extrusion rheometer, 336
Instron Universal Test Machine, 208, 225, 247
Interpenetrating polymer network (IPN), 329
Isocyanate, 136, 242
  index, 124, 138, 236–8, 244

J–U curve, 183–5, 194, 202, 204

K–I curve, 185
Kim–Rudd relationship, 213
Kinetic energy, 180, 182

Lamé constants, 381
Latex foams, 100
Lederman theory, 79
Lined spheres, 380–1
Liner materials, 380–9
Liquid-filled systems, 195, 199, 201–4
Load–deflection relationship, 299
Load–deformation behaviour, 16, 17
Load–unload hysteresis loop, 210
Low-density foams, 7
Low-density polymer foams, 9
Lumped parameter model, 179

Martin Sweets Machine, 228
MDI (4,4'-diphenylmethane di-isocyanate), 110

Mechanical behaviour. See Dynamic mechanical behaviour; Mechanical properties
Mechanical hysteresis, 3, 102
Mechanical properties, 5, 6, 20–5
  cellular matrix, 190–1
  density relationships, 21
  factors influencing, 20
  general considerations, 20–1
  polyurethane flexible foams, 86
  polyurethane rigid foams, 28, 42, 47–9
  poly(vinyl chloride) foams, 28, 47, 67, 68
  rigid plastic foams, 27–72
  thermoplastic structural foam, 265, 287–311, 319, 339
Mechanical testing
  phenolic rigid foams, 245–7
  rigid plastic foams, 47–58
  test procedures, 55–8, 207–10
  test specification, 51–8
Methane foam, fatigue in cushions, 128–32
Microcavitation, 315, 316
Microcracks, 314
Microspheres, 360, 362, 363, 367
MOCA, 210–11, 219–21
Modulus
  ratio, 79, 80
  temperature profile, 331
Molecular orientation, 22
Molecular weight, 111, 112, 114, 118–21, 124, 126, 134
Moment of inertia, 273
Moulded foam, 117, 131
Moulding pressure, 301, 328
Mountings, 169

Newtonian liquid, 174
Nodal points, 370
Non-Newtonian fluids, 174
Notch effect, 65
Notched specimens, 300
NR–SBR latex foam, 147
Numerical analysis of ordered foams, 369–81
Nylon 6, 329

'o' modulus, 281
'o' values, 269
One-shot foams, 116, 118, 219
Open-cell foams, 1, 11, 17–19, 23, 24, 31–42, 76, 150, 172
Ordered foams
  concluding remarks, 389–90
  numerical analysis, 369–81
  ultimate load behaviour, 385–9
Orientation degree, 50
Orientation distribution, 91
Orientation effect, 351
Orientation factor, 38
Oxypropylene, 112

Packing fraction, 336
Pentagonal dodecahedron cell structure, 16, 44, 351
Pentagonal dodecahedron model, 25, 31, 32, 43
Permeability, 154, 161, 170, 197, 198, 247, 256, 258–60
Phase angle, 159, 165
Phenolic foams, 387, 388
Phenolic resin composition, 255–6
Phenolic rigid foams, 244–60
  chemical composition, 245, 247–60
  general considerations, 244
  process conditions, 245
  process variable effects, 247–60
Phosphoric acid concentration, 255
Plane-strain analysis, 375
Plastic
  collapse, 361
  wave, 191
Poisson's ratio, 77, 265, 363–5, 381
Polybutylene terephthalate (PBT), 280
Polyesters, 233, 326, 329
Polyether diol, 115
Polyethers, 233
Polyethylene, 87, 282, 303, 307, 309, 310, 316–17
Polyethylene terephthalate (PET), 280
Polymer
  composition effects, 217–21
  matrix, 145–50
Polymerisation, 111, 120, 248

Polymethylene polyphenylisocyanate (PMPPI), 234, 242–3
Polyol–glass slurry, 352
Polyols, 108, 109, 111, 118–21, 134
Polyoxyisopropylene, 218
Polyoxypropylene triol (POPT), 240
Polyoxytetramethylene–polyoxyisopropylene prepolymer blends, 223
Polypropylene, 119
Polystyrene, 266, 267, 280–2, 301, 302, 329
Polyurethane flexible foams, 224–44
  acceleration efficiency, 183
  air-filled, 164, 189
  chemical composition, 228
  cushioning applications, 100
  deformation, 192
  dynamic mechanical behaviour, 148, 163
  dynamic modulus, 151
  fluid-flow damping peak, 166
  fluidic effects, 195, 198
  formulation, 228–43
  general considerations, 224–5
  liquid-filled, 203
  loss tangent, 151
  mechanical behaviour, 86
  non-reticulated, 81, 145
  properties and their measurement, 225–8
  ratio y_0/y_m, 193
  reticulated, 11, 87, 190, 191
  rheology, 147
  silicone oil filled, 163
  stress–strain diagrams, 74–6, 187
  structures, 13, 79
  transmissibility, 170
Polyurethane foams, 4, 7, 100, 188, 271, 311
  castings, 221
  one-shot, 219
  reaction injection moulding, 328
  reinforced, 324, 346, 351
Polyurethane processing, 210–11
Polyurethane rigid foams
  creep, 69–70
  deformation, 65

Polyurethane rigid foams—*contd.*
  density effects, 21
  elastic behaviour, 52, 61
  load-bearing limit, 66
  mechanical properties, 28, 42, 47–9
  mechanical testing, 55–6
  reinforced, 353
  stress–strain diagrams, 60
  structures, 30, 31
  tensile strength, 45
  tensile testing, 57
Polyurethane sprayed foam, 327
Poly(vinyl chloride) rigid foams
  bending tests, 58
  compressive strength, 43, 45
  creep, 69, 70
  density, 49, 58, 71
  mechanical properties, 28, 47, 67, 68
  stress–strain diagrams, 60
  structure, 29, 42, 51
Porosity, 161, 200
Potential energy, 180
Power-law function, 273
Power-law model, 270, 306
Pressure effects, 387, 388
Principal stresses, 369, 374
Probability distribution function, 94
Progelhof–Eilers model, 294–6
Proportional limit, 75
Pulverised fuel ash, 360
Pump control, 331
Pumps, axial piston, 331
PVT characteristics, 227

Q value, 218, 219, 224

Rate
  constant, 189
  function, 189
  sensitivity, 213, 219, 220
Rayleigh approximation, 158
Reaction Injection Moulding (RIM)
  process, 324, 328, 344
  chemistry of, 328–31
  components, 329
  impingement mixing, 332–3
  processing principles, 331

INDEX

Reaction rate, 248–50
Rebound energy, 193
Rebound resilience, 181
Reduced core density, 265
Reduced density, 265
　relationship, 343
Reduced flexural modulus, 319
Reduced storage modulus, 145
Reinforced foams, 323–58
　boards, 325–7
　closed-mould systems, 324–5
　development, 323
　fabrication, 324–7
　general considerations, 323
　high-density, 340
　low-density, 345–56
　model systems, 339–56
　panels, 325–7
　sprayed, 327
Reinforced Reaction Injection
　Moulding (RRIM) process,
　328–38, 344, 345
　mechanical properties, 334, 337–8
　model systems, 340–5
　reinforcement, 333–4
　slurry viscosities, 334–8
Resilience, 3, 105, 124, 227
Resonance frequency, 170, 171
Reynolds number, 332
Rigid plastic foams, 27–72
　compressive strength, 52
　definition, 3
　elastic behaviour of, 16, 46, 60, 61
　mechanical properties, 27–72
　long-term investigations, 68–71
　measurement of, 47–58
　see also Phenolic rigid foams;
　　Polyurethane rigid foams;
　　Poly(vinyl chloride) rigid
　　foams

Sag factor, 102–3, 122, 124, 125
Sand-drop shock-test machine, 208
SBR latex coated PUR foams, 188
Schwaber–Meinecke rate function, 189
Shape function, 96, 370

Shear
　bands, 311–18
　deformation, 145
　forces, 37
　loading, 38
　modulus, 273, 364
　strength, 57
　stress, 348, 374, 376, 381, 384
Shift factor, 145, 147
Shock
　mitigation, 179–262
　behaviour prediction, 191
　elastic materials, 181
　'ideal' isolating material, 182
　material behaviour prediction, 184–6
　material properties, 181–4
　model representation, 191–3
　non-elastic materials, 181
　structured materials, 207–24
　motion, 179
　response, 179
Silicone
　oil, 163, 164
　surfactants, 234
Skin
　effect, 276–84
　thickness, 265, 266, 273, 291–3, 302, 303, 319
Skin–core interface, 307
Slabstock, 116–17, 131, 132
Slip planes, 311
SnOct (stannous octoate), 234–6
Southwark Tate Emery Testing Machine, 225
Span 80, 256, 258–60
Specimen
　geometry, 52, 57
　preparation, 47
Spherulitic morphology, 317
Stairstep testing method, 303
Stannous octoate (SnOct), 234–6
Stiff liners, 382–5
Stiffness effects, 3, 172, 179–80, 197, 218, 221, 223, 230, 233, 234, 241, 260, 273–4, 276, 291–3
Stiffness equation, 287
Storage modulus, 165

Strain, 84–5, 89, 191, 198, 294
  energy, 305, 364
  rate, 180, 185, 188, 190, 209
Strain-energy release rate, 349
Strength properties, 37
Strength–weight ratio, 367
Stress, 84–5, 89, 129–30, 144, 154, 156,·179, 190, 294
  components, 370
  concentrations, 375
  distribution, 366–7, 383
  relaxation, 9, 23
Stress–strain behaviour, 15, 18, 74, 79–87, 90, 171, 190, 195, 209, 265, 266, 284, 294, 297, 311
Stress–strain diagrams, 60, 62, 74–6, 82–6, 88, 96, 184–6, 193, 194, 296, 361
Stress–strain ratio, 144, 156, 164, 165, 380
Structural foams, 263–322
Structural models, 15–20
Structure–density relationships, 23–5
Structure–property relationships, 19
Structured materials, 207–24
Strut
  angles, 212
  network, 351–6
  shape effect, 212
  thickness
    effect, 213
    function, 214
Strut height/thickness ratio, 217
Styrene–acrylonitrile copolymer, 150
Styrene–butadiene rubber, 100
Sulphuric acid concentration, 253
Superposition principle, 197
Surfactants, 113–14, 140, 234, 256, 356
Syntactic foams, 359–91
  applications, 359
  development, 359
  failure mechanisms, 362–3
  general considerations, 360
  mechanical behaviour, 360–9
  mechanics, 359
  structure, 360
  use of term, 359

Tackiness, 107
Temperature effects, 60, 66–8, 124, 145, 164, 166–9, 221–4, 253
Tensile elastic behaviour, 79
Tensile loading, 35, 60, 61, 66, 74
Tensile modulus, 75–80, 91, 155, 282, 295
Tensile strength, 37, 45, 56, 225, 282, 367
Tensile stress, 347, 380
Tensile tests, 56, 62, 64
Tension, 265, 266, 294–7
Tension–compression test, 62
Thermal conductivity, 306
Thermal expansion, 331
Thermal failure, 307
Thermoplastic structural foam (TSF), 263–322
  cellular morphology, 266–70
  core-density profile, 274–6
  failure mechanisms, 304–11
  flexural bending of beams, 287–93
  I-beam concept, 272–4
  mechanical properties, 265, 287–311, 319, 339, 346
  models, 287–311, 340–5
  processing, 269
  reinforced, 338
  thermal properties, 339
  uniform-density cell behaviour, 270–2
  variable characteristics in foam densities, 284–7
Time–temperature superposition relationship, 227
Tolylene di-isocyanate, 234, 242–3
Torsion, 362
Transcrystalline morphology, 318
Transcrystalline region, 316
Transmissibility, 169–74
Transmissibility curves, 169, 173
Transmissibility response, 170
Transverse deformation, 155, 156, 158–60, 164, 342
1,1,2-Trichloro-1,2,2-trifluoroethane, 251
Trimer acids, 233, 241
Tween 60, 256–9

Urethane foam, 106, 108–14, 117, 126
  design optimisation, 141
  fatigue resistance optimisation, 133–40
  formulation, 134–40
  preparation, 108
Urethane polymers, 114

Vibration isolation, 169–74
Viscoelasticity, 144, 193, 363
Viscoplasticity, 363, 370
Viscosity, 336
  coefficient, 197
Voids
  cubic, 374–5, 389
  spherical
    face-centred cubic arrangement, 377, 389–90

Voids—*contd.*
  spherical—*contd.*
    simple cubic distribution, 376
    unlined, 373
  Von Mises yield criterion, 314, 369

Water
  concentration effects, 240–1, 250–1
  reactions, 136–8
WLF equation, 147

Yield
  point, 362
  stress, 385, 386, 389
Young's modulus, 270, 271, 273, 299, 306, 310, 354, 363, 364
  see also Elastic modulus